Luc Devroye • Bülent Karasözen • Michael Kohler • Ralf Korn
Editors

Recent Developments in Applied Probability and Statistics

Dedicated to the Memory of Jürgen Lehn

Physica-Verlag

Editors
Prof. Dr. Luc Devroye
McGill University
School of Computer Science
University St.
H9X 3V9 Montreal Québec
3480
Canada
lucdevroye@gmail.com

Prof. Dr. Bülent Karasözen
Middle East Technical University
Mathematics & Institute of
Applied Mathematics
Inonu Bulvari
06531 Ankara
Turkey
bulent@metu.edu.tr

Prof. Dr. Michael Kohler
TU Darmstadt
FB Mathematik
Schloßgartenstr. 7
64289 Darmstadt
Germany
kohler@mathematik.tu-darmstadt.de

Prof. Dr. Ralf Korn
University of Kaiserslautern
Mathematics
E.-Schrödinger-Str.
67663 Kaiserslautern
Germany
korn@mathematik.uni-kl.de

ISBN 978-3-7908-2910-5
ISBN 978-3-7908-2598-5 (eBook)
DOI 10.1007/978-3-7908-2598-5
Springer Heidelberg Dordrecht London New York

Cover design: WMXDesign GmbH, Heidelberg

Printed on acid-free paper

Physica-Verlag is a brand of Springer-Verlag Berlin Heidelberg
Springer is part of Springer Science+Business Media (www.springer.com)

What should we do?

I think, we should be grateful to fate

that we safely emerged
from all adventures -
both from the real ones
and from those we dreamed about.

'Dream novel'

Yours Winni, Patrick, Steffen and Annette
with love

Preface

This book is devoted to Professor Jürgen Lehn, who passed away on September 29, 2008, at the age of 67. It contains invited papers that were presented at the *Workshop on Recent Developments in Applied Probability and Statistics Dedicated to the Memory of Professor Jürgen Lehn*, Middle East Technical University (METU), Ankara, April 23–24, 2009, which was jointly organized by the Technische Universität Darmstadt (TUD) and METU. The papers present surveys on recent developments in the area of applied probability and statistics. In addition, papers from the *Panel Discussion: Impact of Mathematics in Science, Technology and Economics* are included.

Jürgen Lehn was born on the 28th of April, 1941 in Karlsruhe. From 1961 to 1968 he studied mathematics in Freiburg and Karlsruhe, and obtained a Diploma in Mathematics from the University of Karlsruhe in 1968. He obtained his Ph.D. at the University of Regensburg in 1972, and his Habilitation at the University of Karlsruhe in 1978. Later in 1978, he became a C3 level professor of Mathematical Statistics at the University of Marburg. In 1980 he was promoted to a C4 level professorship in mathematics at the TUD where he was a researcher until his death.

Jürgen Lehn supervised many theses, wrote three books, published numerous articles and was the editor of proceedings and journals such as *Statistical Papers, Statistics and Decisions, Mathematische Semesterberichte* and *Journal of Computational and Applied Mathematics*. He managed the interdisciplinary Center for Practical Mathematics in Darmstadt, where he founded the stochastic simulation research group and cooperated with structural engineering teams using principles of applied mathematical statistics. His interdisciplinary work in mathematics gave him satisfaction and joy, and not a small degree of success. He also organized an interdisciplinary series of lectures entitled "What lies behind?", and was a member of the German federal selection committee for mathematical research.

Jürgen Lehn was a devoted teacher and a caring person. His ex-students, some of whom are now professors, are thankful for having known him. Jürgen Lehn rose from the rank of assistant to Dean and senator of TUD. He served on several high-level committees within his university. In addition, the community of mathematicians remembers Jürgen Lehn for his selfless dedication and unconditional dependability. He consulted often for the German Research Foundation (DFG), and he was sought after by hiring committees at various universities. Through his work as a founding member at the DFG and his activities in the Stochastik group of the German Mathematician's Union, he helped the general stochastics community in the country. For six years he served as the secretary of the Conference of University Teachers of Mathematical Stochastics. Throughout all this, Jürgen Lehn tried to include and help the younger scientific generation in research and professional life.

In his work for the mathematical community, the Mathematical Research Institute at Oberwolfach played a special role. From 1992 until 2003, Jürgen Lehn was the treasurer of the association *Friends of Oberwolfach*, and from 2003 on, he served as its secretary. As one of the founding fathers of this association, Jürgen Lehn's energy and dedication saw the group through difficult early years, and was instrumental in many emergency relief measures.

He played a most important role in establishing international cooperation between the Department of Mathematics of TUD and several mathematics departments around the globe. In particular, he was one of the leading figures in the foundation of The Institute of Applied Mathematics (IAM) at METU.

As the editors of this book, we would like to express our thanks to all contributors, many of whom were not only his scientific colleagues but also his personal friends. We are specially grateful to Alice Blanck and Dr. Niels Peter Thomas from Springer for their cooperation with the publication of this book.

Ankara, Darmstadt, Kaiserslautern, Montreal
January 2010

Luc Devroye, Bülent Karasözen
Michael Kohler, Ralf Korn

Contents

List of Contributors

Björn Bornkamp Fakultät Statistik, TU Dortmund, Dortmund, Germany, bornkamp@statistik.tu-dortmund.de

Luc Devroye School of Computer Science, McGill University, Montreal, H3A 2K6, Canada, lucdevroye@gmail.com

Roland Fried Fakultät Statistik, Technische Universität Dortmund, Dortmund, Germany, fried@statistik.tu-dortmund.de

Ursula Gather Fakultät Statistik, TU Dortmund, Dortmund, Germany, gather@statistik.tu-dortmund.de

Katja Ickstadt Fakultät Statistik, TU Dortmund, Dortmund, Germany, ickstadt@statistik.tu-dortmund.de

Bülent Karasözen Institute of Applied Mathematics & Department of Mathematics, Middle East Technical University, 06531 Ankara, Turkey, bulent@metu.edu.tr

Michael Kohler Department of Mathematics, Technische Universität Darmstadt, Schloßgartenstraße 7, 64289 Darmstadt, Germany, kohler@mathematik.tu-darmstadt.de

Ralf Korn Center for Mathematical and Computational Modeling (CM)² and Department of Mathematics, University of Kaiserslautern, 67653 Kaiserslautern, Germany, korn@mathematik.uni-kl.de

Hilke Kracker Fakultät Statistik, TU Dortmund, Dortmund, Germany, kracker@statistik.tu-dortmund.de

Sonja Kuhnt Fakultät Statistik, TU Dortmund, Dortmund, Germany, kuhnt@statistik.tu-dortmund.de

Wilfried Meidl MDBF, Sabancı University, Orhanlı, 34956 Tuzla, İstanbul, Turkey, wmeidl@sabanciuniv.edu

Stefanie Müller Center for Mathematical and Computational Modeling (CM)²
and Department of Mathematics, University of Kaiserslautern,
67653 Kaiserslautern, Germany, stefanie@mathematik.uni-kl.de

Andreas Rößler Department Mathematik, Universität Hamburg, Bundesstraße 55,
20146 Hamburg, Germany, andreas.roessler@math.uni-hamburg.de

Michael Schäfer Institute of Numerical Methods in Mechanical Engineering,
Technische Universität Darmstadt, Dolivostr. 15, 64293 Darmstadt, Germany,
schaefer@fnb.tu-darmstadt.de

Alev Topuzoğlu MDBF, Sabancı University, Orhanlı, 34956 Tuzla, İstanbul,
Turkey, alev@sabanciuniv.edu

Daniel Vogel Fakultät Statistik, Technische Universität Dortmund, Dortmund,
Germany, daniel.vogel@tu-dortmund.de

Harro Walk Department of Mathematics, Universität Stuttgart,
Pfaffenwaldring 57, 70569 Stuttgart, Germany,
harro.walk@mathematik.uni-stuttgart.de

On Exact Simulation Algorithms for Some Distributions Related to Brownian Motion and Brownian Meanders

Luc Devroye

Abstract We survey and develop exact random variate generators for several distributions related to Brownian motion, Brownian bridge, Brownian excursion, Brownian meander, and related restricted Brownian motion processes. Various parameters such as maxima and first passage times are dealt with at length. We are particularly interested in simulating process variables in expected time uniformly bounded over all parameters.

1 Introduction

The purpose of this note is to propose and survey efficient algorithms for the exact generation of various functionals of Brownian motion $\{B(t), 0 \le t \le 1\}$. Many applications require the simulation of these processes, often under some restrictions. For example, financial stochastic modeling (Duffie and Glynn 1995; Calvin 2001; McLeish 2002) and the simulation of solutions of stochastic differential equations (Kloeden and Platen 1992; Beskos and Roberts 2005) require fast and efficient methods for generating Brownian motion restricted in various ways. Exact generation of these processes is impossible as it would require an infinite effort. But it is possible to exactly sample the process at a finite number of points that are either fixed beforehand or chosen "on the fly", in an adaptive manner. Exact simulation of various quantities related to the processes, like maxima, first passage times, occupation times, areas, and integrals of functionals, is also feasible. Simulation of the process itself can be achieved by three general strategies.

1. Generate the values of $B(t)$ at $0 = t_0 < t_1 < \cdots < t_n = 1$, where the t_i's are given beforehand. This is a *global* attack of the problem.

Luc Devroye, School of Computer Science, McGill University, Montreal, H3A 2K6, Canada
e-mail: lucdevroye@gmail.com

L. Devroye et al. (eds.), *Recent Developments in Applied Probability and Statistics*,
DOI 10.1007/978-3-7908-2598-5_1, © Springer-Verlag Berlin Heidelberg 2010

2. Simulation by subdivision. In the popular binary division (or "bridge sampling") method (see, e.g., Fox 1999), one starts with $B(0)$ and $B(1)$, then generates $B(1/2)$, then $B(1/4)$ and $B(3/4)$, always refining the intervals dyadically. This can be continued until the user is satisfied with the accuracy. One can imagine other situations in which intervals are selected for sampling based on some criteria, and the sample locations may not always be deterministic. We call these methods *local*. The fundamental problem here is to generate $B(\lambda t + (1-\lambda)s)$ for some $\lambda \in (0,1)$, given the values $B(t)$ and $B(s)$.
3. Generate the values of $B(t)$ sequentially, or by extrapolation. That is, given $B(t)$, generate $B(t+s)$, and continue forward in this manner. We call this a *linear* method, or simply, an *extrapolation* method.

We briefly review the rather well-known theory for all strategies. Related simulation problems will also be discussed. For example, in case (ii), given an interval with certain restrictions at the endpoints, exact simulation of the minimum, maximum, and locations of minima and maxima in the interval becomes interesting. Among the many possible functionals, maxima and minima stand out, as they provide a rectangular cover of the sample path $B(t)$, which may of interest in some applications. Brownian motion may be restricted in various ways, e.g., by being nonnegative (Brownian meander), by staying within an interval (Brownian motion on an interval), or by attaining a fixed value at $t = 1$ (Brownian bridge). This leads to additional simulation challenges that will be discussed in this paper.

We keep three basic principles in mind, just as we did in our book on random variate generation (Devroye 1986). First of all, we are only concerned with exact simulation methods, and to achieve this, we assume that real numbers can be stored on a computer, and that standard algebraic operations, and standard functions such as the trigonometric, exponential and logarithmic functions, are exact. Secondly, we assume that we have a source capable of producing an i.i.d. sequence of uniform $[0,1]$ random variables $U_1, U_2, U_3, \ldots$. Thirdly, we assume that all standard operations, function evaluations, and accesses to the uniform random variate generator take one unit of time. Computer scientists refer to this as the RAM model of computation. Under the latter hypothesis, we wish to achieve uniformly bounded expected complexity (time) for each of the distributions that we will be presented with. The uniformity is with respect to the parameters of the distribution. Users will appreciate not having to worry about bad input parameters. Developing a uniformly fast algorithm is often a challenging and fun exercise. Furthermore, this aspect has often been neglected in the literature, so we hope that this will make many applications more efficient.

We blend a quick survey of known results with several new algorithms that we feel are important in the exact simulation of Brownian motion, and for which we are not aware of uniformly efficient exact methods. The new algorithms apply, for example, to the joint location and value of the maximum of a Brownian bridge, the value of a Brownian meander on a given interval when only the values at its endpoints are given, and the maximum of a Brownian meander with given endpoint.

This paper is a first in a series of papers dealing with the simulation of Brownian processes, focusing mainly on the process itself and simple parameters such as the maximum and location of the maximum in such processes. Further work is needed for the efficient and exact simulation of passage times, occupation times, areas (like the area of the Brownian excursion, which has the Airy distribution, for which no exact simulation algorithm has been published to date), the maximum of Bessel bridges and Bessel processes of all dimensions.

2 Notation

We adopt Pitman's notation (see, e.g., Pitman 1999) and write

$B(t)$	Brownian motion, $B(0) = 0$,
$B_r(t)$	Brownian bridge: same as B conditional on $B(1) = r$,
$B^{\text{br}}(t)$	standard Brownian bridge: same as B_r with $r = 0$,
$B^{\text{me}}(t)$	Brownian meander: same as B conditional on $B(t) \geq 0$ on $[0, 1]$,
$B_r^{\text{me}}(t)$	restricted Brownian meander: same as B^{me} conditional on $B^{\text{me}}(1) = r$,
$B^{\text{ex}}(t)$	Brownian excursion: same as B_r^{me} with $r = 0$.

Conditioning on zero probability events can be rigorously justified either by weak limits of some lattice walks or as weak limits of processes conditioned on ϵ-probability events and letting $\epsilon \downarrow 0$ (see, e.g., Durrett et al. (1977), and consult Bertoin and Pitman (1994) or Borodin and Salminen (2002) for further references). The absolute values of the former three processes, also called reflected Brownian motion and reflected Brownian bridge will only be briefly mentioned.

The maxima of these processes on [0, 1] are denoted, respectively, by

$$M, M_r, M^{\text{br}}, M^{\text{me}}, M_r^{\text{me}}, M^{\text{ex}}.$$

In what follows, we reserve the notation $N, N', N'', N_1, N_2, \ldots$ for i.i.d. standard normal random variables, $E, E', E'', E_1, E_2, \ldots$ for i.i.d. exponential random variables, $U, U', U'', U_1, U_2, \ldots$ for i.i.d. uniform [0, 1] random variables, and G_a for a gamma random variable of shape parameter $a > 0$. All random variables appearing together in an expression are independent. Thus, $U - U = 0$ but $U - U'$ has a triangular density. The symbol $\stackrel{\mathcal{L}}{=}$ denotes equality in distribution. We use ϕ for the normal density, and Φ for its distribution function. Convergence in distribution is denoted by $\stackrel{\mathcal{L}}{\to}$. The notation $\equiv$ means equality in distribution as a process. Also, we use "$X \in dy$" for "$X \in [y, y + dy]$".

3 Brownian Motion: Survey of Global and Local Strategies

We recall that $B(1) \stackrel{\mathcal{L}}{=} N$ and that for $0 \leq t \leq 1$, $\{B(ts), 0 \leq t \leq 1\} \stackrel{\mathcal{L}}{=} \{\sqrt{s}B(t), 0 \leq t \leq 1\}$. Furthermore, there are many constructions that relate the sample paths of the processes. Most useful is the definition, which states that for any $t_0 < t_1 < \cdots, < t_n$, we have that

$$(B(t_1) - B(t_0), \ldots, B(t_n) - B(t_{n-1})) \stackrel{\mathcal{L}}{=} \left(\sqrt{t_1 - t_0} N_1, \ldots, \sqrt{t_n - t_{n-1}} N_n\right).$$

The simplest representation of Brownian bridges is the drift decomposition of B_r: assuming a bridge on [0, 1] with endpoint r, we have

$$B_r(t) \equiv B(t) + t(r - B(1)), 0 \le t \le 1.$$

Thus, given $B(t_i)$ at points $t_0 = 0 < t_1 < \cdots < t_n = 1$, we immediately have $B_r(t_i)$ by the last formula.

As $B(1)$ is a sum of two independent components, $B(t)$ and $B(1) - B(t) \stackrel{\mathcal{L}}{=} B(1-t)$, so that for a fixed t,

$$B_r(t) \stackrel{\mathcal{L}}{=} tr + \sqrt{t}(1-t)N_1 + t\sqrt{1-t}N_2 \stackrel{\mathcal{L}}{=} tr + \sqrt{t(1-t)}N.$$

This permits one to set up a simple local strategy. Given shifted Brownian motion (i.e., Brownian motion translated by a value a) with values $B(0) = a$, $B(1) = b$, then interval splitting can be achieved by the recipe

$$B(t) = a + t(b-a) + \sqrt{t(1-t)}N.$$

Introducing scaling, we have, with $B(0) = a$, $B(T) = b$,

$$B(t) = a + \frac{t}{T}(b-a) + \sqrt{\frac{t}{T}\left(1 - \frac{t}{T}\right)} N\sqrt{T}, \quad 0 \le t \le T.$$

All further splitting can be achieved with fresh independent normal random varieties. Extrapolation beyond t for Brownian motion is trivial, as $B(t+s) \stackrel{\mathcal{L}}{=} B(t) + N\sqrt{s}, s > 0$.

In 1999, Jim Pitman published an important paper on the joint law of the various Brownian motion processes sampled at the order statistics of a uniform [0, 1] cloud of points. These yield various distributional identities but also fast methods of simulation. For the sake of completeness, we briefly recall his results here. The sampling period is [0, 1]. The order statistics of n i.i.d. uniform [0, 1] random variables are denoted by

$$0 = U_{(0)} < U_{(1)} < \cdots < U_{(n)} < U_{(n+1)} = 1.$$

It is well-known that this sample can be obtained from a sequence of i.i.d. exponential random variables $E_1, E_2, \ldots$ in the following manner, denoting $S_i = E_1 + \cdots + E_i$:

$$U_{(i)} = \frac{S_i}{S_{n+1}}, \quad 1 \le i \le n+1.$$

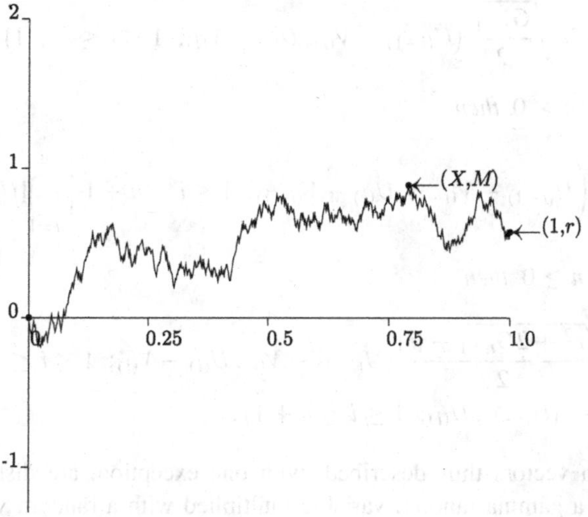

Fig. 1 Simulation of Brownian motion

See, e.g. Shorack and Wellner (1986). Denote by $X(t)$ any of the processes defined at the outset of this paper, and let it be independent of the uniform sample. Let T_i be a time in $[U_{(i-1)}, U_{(i)}]$ when X attains its infimum on that interval. Consider then the $2n+2$-dimensional random vector

$$X_n \stackrel{\text{def}}{=} \left(X(T_1), X(U_{(1)}), X(T_2), X(U_{(2)}), \ldots, X(T_{n+1}), X(U_{(n+1)})\right).$$

Obtain an independent uniform sample

$$0 = V_{(0)} < V_{(1)} < \cdots < V_{(n)} < V_{(n+1)} = 1,$$

which is based on an independent collection of exponentials with partial sums S'_i, $1 \le i \le n+1$, so $V_{(i)} = S'_i / S'_{n+1}$.

Proposition 1 (Pitman 1999). *If $X \equiv B$ and $n \ge 0$, then*

$$X_n \stackrel{\mathcal{L}}{=} \sqrt{2G_{n+3/2}} \left(\frac{S_{i-1} - S'_i}{S_{n+1} + S'_{n+1}}, \frac{S_i - S'_i}{S_{n+1} + S'_{n+1}}; \ 1 \le i \le n+1 \right).$$

If $X \equiv B^{\mathrm{me}}$ and $n \ge 0$, then

$$X_n \stackrel{\mathcal{L}}{=} \sqrt{2G_{n+1}} \left(\frac{S_{i-1} - S'_i}{S_{n+1} + S'_{n+1}}, \frac{S_i - S'_i}{S_{n+1} + S'_{n+1}}; \ 1 \le i \le n+1 \ \middle| \ \bigcap_{i=1}^{n} [S_i > S'_i] \right).$$

If $X \equiv B^{\mathrm{br}}$ and $n \ge 0$, then

$$X_n \stackrel{\mathcal{L}}{=} \sqrt{\frac{G_{n+1}}{2}} \left(U_{(i-1)} - V_{(i)}, U_{(i)} - V_{(i)};\ 1 \le i \le n+1\right).$$

If $X \equiv B^{\text{ex}}$ and $n \ge 0$, then

$$X_n \stackrel{\mathcal{L}}{=} \sqrt{\frac{G_{n+1}}{2}} \left(U_{(i-1)} - V_{(i-1)}, U_{(i)} - V_{(i-1)};\ 1 \le i \le n+1 \,\Bigg|\, \bigcap_{i=1}^{n} [U_{(i)} > V_{(i)}]\right).$$

If $X \equiv B_r$ and $n \ge 0$, then

$$X_n \stackrel{\mathcal{L}}{=} \frac{\sqrt{r^2 + 2G_{n+1}} - |r|}{2} (U_{(i-1)} - V_{(i)}, U_{(i)} - V_{(i)};\ 1 \le i \le n+1) + r\left(U_{(i-1)}, U_{(i)};\ 1 \le i \le n+1\right).$$

The random vectors thus described, with one exception, are distributed as a square root of a gamma random variable multiplied with a random vector that is uniformly distributed on some polytope of $\mathcal{R}^{2n+2}$. Global sampling for all these processes in time $O(n)$ is immediate, provided that one can generate a gamma random varieties G_a in time $O(a)$. Since we need only integer values of a or integer values plus $1/2$, one can achieve this by using $G_n \stackrel{\mathcal{L}}{=} E_1 + \cdots + E_n$ and $G_{n+1/2} \stackrel{\mathcal{L}}{=} E_1 + \cdots + E_n + N^2/2$. However, there are also more sophisticated methods that take expected time $O(1)$ (see, e.g., Devroye 1986).

There are numerous identities that follow from Pitman's proposition. For example,

$$\left(\min_{0 \le t \le U} B^{\text{br}}(t), B^{\text{br}}(U), \min_{U \le t \le 1} B^{\text{br}}(t)\right) \stackrel{\mathcal{L}}{=} \sqrt{\frac{G_2}{2}} \left(-U', U - U', U - 1\right).$$

This implies that $|B^{\text{br}}(U)| \stackrel{\mathcal{L}}{=} \sqrt{G_2/2}|U - U'| \stackrel{\mathcal{L}}{=} U\sqrt{E/2}$.

In the last statement of Pitman's result, we replaced a random variable $L_{n,r}$ (with parameter $r > 0$) in Sect. 8 of Pitman by the equivalent random variable $\sqrt{r^2 + 2G_{n+1}} - r$. It is easy to verify that it has the density

$$\frac{y^n (y+r)(y+2r)^n}{n! 2^n} \times \exp\left(-\frac{y^2}{2} - ry\right), y > 0.$$

4 Brownian Motion: Extremes and Locations of Extremes

The marginal distributions of the maximum M and its location X for B on $[0, 1]$ are well-known. We mention them for completeness (see, e.g., Karatzas and Shreve 1998): X is arc sine distributed, and

$$M \stackrel{\mathcal{L}}{=} |N|.$$

The arc-sine, or beta (1/2, 1/2) distribution, corresponds to random variables that can be represented equivalently in all these forms, where C is standard Cauchy:

$$\frac{G_{1/2}}{G_{1/2}+G'_{1/2}} \stackrel{\mathcal{L}}{=} \frac{N^2}{N^2+N'^2} \stackrel{\mathcal{L}}{=} \frac{1}{1+C^2} \stackrel{\mathcal{L}}{=} \sin^2(2\pi U) \stackrel{\mathcal{L}}{=} \sin^2(\pi U) \stackrel{\mathcal{L}}{=} \sin^2(\pi U/2)$$
$$\stackrel{\mathcal{L}}{=} \frac{1+\cos(2\pi U)}{2} \stackrel{\mathcal{L}}{=} \frac{1+\cos(\pi U)}{2}.$$

In simulation, M is rarely needed on its own. It is usually required jointly with other values of the process. The distribution function of M_r (see Borodin and Salminen 2002, p. 63) is

$$F(x) = 1 - \exp\left(\frac{1}{2}(r^2 - (2x-r)^2)\right), \quad x \geq \max(r, 0).$$

By the inversion method, this shows that

$$M_r \stackrel{\mathcal{L}}{=} \frac{1}{2}\left(r + \sqrt{r^2+2E}\right). \tag{1}$$

This was used by McLeish (2002) in simulations. Therefore, replacing r by N, we have the following joint law:

$$(M, B(1)) \stackrel{\mathcal{L}}{=} \left(\frac{1}{2}\left(N + \sqrt{N^2+2E}\right), N\right).$$

Putting $r = 0$ in (1), we observe that $M^{\text{br}} \stackrel{\mathcal{L}}{=} \sqrt{E/2}$, a result due to Lévy (1939). It is also noteworthy that

$$M \stackrel{\mathcal{L}}{=} |N| \stackrel{\mathcal{L}}{=} M - B(1).$$

The rightmost result is simply due to Lévy's observation (1948) that $|B(t)|$ is equivalent as a process to $M(t) - B(t)$ where $M(t)$ is the maximum of B over $[0, t]$.

Pitman's Proposition together with the observation that $2G_{3/2} \stackrel{\mathcal{L}}{=} N^2 + 2E''$, show that

$$(M, B(1)) \stackrel{\mathcal{L}}{=} \sqrt{2G_{3/2}} \times \left(\frac{E}{E+E'}, \frac{E-E'}{E+E'}\right)$$
$$\stackrel{\mathcal{L}}{=} \sqrt{N^2+2E''} \times \left(\frac{E}{E+E'}, \frac{E-E'}{E+E'}\right)$$
$$\stackrel{\mathcal{L}}{=} \left(U\sqrt{N^2+2E}, (2U-1)\sqrt{N^2+2E}\right).$$

Furthermore, Pitman's results allow us to rediscover Lévy's result $M^{\text{br}} \stackrel{\mathcal{L}}{=} \sqrt{E/2}$. Using $E/(E+E') \stackrel{\mathcal{L}}{=} U$, we also have

$$M \stackrel{\mathcal{L}}{=} U\sqrt{2G_{3/2}} \stackrel{\mathcal{L}}{=} U\sqrt{N^2+2E} \stackrel{\mathcal{L}}{=} \frac{1}{2}\left(N+\sqrt{N^2+2E}\right).$$

For $x > 0$, we define the first passage time (also called hitting time)

$$T_x = \min\{t : B(t) = x\}.$$

For $t > 0$,

$$\begin{aligned}
\mathcal{P}\{T_x > t\} &= \mathcal{P}\left\{\max_{0\le s\le t} B(s) < x\right\} \\
&= \mathcal{P}\left\{\max_{0\le s\le 1} B(s) < x/\sqrt{t}\right\} \\
&= \mathcal{P}\left\{\frac{1}{2}\left(N+\sqrt{N^2+2E}\right) < x/\sqrt{t}\right\} \\
&= \mathcal{P}\left\{\left(\frac{2x}{N+\sqrt{N^2+2E}}\right)^2 > t\right\},
\end{aligned}$$

and therefore,

$$T_x \stackrel{\mathcal{L}}{=} \left(\frac{x}{M}\right)^2.$$

Simulating hitting times and maxima are in fact equivalent computational questions. The same argument can be used for Brownian meanders: the hitting time of $x > 0$ for a Brownian meander is distributed as

$$\left(\frac{x}{M^{\mathrm{me}}}\right)^2.$$

Consider now the joint density of the triple $(X, M, B(1))$. Using (x, m, y) as the running coordinates for $(X, M, B(1))$, Shepp (1979) [see also Karatzas and Shreve 1998, p. 100] showed that this density is

$$\frac{m(m-y)}{\pi x^{3/2}(1-x)^{3/2}} \times \exp\left(-\frac{m^2}{2x} - \frac{(m-y)^2}{2(1-x)}\right), \qquad m \ge y \in \mathcal{R},\ x \in (0,1).$$

This suggests a simple method for their joint generation:

$$(X, M, B(1)) \stackrel{\mathcal{L}}{=} \left(X \stackrel{\text{def}}{=} \frac{1+\cos(2\pi U)}{2}, \sqrt{2XE}, \sqrt{2XE} - \sqrt{2(1-X)E'}\right).$$

This is easily seen by first noting that if $(X, M) = (x, m)$, then $B(1) \stackrel{\mathcal{L}}{=} m - \sqrt{2(1-x)E'}$. Then, given $X = x$, $M \stackrel{\mathcal{L}}{=} \sqrt{2xE}$.

Finally, we consider the joint law of (X, M_r) for B_r. This is a bit more cumbersome, especially if we want to simulate it with expected complexity uniformly bounded over all r. The joint density can be written as

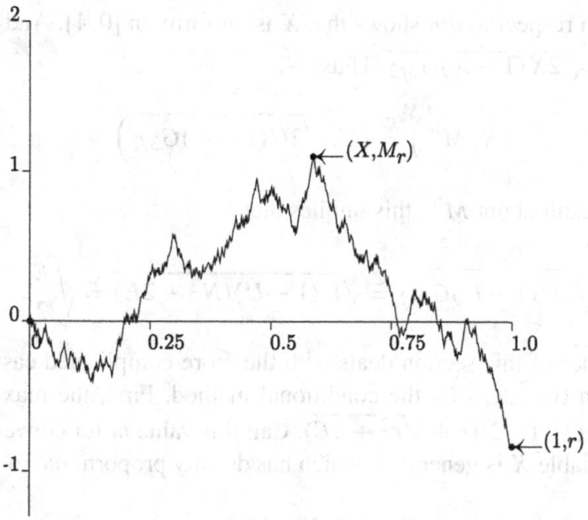

Fig. 2 A simulation of a Brownian bridge

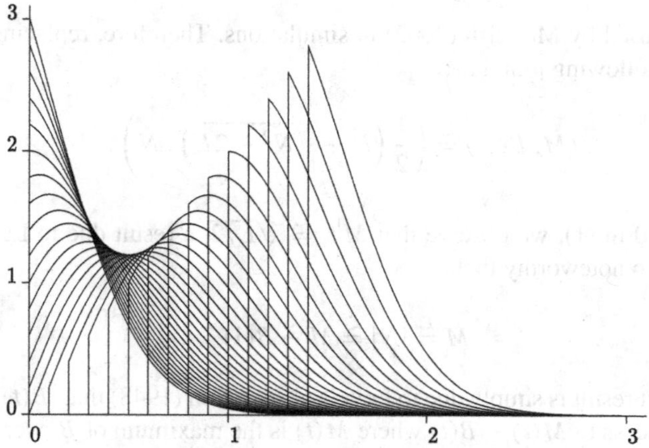

Fig. 3 The density of a Brownian bridge for endpoint values varying from -1.5 to 1.4 in increments of 0.1. For all values of r, 0 excepted, there is a discontinuity. For the standard Brownian bridge, we recover the scaled Rayleigh density

$$\frac{m(m-r)\sqrt{2\pi}e^{r^2/2}}{\pi x^{3/2}(1-x)^{3/2}} \times \exp\left(-\frac{m^2}{2x} - \frac{(m-r)^2}{2(1-x)}\right), \quad 0 \le x \le 1,\ m \ge \max(r, 0).$$

The standard Brownian bridge: $r = 0$. The special case of the standard Brownian bridge ($r = 0$) has a simple solution. Indeed, the joint density reduces to

$$\frac{2m^2}{\sqrt{2\pi}x^{3/2}(1-x)^{3/2}} \times \exp\left(-\frac{m^2}{2x(1-x)}\right), \quad 0 \le x \le 1,\ m \ge 0.$$

Integrating with respect to dm shows that X is uniform on $[0, 1]$. And given X, we see that $M^{\text{br}} \stackrel{\mathcal{L}}{=} \sqrt{2X(1-X)G_{3/2}}$. Thus,

$$\left(X, M^{\text{br}}\right) \stackrel{\mathcal{L}}{=} \left(U, \sqrt{2U(1-U)G_{3/2}}\right).$$

Using Lévy's result about M^{br}, this implies that

$$\sqrt{2U(1-U)G_{3/2}} \stackrel{\mathcal{L}}{=} \sqrt{U(1-U)(N^2+2E)} \stackrel{\mathcal{L}}{=} \sqrt{\frac{E}{2}}.$$

The remainder of this section deals with the more complicated case $r \neq 0$. We will simulate in two steps by the conditional method. First, the maximum M_r is generated as in (1): $(1/2)(r + \sqrt{r^2+2E})$. Call this value m for convenience. Then the random variable X is generated, which has density proportional to

$$\frac{\exp(-\frac{m^2}{2x} - \frac{(m-r)^2}{2(1-x)})}{x^{3/2}(1-x)^{3/2}}, \quad 0 < x < 1. \tag{2}$$

This was used by McLeish (2002) in simulations. Therefore, replacing r by N, we have the following joint law:

$$(M, B(1)) \stackrel{\mathcal{L}}{=} \left(\frac{1}{2}\left(N + \sqrt{N^2+2E}\right), N\right).$$

Putting $r = 0$ in (1), we observe that $M^{\text{br}} \stackrel{\mathcal{L}}{=} \sqrt{E/2}$, a result due to Lévy (1939, (20)). It is also noteworthy that

$$M \stackrel{\mathcal{L}}{=} |N| \stackrel{\mathcal{L}}{=} M - B(1).$$

The rightmost result is simply due to Lévy's observation (1948) that $|B(t)|$ is equivalent as a process to $M(t) - B(t)$ where $M(t)$ is the maximum of B over $[0, t]$.

Pitman's Proposition together with the observation that $2G_{3/2} \stackrel{\mathcal{L}}{=} N^2 + 2E''$, show that

$$\begin{aligned}
(M, B(1)) &\stackrel{\mathcal{L}}{=} \sqrt{2G_{3/2}} \times \left(\frac{E}{E+E'}, \frac{E-E'}{E+E'}\right) \\
&\stackrel{\mathcal{L}}{=} \sqrt{N^2+2E''} \times \left(\frac{E}{E+E'}, \frac{E-E'}{E+E'}\right) \\
&\stackrel{\mathcal{L}}{=} \left(U\sqrt{N^2+2E}, (2U-1)\sqrt{N^2+2E}\right).
\end{aligned}$$

Furthermore, Pitman's results allow us to rediscover Lévy's result $M^{\text{br}} \stackrel{\mathcal{L}}{=} \sqrt{E/2}$. Using $E/(E+E') \stackrel{\mathcal{L}}{=} U$, we also have

$$M \stackrel{\mathcal{L}}{=} U\sqrt{2G_{3/2}} \stackrel{\mathcal{L}}{=} U\sqrt{N^2+2E} \stackrel{\mathcal{L}}{=} \frac{1}{2}\left(N+\sqrt{N^2+2E}\right).$$

For $x > 0$, we define the first passage time (also called hitting time)

$$T_x = \min\{t : B(t) = x\}.$$

For $t > 0$,

$$\begin{aligned}
\mathcal{P}\{T_x > t\} &= \mathcal{P}\left\{\max_{0\le s\le t} B(s) < x\right\} \\
&= \mathcal{P}\left\{\max_{0\le s\le 1} B(s) < x/\sqrt{t}\right\} \\
&= \mathcal{P}\left\{\frac{1}{2}\left(N+\sqrt{N^2+2E}\right) < x/\sqrt{t}\right\} \\
&= \mathcal{P}\left\{\left(\frac{2x}{N+\sqrt{N^2+2E}}\right)^2 > t\right\},
\end{aligned}$$

and therefore,

$$T_x \stackrel{\mathcal{L}}{=} \left(\frac{x}{M}\right)^2.$$

Simulating hitting times and maxima are in fact equivalent computational questions. The same argument can be used for Brownian meanders: the hitting time of $x > 0$ for a Brownian meander is distributed as

$$\left(\frac{x}{M^{\text{me}}}\right)^2.$$

For this, we propose a rejection algorithm with rejection constant (the expected number of iterations before halting, or, equivalently, one over the acceptance probability) $R(m, r)$ depending upon m and r, uniformly bounded in the following sense:

$$\sup_r \mathcal{E}\{R(M_r, r)\} < \infty. \tag{3}$$

Note that $\sup_{r,m\ge\max(r,0)} R(m, r) = \infty$, but this is of secondary importance. In fact, by insisting only on (3), we can design a rather simple algorithm. Since we need to refer to it, and because it is fashionable to do so, we will give this algorithm a name, MAXLOCATION.

ALGORITHM "MAXLOCATION"
Case I ($m \geq \sqrt{2}$)
Repeat Generate U, N. Set $Y \leftarrow 1 + \frac{(m-r)^2}{N^2}$
Until $U \exp(-m^2/2) \leq Y \exp(-Ym^2/2)$
Return $X \leftarrow 1/Y$
Case II ($m - r \geq \sqrt{2}$)
Repeat Generate U, N. Set $Y \leftarrow 1 + \frac{m^2}{N^2}$
Until $U \exp(-(m-r)^2/2) \leq Y \exp(-Y(m-r)^2/2)$
Return $X \leftarrow 1 - 1/Y$
Case III ($m - r \leq \sqrt{2}, m \leq \sqrt{2}$)
Repeat Generate U, N. Set $X \leftarrow$ beta $(1/2, 1/2)$
Until $U \frac{4}{\sqrt{X(1-X)}e^2m^2(m-r)^2} \leq \frac{\exp(-m^2/2X-(m-r)^2/2(1-X))}{(X(1-X))^{3/2}}$
Return X

No attempt was made to optimize the algorithm with respect to its design parameters like the cut-off points. Our choices facilitate easy design and analysis. Note also that the three cases in MAXLOCATION overlap. In overlapping regions, any choice will do. Gou (2009) has another algorithm for this, but it is not uniformly fast. However, for certain values of the parameters, it may beat MAXLOCATION in given implementations.

Theorem 1. *Algorithm* MAXLOCATION *generates a random variable X with density proportional to* (2). *Furthermore, if m is replaced by $M_r = (1/2)(r + \sqrt{r^2 + 2E})$, then (X, M_r) is distributed as the joint location and value of the maximum of a Brownian bridge B_r. Finally, the complexity is uniformly bounded over all values of r in the sense of* (3).

Proof. The first two cases are symmetric—indeed, X for given input values m, r is distributed as $1 - X'$, where X' has input parameters $m - r$ and $-r$. This follows from considering the Brownian motion backwards. Case I: Let X have density proportional to (2), and let $Y = 1/X$. Then Y has density proportional to

$$y \exp\left(-\frac{m^2 y}{2}\right) \times (y-1)^{-\frac{3}{2}} \exp\left(-\frac{(m-r)^2 y}{2(y-1)}\right), \quad y > 1.$$

If $m \geq \sqrt{2}$, then the leftmost of the two factors is not larger than $\exp(-m^2/2)$, while the rightmost factor is proportional to the density of $1 + (m-r)^2/N^2$, as is readily verified. This confirms the validity of the rejection method for cases I and II. Case III: note that (2) is bounded by

$$\frac{4}{\sqrt{x(1-x)}e^2m^2(m-r)^2},$$

which is proportional to the beta (1/2, 1/2) density. To see this, observe that $(m^2/(2x))\exp(-(m^2/(2x))) \le 1/e$, and $((m-r)^2/(2(1-x)))\exp(-((m-r)^2/(2(1-x)))) \le 1/e$.

Finally, we verify (3) when the supremum is taken over the parameter ranges that correspond to the three cases. It is helpful to note that m is now random and equal to $(1/2)(r+\sqrt{r^2+2E})$. Thus, $m(m-r)=E/2$, a property that will be very useful. The acceptance rate in case I (using the notation of the algorithm) is

$$\begin{aligned}\mathcal{P}\{U\exp(-m^2/2)\le Y\exp(-Ym^2/2)\} &= \mathcal{E}\{Y\exp((1-Y)m^2/2)\}\\ &= \mathcal{E}\{Y\exp(-m^2(m-r)^2/2N^2)\}\\ &\ge \mathcal{E}\{\exp(-m^2(m-r)^2/2N^2)\}\\ &= \mathcal{E}\{\exp(-E^2/8N^2)\}\\ &\stackrel{\text{def}}{=} \delta>0.\end{aligned}$$

The acceptance rate for case II is dealt with in precisely the same manner—it is also at least δ. Finally, in case III, the acceptance rate is

$$\begin{aligned}&\mathcal{P}\left\{U\frac{4}{\sqrt{X(1-X)}e^2m^2(m-r)^2}\le\frac{\exp(-m^2/2X-(m-r)^2/2(1-X))}{(X(1-X))^{3/2}}\right\}\\ &=\mathcal{E}\left\{\frac{e^2m^2(m-r)^2\exp(-m^2/2X-(m-r)^2/2(1-X))}{4X(1-X)}\right\}\\ &\ge\mathcal{E}\left\{\frac{e^2m^2(m-r)^2\exp(-1/X-1/(1-X))}{4X(1-X)}\right\}\\ &=\mathcal{E}\left\{\frac{e^2E^2\exp(-1/X(1-X))}{16X(1-X)}\right\}\\ &=\mathcal{E}\left\{\frac{e^2\exp(-1/X(1-X))}{8X(1-X)}\right\}\\ &\ge\mathcal{E}\left\{\frac{e^2\exp(-16/3)}{8(3/16)}1_{[X\in[1/4,3/4]]}\right\}\\ &=\frac{1}{3e^{10/3}}.\end{aligned}$$

Therefore,

$$\mathcal{E}\{R(M_r,r)\}\le\max(1/\delta,3e^{10/3}).$$ □

The joint maximum and minimum of Brownian bridge and Brownian motion. The joint maximum and minimum of B_r can be done in two steps. First, we generate $M_r=(1/2)(r+\sqrt{r^2+2E})$ and then apply MAXLOCATION to generate the location X of the maximum. Using a decomposition of Williams (1974) and Denisov (1984), we note that the process cut at X consists of two Brownian meanders, back to back. More specifically,

$$M_r - B_r(X+t), \quad 0 \le t \le 1-X,$$

is a Brownian meander with endpoint $B^{\text{me}}(1-X) = M_r - r$. The maximum Z_1 of this process is distributed as

$$\sqrt{1-X} \times M_s^{\text{me}} \quad \text{with } s = \frac{M_r - r}{\sqrt{1-X}}.$$

The value M_s^{me} is generated by our algorithm MAXMEANDER, which will be developed further on in the paper. Similarly, the process

$$M_r - B_r(X-t), \quad 0 \le t \le X,$$

is a Brownian meander with endpoint $B^{\text{me}}(X) = M_r$. The maximum Z_2 of this process is distributed as

$$\sqrt{X} \times M_s^{\text{me}} \quad \text{with } s = \frac{M_r}{\sqrt{X}}.$$

The value M_s^{me} is again generated by our algorithm MAXMEANDER. Putting things together, and using the Markovian nature of Brownian motion, we see that the minimum of B_r on $[0, 1]$ is equal to

$$M_r - \max(Z_1, Z_2).$$

The joint maximum and minimum for B is dealt with as above, for B_r, when we start with $r = N$.

5 Brownian Meander: Global Methods

Simple computations involving the reflection principle show that the density of $B(1)$ for Brownian motion started at $a > 0$ and restricted to remain positive on $[0, 1]$ is

$$f(x) \stackrel{\text{def}}{=} \frac{\exp(-\frac{(x-a)^2}{2}) - \exp(-\frac{(x+a)^2}{2})}{\sqrt{2\pi}\mathcal{P}\{|N| \le a\}}, \quad x > 0.$$

The limit of this as $a \downarrow 0$ is the Rayleigh density $x\exp(-x^2/2)$, i.e., the density of $\sqrt{2E}$. An easy scaling argument then shows that

$$B^{\text{me}}(t) \stackrel{\mathcal{L}}{=} \sqrt{2tE}.$$

This permits simulation at a single point, but cannot be used for a sequence of points.

A useful property (see Williams 1970 or Imhof 1984) of the Brownian meander permits carefree simulation: a restricted Brownian meander B_r^{me} can be represented as a sum of three independent standard Brownian bridges:

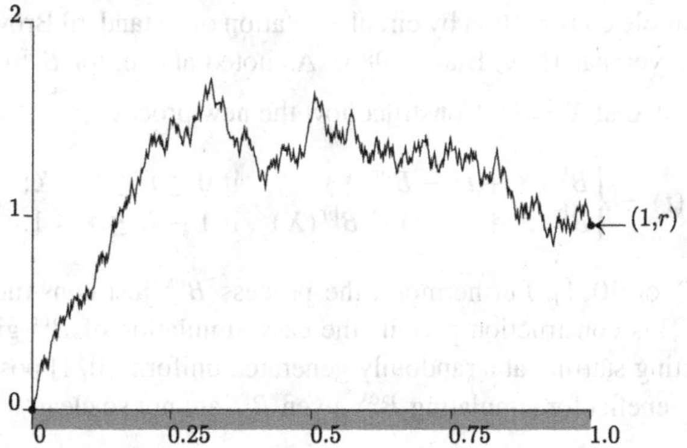

Fig. 4 Simulation of Brownian meander with end value r

$$B_r^{\mathrm{me}}(t) \equiv \sqrt{(rt + B^{\mathrm{br}}(t))^2 + (B^{\mathrm{br}'}(t))^2 + (B^{\mathrm{br}''}(t))^2}.$$

This is called the three-dimensional Bessel bridge from 0 to r. We obtain B^{me} as B_r^{me} with $r = \sqrt{2E}$. B^{me} can also be obtained from the sample path of B directly: let $\tau = \sup\{t \in [0,1] : B(t) = 0\}$. Then

$$B^{\mathrm{me}}(t) \equiv \frac{|B(\tau + t(1-\tau))|}{\sqrt{1-\tau}}, \quad 0 \le t \le 1.$$

This is not very useful for simulating B^{me} though.

For the standard Brownian bridge, $B^{\mathrm{br}}(t) \equiv B(t) - tB(1)$. Maintaining three independent copies of such bridges gives a simple global algorithm for simulating B_r^{me} at values $0 = t_0 < t_1 < \cdots < t_n = 1$, based on the values $B^{\mathrm{br}}(t_i)$.

There is also a way of simulating B_r^{me} inwards, starting at $t = 1$, and then obtaining the values at points $1 = t_0 > t_1 > t_2 > \cdots > 0$. Using $B^{\mathrm{br}}(t) \stackrel{\mathcal{L}}{=} \sqrt{t(1-t)}N$, we have

$$\begin{aligned} B_r^{\mathrm{me}}(t) &\stackrel{\mathcal{L}}{=} \sqrt{(rt + \sqrt{t(1-t)}N)^2 + t(1-t)(N_2^2 + N_3^2)} \\ &\stackrel{\mathcal{L}}{=} \sqrt{(rt + \sqrt{t(1-t)}N)^2 + 2Et(1-t)} \\ &\stackrel{\mathrm{def}}{=} Z(t,r). \end{aligned}$$

So, we have $B(t_0) = B(1) = r$. Then

$$B(t_{n+1}) \stackrel{\mathcal{L}}{=} Z\left(\frac{t_{n+1}}{t_n}, B(t_n)\right), \quad n \ge 0,$$

where the different realizations of $Z(\cdot,\cdot)$ can be generated independently, so that $B(t_n), n \ge 0$ forms a Markov chain imploding towards zero.

For B^{ex}, a simple construction by circular rotation of a standard Brownian bridge B^{br} is possible (Vervaat 1979; Biane 1986). As noted above, for B^{br} on [0, 1], the minimum is located at $X \stackrel{\mathcal{L}}{=} U$. Construct now the new process

$$Y(t) = \begin{cases} B^{br}(X+t) - B^{br}(X) & \text{if } 0 \le t \le 1-X; \\ B^{br}(X+t-1) - B^{br}(X) & \text{if } 1-X \le t \le 1. \end{cases}$$

Then $Y \equiv B^{ex}$ on [0, 1]. Furthermore, the process B^{ex} just constructed is independent of X. This construction permits the easy simulation of B^{br} given B^{ex}, by cutting and pasting starting at a randomly generated uniform [0, 1] position U. But vice versa, the benefits for simulating B^{ex} given B^{br} are not so clear.

6 Brownian Meander: Local Methods

The local simulation problem for Brownian meanders can be summarized as follows: given $a, b \ge 0$, and $B^{me}(0) = a$, $B^{me}(1) = b$, generate the value of $B^{me}(t)$ for given $t \in (0, 1)$ in expected time bounded uniformly over a, b, t. Armed with such a tool, we can continue subdividing intervals at unit expected cost per subdivision. We may need to rescale things. Let us denote by $B^{me}(t; a, b, s)$ the value $B^{me}(t)$ when $0 \le t \le s$, given that $B^{me}(0) = a$, $B^{me}(s) = b$. Then

$$B^{me}(t; a, b, s) \stackrel{\mathcal{L}}{=} \sqrt{s} B^{me}\left(\frac{t}{s}; \frac{a}{\sqrt{s}}, \frac{b}{\sqrt{s}}, 1\right).$$

Random variate generation can be tackled by a variant of the global method if one is willing to store and carry through the values of all three Brownian motions in the three-dimensional Bessel bridge approach. However, if this is not done, and the boundaries of an interval are fixed, then one must revert to a truly local method. This section discusses the simulation of $B^{me}(t; a, b, 1)$. In the remainder of this section, we will write $B^{me}(t)$ instead of $B^{me}(t; a, b, 1)$.

The (known) density of $B^{me}(t)$ can be derived quite easily. We repeat the easy computations because some intermediate results will be used later. Let us start from the well-known representation for Brownian motion $X(t)$ restricted to $X(0) = a$, $X(1) = b$ with $0 \le a < b$:

$$X(t) \stackrel{\mathcal{L}}{=} a + B(t) + t(b - a - B(1)), \quad 0 \le t \le 1.$$

Writing $B(1) = B(t) + B'(1-t)$ (B' being independent of B), and replacing $B(t) = \sqrt{t}N$, $B'(1-t) = \sqrt{1-t}N'$, we have

$$\begin{aligned} X(t) &\stackrel{\mathcal{L}}{=} a + t(b-a) + \sqrt{t}(1-t)N - \sqrt{1-t}tN' \\ &\stackrel{\mathcal{L}}{=} a + t(b-a) + \sqrt{t(1-t)}N. \end{aligned}$$

For the Brownian bridge B_r on [0, 1], we know that

$$M_r \stackrel{\mathcal{L}}{=} \frac{1}{2}\left(r + \sqrt{r^2 + 2E}\right),$$

and thus, the minimum is distributed as

$$\frac{1}{2}\left(r - \sqrt{r^2 + 2E}\right).$$

Since $X(t)$ is just $a + B_r(t)$ with $r = b - a$,

$$\begin{aligned}
&\mathcal{P}\left\{\min_{0 \le t \le 1} X(t) \ge 0\right\} \\
&\quad = \mathcal{P}\left\{a + \frac{1}{2}\left(b - a - \sqrt{(b-a)^2 + 2E}\right) \ge 0\right\} \\
&\quad = \mathcal{P}\left\{\sqrt{(b-a)^2 + 2E} \le a + b\right\} \\
&\quad = \mathcal{P}\{(b-a)^2 + 2E \le (a+b)^2\} \\
&\quad = \mathcal{P}\{E \le 2ab\} \\
&\quad = 1 - \exp(-2ab).
\end{aligned}$$

For $x > 0$, assuming $ab > 0$,

$$\begin{aligned}
&\mathcal{P}\{B^{\text{me}}(t) \in dx\} \\
&\quad = \mathcal{P}\left\{X(t) \in dx \,\middle|\, \min_{0 \le s \le 1} X(s) \ge 0\right\} \\
&\quad = \frac{\mathcal{P}\{X(t) \in dx, \min_{0 \le s \le 1} X(s) \ge 0\}}{\mathcal{P}\{\min_{0 \le s \le 1} X(s) \ge 0\}} \\
&\quad = \frac{\mathcal{P}\{X(t) \in dx\}\mathcal{P}\{\min_{0 \le s \le t} Y(s) \ge 0\}\mathcal{P}\{\min_{t \le s \le 1} Z(s) \ge 0\}}{\mathcal{P}\{\min_{0 \le s \le 1} X(s) \ge 0\}}
\end{aligned}$$

where $Y(s)$ is Brownian motion on $[0, t]$ with endpoint values a, x, and $Z(s)$ is Brownian motion on $[t, 1]$ with endpoint values x, b. The decomposition into a product in the numerator follows from the Markovian nature of Brownian motion. Using scaling, we see that

$$\mathcal{P}\left\{\min_{0 \le s \le t} Y(s) \ge 0\right\} = 1 - \exp(-2(a/\sqrt{t})(x/\sqrt{t})) = 1 - \exp(-2ax/t),$$

and similarly,

$$\mathcal{P}\left\{\min_{t \le s \le 1} Z(s) \ge 0\right\} = 1 - \exp(-2bx/(1-t)).$$

Therefore, putting $\mu = a + t(b - a)$,

$$\mathcal{P}\left\{B^{\text{me}}(t) \in dx\right\} = \mathcal{P}\left\{a + t(b-a) + \sqrt{t(1-t)}N \in dx\right\} \times \frac{(1-\exp(-2ax/t))(1-\exp(-2bx/(1-t)))}{1-\exp(-2ab)}.$$

The density of $B^{\text{me}}(t)$ is

$$f(x) = g(x) \times h(x),$$

where, for $x > 0$,

$$g(x) \stackrel{\text{def}}{=} \frac{1}{\sqrt{2\pi t(1-t)}} \exp\left(-\frac{(x-\mu)^2}{2t(1-t)}\right),$$

$$h(x) \stackrel{\text{def}}{=} \begin{cases} \frac{(1-\exp(-2ax/t))(1-\exp(-2bx/(1-t)))}{1-\exp(-2ab)} & \text{if } ab > 0, \\ \frac{x}{bt}(1-\exp(-2bx/(1-t))) & \text{if } a = 0, b > 0, \\ \frac{x}{a(1-t)}(1-\exp(-2ax/t)) & \text{if } a > 0, b = 0, \\ \frac{2x^2}{t(1-t)} & \text{if } a = b = 0. \end{cases}$$

When $a = 0$ or $b = 0$ or both, the density was obtained by a continuity argument. The case $a = 0, b > 0$ corresponds to Brownian meander started at the origin and ending at b, and the case $a = b = 0$ is just Brownian excursion. In the latter case, the density is

$$\frac{2x^2}{\sqrt{2\pi(t(1-t))^3}} \exp\left(-\frac{x^2}{2t(1-t)}\right), \quad x > 0,$$

which is the density of

$$\sqrt{2t(1-t)G_{3/2}} \stackrel{\mathcal{L}}{=} \sqrt{t(1-t)(N^2+2E)}.$$

More interestingly, we already noted the 3d representation of Brownian meanders, which gives for $a = 0$ the recipe

$$B^{\text{me}}(t) \stackrel{\mathcal{L}}{=} \sqrt{\left(bt + \sqrt{t(1-t)}N\right)^2 + 2Et(1-t)},$$

and, by symmetry, for $b = 0$,

$$B^{\text{me}}(t) \stackrel{\mathcal{L}}{=} \sqrt{\left(a(1-t) + \sqrt{t(1-t)}N\right)^2 + 2Et(1-t)}.$$

We rediscover the special case $a = b = 0$. We do not know a simple generalization of these sampling formulae for $ab > 0$. In the remainder of this section, we therefore develop a uniformly fast rejection method for f.

If $ab \geq 1/2$, we have $1 - \exp(-2ab) \geq 1 - 1/e$, and thus

$$h(x) \le \frac{e}{e-1}.$$

Since g is a truncated normal density, the rejection method is particularly simple:

Repeat Generate U, N. Set $X \leftarrow \mu + \sqrt{t(1-t)}N$
Until $X \ge 0$ and $Ue/(e-1) \le h(X)$
Return X

The expected number of iterations is the integral under the dominating curve, which is $e/(e-1)$.

Consider now the second case, $ab \le 1/2$. Using the general inequality $1-e^{-u} \le u$, and $1-e^{-u} \ge u(1-1/e)$ for $u \le 1$, we have

$$h(x) \le \frac{e}{e-1} \times \frac{x}{bt}(1-\exp(-2bx/(1-t))),$$

where on the right hand side, we recognize the formula for h when $a = 0, b > 0$ discussed above. Thus, using the sampling formula for that case, we obtain a simple rejection algorithm with expected number of iterations again equal to $e/(e-1)$.

Repeat Generate U, N, E. Set $X \leftarrow \sqrt{\left(bt + \sqrt{t(1-t)}N\right)^2 + 2Et(1-t)}$
Until $\frac{e}{e-1}\frac{UX}{bt} \le \frac{1-\exp(-2aX/t)}{1-\exp(-2ab)}$
Return X

7 Brownian Meander: Extrapolation

Given $B^{\mathrm{me}}(t) = a$, we are asked to simulate $B^{\mathrm{me}}(t+s)$. We recall first that Brownian meanders are translation invariant, i.e., $B^{\mathrm{me}}(t; a, b, t'), 0 \le t \le t'$ is equivalent to $B^{\mathrm{me}}(t+s; a, b, t'+s), 0 \le t \le t'$ for all $s > 0$. Also, it is quickly verified that $B^{\mathrm{me}}(t; a, b, t'), 0 \le t \le t'$ is equivalent to Brownian motion on $[0, t']$ starting from a and ending at b, conditional on staying positive (if a or b are zero, then limits must be taken). Finally, scaling is taken care of by noting that given $B^{\mathrm{me}}(t) = a$, $B^{\mathrm{me}}(t+s)$ is distributed as $\sqrt{s}B^{\mathrm{me}}(t+1)$ started at $a/\sqrt{s}$. These remarks show that we need only be concerned with the simulation of Brownian motion $B(1)$ on [0, 1], given $B(0) = a > 0$ and conditional on $\min_{0 \le t \le 1} B(t) > 0$. The case $a = 0$ reduces to standard Brownian meander $B^{\mathrm{me}}(1)$, which we know is distributed as $\sqrt{2E}$.

As we remarked earlier, simple computations involving the reflection principle show that the density of $B(1)$ under the above restrictions is

$$f(x) \stackrel{\text{def}}{=} \frac{\exp(-\frac{(x-a)^2}{2}) - \exp(-\frac{(x+a)^2}{2})}{\sqrt{2\pi}\mathcal{P}\{|N| \le a\}}, \quad x > 0.$$

The limit of this as $a \downarrow 0$ is $x \exp(-x^2/2)$, the density of $\sqrt{2E}$. The distribution function is given by

$$F(x) = 1 - \frac{\Phi(x+a) - \Phi(x-a)}{\Phi(a) - \Phi(-a)},$$

where we recall that Φ is the distribution function of N. This is not immediately helpful for random variate generation. We propose instead the following simple rejection algorithm, which has uniformly bounded expected time.

(Case $a \geq 1$)
  Repeat Generate U uniform $[0, 1]$, N standard normal
    $X \leftarrow a + N$
  Until $X > 0$ and $U \geq \exp(-2aX)$
  Return X
(Case $0 < a \leq 1$)
  Repeat Generate U uniform $[0, 1]$, E exponential
    $X \leftarrow \sqrt{2E/(1-a^2/3)}$
  Until $2aUX \exp(a^2X^2/6) \leq \exp(aX) + \exp(-aX)$
  Return X

For $a \geq 1$, we apply rejection with as dominating curve the first term in the expression of f, which is nothing but the normal density with mean a. The acceptance condition is simply $U \geq \exp(-2aX)$, which we leave as a simple exercise. The probability of rejection is

$$\begin{aligned}\mathcal{P}\{[a+N<0] \cup [U \leq \exp(-2a(a+N))]\} &\leq \mathcal{P}\{N > a\} + \mathcal{E}\{\exp(-2a(a+N))\} \\ &= 1 - \Phi(a) + \exp(-4a^2) \\ &\leq 1 - \Phi(1) + e^{-4} \\ &< 0.18.\end{aligned}$$

This method applies for all a, but as $a \downarrow 0$, we note with disappointment that the rejection probability approaches 1. For $0 < a \leq 1$, we rewrite the numerator in f as

$$\exp\left(-\frac{x^2}{2} - \frac{a^2}{2}\right) \times \left(e^{ax} + e^{-ax}\right),$$

and bound

$$e^{ax} + e^{-ax} \leq 2axe^{(ax)^2/6},$$

which is easily verified by comparing Taylor series on both sides. This explains the rejection condition. Furthermore, the dominating curve is proportional to

$$x \exp\left(-\frac{x^2(1-a^2/3)}{2}\right),$$

which in turn is proportional to the density of $\sqrt{2E/(1-a^2/3)}$. The probability of acceptance is one over the integral of the dominating curve, which is

$$\frac{2ae^{-a^2/2}}{\sqrt{2\pi}\mathcal{P}\{|N| \le a\}(1-a^2/3)} \le \frac{1}{1-a^2/3} \le \frac{3}{2}.$$

Thus, the rejection probability is less than $1/3$. Therefore, the expected time taken by the algorithm above is uniformly bounded over all choices of a.

8 Brownian Meander: Extremes

The maxima related to B^{me} are slightly more complicated to describe:

$$\mathcal{P}\{M^{\text{me}} \le x\} = 1 + 2\sum_{k=1}^{\infty}(-1)^k \exp\left(-k^2x^2/2\right), \quad x > 0$$

(Chung 1975, 1976; Durrett and Iglehart 1977). This is also known as the (scaled) Kolmogorov-Smirnov limit distribution. For this distribution, fast exact algorithms exist (Devroye 1981)—more about this in the last section. Furthermore,

$$\mathcal{P}\{M^{\text{ex}} \le x\} = 1 + 2\sum_{k=1}^{\infty}(1-4k^2x^2)\exp\left(-2k^2x^2\right), \quad x > 0$$

(Chung 1975, 1976; Durrett and Iglehart 1977). This is also called the theta distribution. For this too, we have fast exact methods (Devroye 1997).

The remainder of this section deals with M_r^{me}. Once we can simulate this, we also have

$$\begin{aligned} \left(M^{\text{me}}, B^{\text{me}}(1)\right) &\equiv \left(M_r^{\text{me}}, r\right) \quad \text{with } r = \sqrt{2E}, \\ \left(M^{\text{ex}}, B^{\text{ex}}(1)\right) &\equiv \left(M_r^{\text{me}}, r\right) \quad \text{with } r = 0. \end{aligned}$$

The starting point is the following joint law,

$$\begin{aligned} &\mathcal{P}\{M^{\text{me}} \le x, B^{\text{me}}(1) \le y\} \\ &\quad = \sum_{k=-\infty}^{\infty} \left[\exp\left(-(2kx)^2/2\right) - \exp\left(-(2kx+y)^2/2\right)\right], \quad x \ge y \ge 0, \end{aligned}$$

as obtained by Durrett and Iglehart (1977) and Chung (1976). Straightforward calculations then show

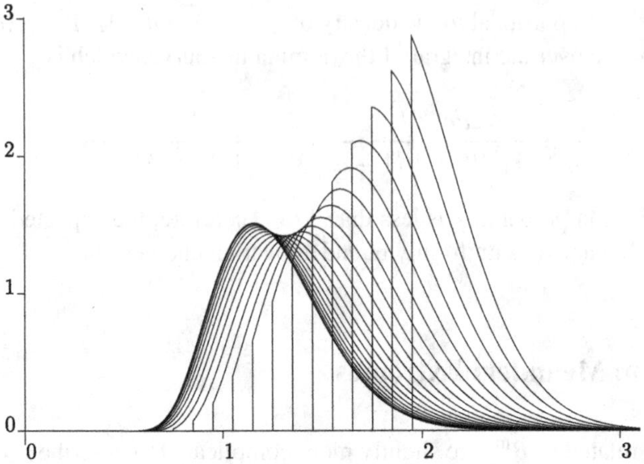

Fig. 5 The density of the maximum of a Brownian meander for r varying from 0.05 to 1.95 in steps of 0.1. Note the accumulation at $r = 0$, which corresponds to the maximum of a Brownian excursion, i.e., the theta distribution

$$\mathcal{P}\{M^{\mathrm{me}} \le x, B^{\mathrm{me}}(1) \in dy\} = \sum_{k=-\infty}^{\infty} (2kx + y) \exp\big(-(2kx + y)^2/2\big) dy,$$

$$x \ge y \ge 0.$$

Because $B^{\mathrm{me}}(1)$ has density $y \exp(-y^2/2)$, we see that the distribution function of M_r^{me} is

$$\sum_{k=-\infty}^{\infty} \frac{2kx + r}{r} \exp\big(r^2/2 - (2kx + r)^2/2\big), \quad x \ge r > 0.$$

Its density is

$$f(x) \stackrel{\mathrm{def}}{=} r^{-1} e^{r^2/2} \sum_{k=-\infty}^{\infty} 2k\big(1 - (2kx + r)^2\big) \exp\big(-(2kx + r)^2/2\big), \quad x \ge r > 0. \tag{4}$$

It helps to rewrite the density (4) of M_r^{me} by grouping the terms:

$$f(x) = \sum_{k=1}^{\infty} f_k(x) \stackrel{\mathrm{def}}{=} \sum_{k=1}^{\infty} 2k e^{-2k^2x^2} g(r, k, x),$$

with

$$\begin{aligned} g(r, k, x) &= \frac{1}{r} \times \Big(\big(1 - (r + 2kx)^2\big) e^{-2kxr} - \big(1 - (r - 2kx)^2\big) e^{2kxr}\Big) \\ &= \frac{1}{r} \times \Big((r^2 + 4k^2x^2 - 1) \sinh(2kxr) - 4kxr \cosh(2kxr)\Big). \end{aligned}$$

The Jacobi theta function

$$\theta(x) = \sum_{n=-\infty}^{\infty} \exp(-n^2\pi x), \quad x > 0,$$

has the remarkable property that $\sqrt{x}\theta(x) = \theta(1/x)$, which follows from the Poisson summation formula, and more particularly from Jacobi's theta function identity

$$\frac{1}{\sqrt{\pi x}} \sum_{n=-\infty}^{\infty} \exp\left(-\frac{(n+y)^2}{x}\right) = \sum_{n=-\infty}^{\infty} \cos(2\pi n y) \exp(-n^2\pi^2 x),$$
$$y \in \mathcal{R},\ x > 0.$$

Taking derivatives with respect to y then shows the identity

$$\frac{1}{\sqrt{\pi x^3}} \sum_{n=-\infty}^{\infty} (n+y) \exp\left(-\frac{(n+y)^2}{x}\right)$$
$$= \sum_{n=-\infty}^{\infty} \pi n \sin(2\pi n y) \exp(-n^2\pi^2 x), \quad y \in \mathcal{R},\ x > 0.$$

A term by term comparison yields the alternative representation

$$\begin{aligned}
&\mathcal{P}\{M^{\text{me}} \le x, B^{\text{me}}(1) \in dy\} \\
&= 2x \sum_{k=-\infty}^{\infty} \left(k + \frac{y}{2x}\right) \exp\left(-\frac{(k+\frac{y}{2x})^2}{1/(2x^2)}\right) dy \\
&= 2x\sqrt{\frac{\pi}{8x^6}} \sum_{n=-\infty}^{\infty} \pi n \sin\left(2\pi n \frac{y}{2x}\right) \exp\left(-\frac{n^2\pi^2}{2x^2}\right) dy \\
&= \sqrt{\frac{\pi}{2x^4}} \sum_{n=-\infty}^{\infty} \pi n \sin\left(2\pi n \frac{y}{2x}\right) \exp\left(-\frac{n^2\pi^2}{2x^2}\right) dy, \quad x \ge y \ge 0.
\end{aligned}$$

The distribution function of M_r^{me} can also be written as

$$F(x) = \sum_{n=1}^{\infty} F_n(x) \sin\left(\frac{\pi n r}{x}\right), \quad x \ge r \ge 0, \tag{5}$$

where

$$F_n(x) = \sqrt{2\pi} x^{-2} r^{-1} e^{r^2/2} \pi n \exp\left(-\frac{n^2\pi^2}{2x^2}\right).$$

This yields the density

$$\begin{aligned} f(x) &= \sum_{n=1}^{\infty} F_n'(x)\sin(\pi nr/x) - \sum_{n=1}^{\infty} F_n(x)(\pi nr/x^2)\cos(\pi nr/x) \\ &= \sum_{n=1}^{\infty} F_n(x)\left(\left(\frac{n^2\pi^2 - 2x^2}{x^3}\right)\sin(\pi nr/x) - \left(\frac{\pi nr}{x^2}\right)\cos(\pi nr/x)\right) \\ &\stackrel{\text{def}}{=} \sum_{n=1}^{\infty} \psi_n(x). \end{aligned} \tag{6}$$

Armed with the dual representations (4) and (5), we develop an algorithm called MAXMEANDER. It is based upon rejection combined with the series method developed by the author in 1981 (see also Devroye 1986). The challenge here is to have an expected time uniformly bounded over all choices of r. For rejection, one should make use of the properties of the family when r approaches its extremes. as $r \downarrow 0$, the figure above suggests that $M_r \stackrel{\mathcal{L}}{\to} M_0$, and that bounds on the density for M_0 should help for small r. As $r \to \infty$, the distribution "escapes to infinity". In fact, $2r(M_r - r) \stackrel{\mathcal{L}}{\to} E$, a fact that follows from our bounds below. Thus, we should look for exponential tail bounds that hug the density tightly near $x = r$. We have two regimes, $r \geq 3/2$, and $r \leq 3/2$.

REGIME I: $r \geq 3/2$.

Lemma 1. *Assume $r \geq 3/2$. For every $K \geq 1$, $x \geq r \geq 3/2$,*

$$\begin{aligned} &-\frac{1}{1-\zeta} \times \frac{4K(1+4Kxr)}{r} \times e^{-2K^2x^2+2Kxr} \\ &\leq \sum_{k=K}^{\infty} f_k(x) \\ &\leq \frac{1}{1-\xi} \times 2K(r + 4K^2x^2/r) \times e^{-2K^2x^2+2Kxr}, \end{aligned}$$

where $\xi = 6.8e^{-9}$ and $\zeta = 2.2e^{-9}$. Next, when $r \leq 3/2$, $x \geq 3/2$, we have

$$-\frac{8K^2xe^{2Kxr-2K^2x^2}}{1-\tau} \leq \sum_{k=K}^{\infty} f_k(x) \leq \frac{164}{9(1-\nu)}K^4x^3e^{2Kxr-2K^2x^2},$$

where $\nu = 16e^{-9}$ and $\tau = 4e^{-9}$.

This leads to an algorithm for $r \geq 3/2$. Note that for this, we need a bound on f. By Lemma 1,

$$f(x) \leq s(x) \stackrel{\text{def}}{=} \frac{2r^2 + 8x^2}{(1-\xi)r} \times e^{-2x^2+2xr}$$

and the upper bound is easily checked to be log-concave for $x \geq r \geq 3/2$. Thus,

$$s(x) \leq s(r) \exp((\log s)'(r)(x-r)), \quad x \geq r,$$

and since $(\log s)'(r) = (8 - 10r^2)/(5r)$, we have

$$f(x) \leq g(x) \stackrel{\text{def}}{=} \frac{10r}{(1-\xi)} \times \exp\left(-\frac{10r^2-8}{5r}(x-r)\right), \quad x \geq r.$$

We have

$$\begin{aligned}\int_r^\infty g(x)dx &= \frac{10r \times 5r}{(1-\xi)(10r^2-8)} \\ &= \frac{5}{(1-\xi)\left(1-8/(10r^2)\right)} \\ &\leq \frac{5}{(1-\xi)(1-32/90)} \\ &< 7.77.\end{aligned}$$

This suggests that using g as a dominating curve for rejection yields an algorithm that is uniformly fast when $r \geq 3/2$. Also, the function g is proportional to the density of $r + cE$ with $c = 5r/(10r^2 - 8)$.

ALGORITHM "MAXMEANDER" (for $r \geq 3/2$)
Repeat
  Generate $X = r + cE$ where $c = 5r/(10r^2 - 8)$
  Generate V uniformly on $[0, 1]$, and set $Y \leftarrow Vg(X) = \frac{10rVe^{-E}}{(1-\xi)}$
  $k \leftarrow 2$, $S \leftarrow f_1(X)$
  Decision $\leftarrow$ "Undecided"
  Repeat If $Y \leq S - \frac{1}{1-\xi} \times \frac{4k(1+4kXr)}{r} \times e^{-2k^2X^2+2kXr}$ then Decision $\leftarrow$ "Accept"
    If $Y \geq S + \frac{2k}{1-\xi}\left(r + \frac{4k^2X^2}{r}\right) \times e^{-2k^2X^2+2kXr}$ then Decision $\leftarrow$ "Reject"
    $S \leftarrow S + f_k(X)$
    $k \leftarrow k + 1$
  Until Decision $\neq$ "Undecided"
Until Decision = "Accept"
Return X

REGIME II: $r \leq 3/2$. The next Lemma provides approximation inequalities for small values of x, thanks to the Jacobi-transformed representation (5).

Lemma 2. *For every $K \geq 1$, $x \leq 3/2$, $r \leq 3/2$,*

$$\left|\sum_{k=K}^{\infty} \psi_k(x)\right| \leq \frac{1}{1-\mu} F_K(x) \times \frac{K^3\pi^3 r}{x^4},$$

where $\mu = 16\exp(-\frac{2\pi^2}{3}) = 0.0222\ldots$.

Consider first $x \le 3/2$. Rejection can be based on the inequality

$$f(x) \le \frac{1}{1-\mu} F_1(x) \left(\frac{\pi^3 r}{x^4}\right)$$
$$= \frac{\sqrt{2\pi}\, e^{r^2/2} \pi^4}{(1-\mu)x^6} \times \exp\left(-\frac{\pi^2}{2x^2}\right)$$
$$\le g(x) \stackrel{\text{def}}{=} \frac{\sqrt{2\pi}\, e^{9/8} \pi^4}{(1-\mu)x^6} \times \exp\left(-\frac{\pi^2}{2x^2}\right). \tag{7}$$

It is remarkable, but not surprising, that the dominating function does not depend upon r. It is uniformly valid over the range. We will apply it over $\mathcal{R}^+$. The random variable $\pi/\sqrt{2G_{5/2}} = \pi/\sqrt{N^2 + 2E_1 + 2E_2}$ has density

$$\frac{\sqrt{2\pi}\,\pi^4}{3x^6} \exp\left(-\frac{\pi^2}{2x^2}\right), \quad x > 0,$$

which is $g(x)/p$ with $p = 3e^{9/8}/(1-\mu)$. Thus, $\int_0^\infty g(x)dx = p$, and rejection is universally efficient.

Consider next $x \ge 3/2$, a situation covered by the inequalities of Lemma 1. Here we first need an upper bound to be able to apply rejection. Once again, an exponential bound is most appropriate. To see this, not that

$$f(x) \le \frac{164}{9(1-\nu)} x^3 e^{2xr - 2x^2}, \quad x \ge 3/2.$$

The upper bound is log-concave in x, and we can apply the exponential tail technique for log-concave densities, which yields the further bound

$$f(x) \le g^*(x) \stackrel{\text{def}}{=} q \times (4-2r)e^{-(4-2r)(x-3/2)}, \quad x \ge 3/2, \tag{8}$$

where

$$q \stackrel{\text{def}}{=} \int_{3/2}^\infty g^*(x)dx = \frac{123 \times e^{3r-9/2}}{2(1-\nu)(4-2r)}.$$

The function g^* is proportional to the density of $3/2 + E/(4-2r)$. We are thus set up to apply rejection with a choice of dominating curves, one having weight p for $x \le 3/2$, and one of weight q for $x \ge 3/2$. The algorithm, which has an expected time uniformly bounded over the range $r \le 3/2$ (since $p+q$ is uniformly bounded) can be summarized as follows:

ALGORITHM "MAXMEANDER" (for $r \leq 3/2$)
Set $p = \frac{3e^{9/8}}{1-\mu}$, $q = \frac{123 \times e^{3r-9/2}}{2(1-\nu)(4-2r)}$.
Repeat Generate U, V uniformly on $[0, 1]$
If $U \leq \frac{p}{p+q}$ then $X \leftarrow \frac{\pi}{\sqrt{N^2+2E_1+2E_2}}$
$Y \leftarrow Vg(X)$ [g is as in (7)]
$k \leftarrow 2$, $S \leftarrow \psi_1(X)$
Decision $\leftarrow$ "Undecided"
Repeat If $X \geq 3/2$ then Decision $\leftarrow$ "Reject"
Set $T \leftarrow \frac{1}{1-\mu} F_k(X) \times \frac{k^3 \pi^3 r}{X^4}$
If $Y \leq S - T$ then Decision $\leftarrow$ "Accept"
If $Y \geq S + T$ then Decision $\leftarrow$ "Reject"
$S \leftarrow S + \psi_k(X)$
$k \leftarrow k + 1$
Until Decision $\neq$ "Undecided"
else $X \leftarrow \frac{3}{2} + \frac{E}{4-2r}$
$Y \leftarrow Vg^*(X)$ [g^* is as in (8)]
$k \leftarrow 2$, $S \leftarrow f_1(X)$
Decision $\leftarrow$ "Undecided"
Repeat If $Y \leq S - \frac{8k^2 X e^{2kXr-2k^2X^2}}{1-\tau}$
then Decision $\leftarrow$ "Accept"
If $Y \geq S + \frac{164}{9(1-\nu)} k^4 X^3 e^{2kXr-2k^2X^2}$
then Decision $\leftarrow$ "Reject"
$S \leftarrow S + f_k(X)$
$k \leftarrow k + 1$
Until Decision $\neq$ "Undecided"
Until Decision = "Accept"
Return X

Extensions. Using the ideas of this section, it is possible to develop a uniformly fast generator for Mme when both endpoints are fixed and nonzero: $B^{me}(0) = a$ and $B^{me}(1) = b$. Majumdar et al. (2008) describe the distributions of the locations of maxima in several constrained Brownian motions, including Brownian meanders. It is also possible to develop uniformly fast exact simulation algorithms for them.

9 Notes on the Kolmogorov-Smirnov and Theta Distributions

The Kolmogorov-Smirnov statistic has the limit distribution function

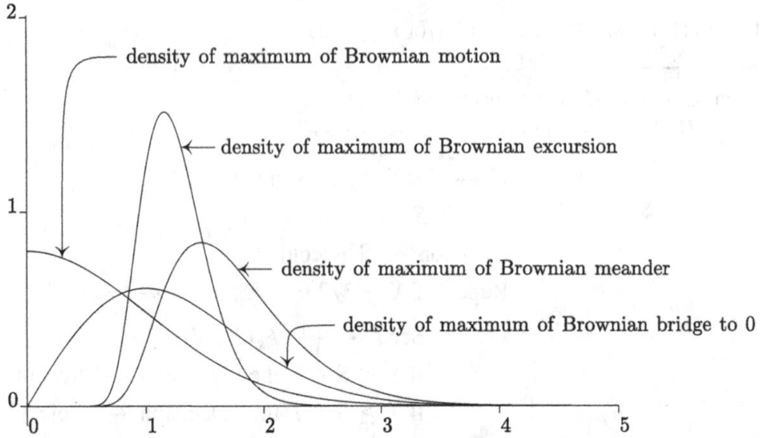

Fig. 6 The four main densities dealt with in this paper. The maximum of Brownian motion has the half-normal density. the maximum of a Brownian excursion has the theta distribution. The maximum of Brownian meander is distributed as $2K$, where K has the Kolmogorov-Smirnov law. Finally, the maximum of a standard Brownian bridge has the Rayleigh density

$$F(x) = \sum_{n=-\infty}^{\infty} (-1)^n e^{-2n^2x^2}, \quad x > 0$$

(Kolmogorov 1933). We call this the Kolmogorov-Smirnov distribution and denote its random variable by K. It is known that

$$2K \stackrel{\mathcal{L}}{=} M^{\text{me}}.$$

Exact random variate generation for the Kolmogorov-Smirnov law was first proposed by Devroye (1981), who used the so-called alternating series method, which is an extension of von Neumann's (1963) rejection method. This method is useful whenever densities can be written as infinite sums,

$$f(x) = \sum_{n=0}^{\infty} (-1)^n a_n(x),$$

where $a_n(x) \geq 0$ and for fixed x, $a_n(x)$ is eventually decreasing in n. Jacobi functions are prime examples of such functions. In the present paper, we proposed an algorithm for M_r^{me} that is uniformly fast over all r, and is thus more general. Replacing r by $\sqrt{E/2}$ yields a method for simulating $2K$.

We say that T is theta distributed if it has distribution function

$$G(x) = \sum_{n=-\infty}^{\infty} \left(1 - 2n^2x^2\right) e^{-n^2x^2}, \quad x > 0.$$

We warn that some authors use a different scaling: we call a random variable with distribution function G a theta random variable, and denote it by T. It appears as the limit law of the height of random conditional Galton-Watson trees (see, e.g., Renyi and Szekeres 1967; de Bruijn et al. 1972; Chung 1975; Kennedy 1976; Meir and Moon 1978; Flajolet and Odlyzko 1982). Furthermore,

$$\frac{T}{\sqrt{2}} \stackrel{\mathcal{L}}{=} M^{\text{ex}}$$

(see, e.g., Pitman and Yor 2001). Devroye (1997) published an exact algorithm for T that uses the principle of a converging series representation for the density. The algorithm presented in this paper for M_r^{me} with $r = 0$ can also be used.

Both T and K are thus directly related to the maxima dealt with in this paper. But they are connected in a number of other ways that are of independent interest. To describe the relationships, we introduce the random variables J and J^* where the symbol J refers to Jacobi. The density of J is

$$f(x) = \frac{d}{dx} \sum_{n=-\infty}^{\infty} (-1)^n \exp\left(-\frac{n^2\pi^2 x}{2}\right) = \sum_{n=1}^{\infty} (-1)^{n+1} n^2\pi^2 \exp\left(-\frac{n^2\pi^2 x}{2}\right).$$

The density of J^* is

$$f^*(x) = \pi \sum_{n=0}^{\infty} (-1)^n \left(n + \frac{1}{2}\right) \exp\left(-\frac{(n+1/2)^2\pi^2 x}{2}\right).$$

We note that all moments are finite, and are expressible in terms of the Riemann zeta function. The properties of these laws are carefully laid out by Biane et al. (2001). Their Laplace transforms are given by

$$\mathcal{E}\left\{e^{-\lambda J}\right\} = \frac{\sqrt{2\lambda}}{\sinh(\sqrt{2\lambda})}, \qquad \mathcal{E}\left\{e^{-\lambda J^*}\right\} = \frac{1}{\cosh(\sqrt{2\lambda})}.$$

Using Euler's formulae

$$\sinh z = z \prod_{n=1}^{\infty} \left(1 + \frac{z^2}{n^2\pi^2}\right), \qquad \cosh z = \prod_{n=1}^{\infty} \left(1 + \frac{z^2}{(n-1/2)^2\pi^2}\right),$$

it is easy to see that J and J^* are indeed positive random variables, and that they have the following representation in terms of i.i.d. standard exponential random variables $E_1, E_2, \ldots$:

$$J \stackrel{\mathcal{L}}{=} \frac{2}{\pi^2} \sum_{n=1}^{\infty} \frac{E_n}{n^2}, \qquad J^* \stackrel{\mathcal{L}}{=} \frac{2}{\pi^2} \sum_{n=1}^{\infty} \frac{E_n}{(n-1/2)^2}.$$

It is known that J^* is the first passage time of Brownian motion started at the origin for absolute value 1, and J is similarly defined for the Bessel process of dimension 3 (which is the square root of the sum of the squares of three independent Brownian motions). See, e.g., Yor (1992, 1997). An exact algorithm for J^* is given by Devroye (2009).

Watson (1961) first observed that $K \stackrel{\mathcal{L}}{=} (\pi/2)\sqrt{J}$, and so we have

$$M^{\text{me}} \stackrel{\mathcal{L}}{=} \pi\sqrt{J} \stackrel{\mathcal{L}}{=} 2K.$$

In addition, Mme is distributed as twice the maximum absolute value of a Brownian bridge on [0, 1] (Durrett et al. 1977; Kennedy 1976; Biane and Yor 1987; Borodin and Salminen 2002).

Let us write $K(1), K(2), \ldots$ for a sequence of i.i.d. copies of a Kolmogorov-Smirnov random variable K. As noted by Biane et al. (2001), the distribution function of the sum $J(1) + J(2)$ of two independent copies of J is given by

$$\sum_{n=-\infty}^{\infty} \left(1 - n^2\pi^2 x\right) e^{-n^2\pi^2 x/2}, \quad x > 0.$$

Thus, we have the distributional identity

$$\frac{\pi^2}{2}(J(1) + J(2)) \stackrel{\mathcal{L}}{=} T^2.$$

Using $J \stackrel{\mathcal{L}}{=} (4/\pi^2)K^2$, we deduce

$$T \stackrel{\mathcal{L}}{=} \sqrt{2(K(1)^2 + K(2)^2)}.$$

This provides a route to the simulation of T via a generator for K.

It is also noteworthy that

$$J \stackrel{\mathcal{L}}{=} \frac{J(1) + J(2)}{(1+U)^2}$$

where U is uniform [0, 1] and independent of the $J(i)$'s (Biane et al. 2001, Sect. 3.3). Thus we have the further identities

$$J \stackrel{\mathcal{L}}{=} \frac{2T^2}{\pi^2(1+U)^2} \stackrel{\mathcal{L}}{=} \frac{4K^2}{\pi^2}.$$

Finally,

$$K \stackrel{\mathcal{L}}{=} \frac{T}{(1+U)\sqrt{2}}.$$

Further properties of K and of maxima of Bessel bridges are given by Pitman and Yor (1999).

Acknowledgements

I would like to thank Tingting Gou and Duncan Murdoch (University of Western Ontario) for allowing me to study their work and for posing the problem that led to the uniformly fast algorithm for the joint location and value of the maximum of a Brownian bridge.

Appendix

Proof (of Lemma 1). We deal with $r \geq 3/2$ first. Clearly, using $x \geq r$,

$$g(r,k,x) \leq (r + 4k^2x^2/r)\exp(2kxr).$$

Define

$$h(r,k,x) = 2k(r + 4k^2x^2/r) \times e^{-2k^2x^2+2kxr}.$$

Also,

$$g(r,k,x) \geq -(2/r)(1+4kxr)\exp(2kxr).$$

Define

$$h^*(r,k,x) = \frac{4k(1+4kxr)}{r} \times e^{-2k^2x^2+2kxr}.$$

We have

$$-h^*(r,k,x) \leq f_k(x) \leq h(r,k,x).$$

For $k \geq K \geq 1$,

$$\begin{aligned}
\frac{h(r,k+1,x)}{h(r,k,x)} &= (1+1/k)\,\frac{1+4(k+1)^2(x/r)^2}{1+4k^2(x/r)^2} \times e^{-(4k+2)x^2+2xr} \\
&\leq 2 \times \frac{1+16}{1+4} \times e^{-6x^2+2xr} \\
&\leq \frac{34}{5} \times e^{-4x^2} \leq \frac{34}{5} \times e^{-4r^2} \leq \frac{34e^{-9}}{5} \stackrel{\text{def}}{=} \xi.
\end{aligned}$$

Therefore,

$$\begin{aligned}
\sum_{k=K}^{\infty} f_k(x) \leq \sum_{k=K}^{\infty} h(r,k,x) &\leq \frac{1}{1-\xi} \times h(r,K,x) \\
&= \frac{1}{1-\xi} \times 2K(r + 4K^2x^2/r) \times e^{-2K^2x^2+2Kxr}.
\end{aligned}$$

Reasoning in a similar way,

$$\begin{aligned}\frac{h^*(r,k+1,x)}{h^*(r,k,x)} &= \left(1+\frac{1}{k}\right)\times\left(1+\frac{4xr}{1+4kxr}\right)\times e^{-(4k+2)x^2+2xr}\\&\le 2\left(\frac{2+4r^2}{1+4r^2}\right)\times e^{-6x^2+2xr}\le 2\times\frac{11}{10}e^{-4x^2}\\&\le 2.2e^{-4r^2}\le 2.2e^{-9}\stackrel{\text{def}}{=}\zeta.\end{aligned}$$

Therefore,

$$\begin{aligned}\sum_{k=K}^{\infty} f_k(x) &\ge -\sum_{k=K}^{\infty} h^*(r,k,x)\ge -\frac{1}{1-\zeta}\times h^*(r,K,x)\\&\ge -\frac{1}{1-\zeta}\times\frac{4K(1+4Kxr)}{r}\times e^{-2K^2x^2+2Kxr}.\end{aligned}$$

Consider next the case $r \le 3/2$ but $x \ge 3/2$. Observe that in this range, $r^2 + 4k^2x^2 - 1 \in [8, 5/4+4k^2x^2] \subseteq [8, (41/9)k^2x^2]$. Also, for $\theta \ge 0$, $\sinh\theta \in [\theta, \theta e^{\theta}]$. Thus,

$$-r_k(x) \stackrel{\text{def}}{=} -8k^2xe^{2kxr-2k^2x^2} \le f_k(x) \le \frac{4\times 41}{9}k^4x^3e^{2kxr-2k^2x^2} \stackrel{\text{def}}{=} R_k(x).$$

For $k \ge 1$,

$$\frac{R_{k+1}(x)}{R_k(x)} = (1+1/k)^4e^{2xr-2(2k+1)x^2} \le 16e^{2xr-6x^2} \le 16e^{3r-27/2} \le 16e^{-9} \stackrel{\text{def}}{=} \nu.$$

Thus, $\sum_{k\ge K} f_k(x) \le R_K(x)/(1-\nu)$ for all $K \ge 1$. Similarly,

$$\frac{r_{k+1}(x)}{r_k(x)} = (1+1/k)^2e^{2xr-2(2k+1)x^2} \le 4e^{-9} \stackrel{\text{def}}{=} \tau.$$

Thus, $\sum_{k\ge K} f_k(x) \ge -r_K(x)/(1-\tau)$ for all $K \ge 1$.

Proof (of Lemma 2). For the first part, assume $x \le 3/2$, and let ψ_k and F_k be as defined in (6). For $x \le \pi/\sqrt{2}$ (which is $\ge \Delta$), using $|\sin x| \le |x|$, and, for $\alpha, \beta, \theta \ge 0$, $|\alpha\sin\theta - \beta\cos\theta| \le \alpha\theta + \beta$,

$$\psi_k(x) \le F_k(x)\left(\frac{k^3\pi^3r}{x^4} - \frac{\pi kr}{x^2}\right) \le F_k(x)\left(\frac{k^3\pi^3r}{x^4}\right) \stackrel{\text{def}}{=} H_k(x),$$

and $\psi_k(x) \ge -H_k(x)$. For $k \ge 1$,

$$\frac{H_{k+1}(x)}{H_k(x)} = \left(1+\frac{1}{k}\right)^4 e^{-\frac{(2k+1)\pi^2}{2x^2}} \le 16e^{-\frac{3\pi^2}{2x^2}} \le 16e^{-\frac{2\pi^2}{3}} \stackrel{\text{def}}{=} \mu.$$

We conclude that

$$\sum_{k=K}^{\infty} \psi_k(x) \le \sum_{k=K}^{\infty} H_k(x) \le \frac{1}{1-\mu} H_K(x).$$

Similarly, on the bottom side,

$$\sum_{k=K}^{\infty} \psi_k(x) \ge -\sum_{k=K}^{\infty} H_k(x) \ge -\frac{1}{1-\mu} H_K(x).$$

Further references relevant to the material in this paper include Asmussen et al. (1995), Bertoin et al. (1999), Bonaccorsi and Zambotti (2004), Calvin (2004), Ciesielski and Taylor (1962), Fujita and Yor (2007), Lévy (1948), Pitman and Yor (2003), Revuz and Yor (1991), and Zambotti (2003).

References

Asmussen, S., Glynn, P., Pitman, J.: Discretization error in simulation of one-dimensional reflecting Brownian motion. Ann. Appl. Probab. **5**, 875–896 (1995)

Bertoin, J., Pitman, J.: Path transformations connecting Brownian bridge, excursion and meander. Bull. Sci. Math. **2**(118), 147–166 (1994)

Bertoin, J., Pitman, J., Ruiz de Chavez, J.: Constructions of a Brownian path with a given minimum. Electron. Commun. Probab. **4**, 31–37 (1999)

Beskos, A., Roberts, G.O.: Exact simulation of diffusions. Ann. Appl. Probab. **15**, 2422–2444 (2005)

Biane, Ph.: Relations entre pont et excursion du mouvement Brownien. Ann. Inst. H. Poincaré **22**, 1–7 (1986)

Biane, P., Yor, M.: Valeurs principales associées aux temps locaux Browniens. Bull. Sci. Math. **111**, 23–101 (1987)

Biane, P., Pitman, J., Yor, M.: Probability laws related to the Jacobi theta and Riemann zeta functions, and Brownian excursions. Bull. Am. Math. Soc. **38**, 435–465 (2001)

Bonaccorsi, S., Zambotti, L.: Integration by parts on the Brownian meander. Proc. Am. Math. Soc. **132**, 875–883 (2004)

Borodin, A.N., Salminen, P.: Handbook of Brownian Motion: Facts and Formulae. Birkhäuser, Basel (2002)

Calvin, J.M.: Efficient simulation for discrete path-dependent option pricing. In: Peters, B.A., Smith, J.S., Medeiros, D.J., Rohrer, M.W. (eds.) Proceedings of the 33rd Winter Simulation Conference, pp. 325–328. IEEE Computer Society, Washington (2001)

Calvin, J.M.: Simulation output analysis based on excursions. In: Proceedings of the 36th Winter Simulation Conference, pp. 681–684. IEEE Computer Society, Washington (2004)

Chung, K.L.: Maxima in Brownian excursions. Bull. Am. Math. Soc. **81**, 742–744 (1975)

Chung, K.L.: Excursions in Brownian motion. Ark. Mat. 155–177 (1976)

Ciesielski, Z., Taylor, S.J.: First passage times and sojourn density for Brownian motion in space and the exact Hausdorff measure of the sample path. Trans. Am. Math. Soc. **103**, 434–450 (1962)

de Bruijn, N.G., Knuth, D.E., Rice, S.O.: The average height of planted plane trees. In: Read, R.-C. (ed.) Graph Theory and Computing, pp. 15–22. Academic Press, New York (1972)

Denisov, I.V.: A random walk and a Wiener process near a maximum. Theory Probab. Appl. **28**, 821–824 (1984)

Devroye, L.: The series method in random variate generation and its application to the Kolmogorov-Smirnov distribution. Am. J. Math. Manag. Sci. **1**, 359–379 (1981)

Devroye, L.: Non-Uniform Random Variate Generation. Springer, New York (1986)
Devroye, L.: Simulating theta random variates. Stat. Probab. Lett. **31**, 2785–2791 (1997)
Devroye, L.: On exact simulation algorithms for some distributions related to Jacobi theta functions. Stat. Probab. Lett. **29**, 2251–2259 (2009)
Duffie, D., Glynn, P.: Efficient Monte Carlo simulation of security prices. Ann. Appl. Probab. **5**, 897–905 (1995)
Durrett, R., Iglehart, D.L.: Functionals of Brownian meander and Brownian excursion. Ann. Probab. **5**, 130–135 (1977)
Durrett, R., Iglehart, D.L., Miller, D.R.: Weak convergence to Brownian meander and Brownian excursion. Ann. Probab. **5**, 117–129 (1977)
Flajolet, P., Odlyzko, A.: The average height of binary trees and other simple trees. J. Comput. Syst. Sci. **25**, 171–213 (1982)
Fox, B.L.: Strategies for Quasi-Monte Carlo. Kluwer, Norwell (1999)
Fujita, T., Yor, M.: On the remarkable distributions of maxima of some fragments of the standard reflecting random walk and Brownian motion. Probab. Math. Stat. **27**, 89–104 (2007)
Gou, T.: Two recursive sampling methods. PhD dissertation, The University of Western Ontario (2009)
Imhof, J.P.: Density factorizations for Brownian motion, meander and the three-dimensional Bessel process, and applications. J. Appl. Probab. **21**, 500–510 (1984)
Karatzas, I., Shreve, Ş.E.: Brownian Motion and Stochastic Calculus, 2nd edn. Springer, New York (1998)
Kennedy, D.P.: The distribution of the maximum Brownian excursion. J. Appl. Probab. **13**, 371–376 (1976)
Kloeden, P.E., Platen, E.: Numerical Solution of Stochastic Differential Equations. Springer, New York (1992)
Kolmogorov, A.N.: Sulla determinazione empirica delle leggi di probabilita. G. Ist. Ital. Attuari **4**, 1–11 (1933)
Lévy, P.: Sur certains processus stochastiques homogènes. Compos. Math. **7**, 283–339 (1939)
Lévy, P.: Processus Stochastiques et Mouvement Brownien. Gauthier-Villars, Paris (1948)
Majumdar, S.N., Randon-Furling, J., Kearney, M.J., Yor, M.: On the time to reach maximum for a variety of constrained Brownian motions. J. Phys. A, Math. Theor. **41** (2008)
McLeish, D.L.: Highs and lows: Some properties of the extremes of a diffusion and applications in finance. Can. J. Stat. **30**, 243–267 (2002)
Meir, A., Moon, J.W.: On the altitude of nodes in random trees. Can. J. Math. **30**, 997–1015 (1978)
Pitman, J.: Brownian motion, bridge, excursion, and meander characterized by sampling at independent uniform times. Electron. J. Probab. **4**(11), 1–33 (1999)
Pitman, J., Yor, M.: The law of the maximum of a Bessel bridge. Electron. J. Probab. **4**, 1–15 (1999)
Pitman, J., Yor, M.: On the distribution of ranked heights of excursions of a Brownian bridge. Ann. Probab. **29**, 361–384 (2001)
Pitman, J., Yor, M.: Infinitely divisible laws associated with hyperbolic functions. Can. J. Math. **53**, 292–330 (2003)
Renyi, A., Szekeres, G.: On the height of trees. J. Aust. Math. Soc. **7**, 497–507 (1967)
Revuz, D., Yor, M.: Continuous Martingales and Brownian Motion. Grundlehren der mathematischen Wissenschaften, vol. 293. Springer, Berlin (1991)
Shepp, L.A.: The joint density of the maximum and its location for a Wiener process with drift. J. Appl. Probab. **16**, 423–427 (1979)
Shorack, G.R., Wellner, J.A.: Empirical Processes with Applications to Statistics. Wiley, New York (1986)
Vervaat, W.: A relation between Brownian bridge and Brownian excursion. Ann. Probab. **7**, 143–149 (1979)
von Neumann, J.: Various techniques used in connection with random digits. Collect. Works **5**, 768–770 (1963). Also in Monte Carlo Method, National Bureau of Standards Series, vol. 12, pp. 36–38 (1951)

Watson, G.S.: Goodness-of-fit tests on a circle. Biometrika **48**, 109–114 (1961)
Williams, D.: Decomposing the Brownian path. Bull. Am. Math. Soc. **76**, 871–873 (1970)
Williams, D.: Path decomposition and continuity of local time for one dimensional diffusions I. Proc. Lond. Math. Soc. **28**, 738–768 (1974)
Yor, M.: Some Aspects of Brownian Motion, Part I: Some Special Functionals. Birkhäuser, Basel (1992)
Yor, M.: Some Aspects of Brownian Motion, Part II: Some Recent Martingale Problems. Birkhäuser, Basel (1997)
Zambotti, L.: Integration by parts on δ-Bessel bridges, $\delta > 3$, and related SPDEs. Ann. Probab. **31**, 323–348 (2003)

A Review on Regression-based Monte Carlo Methods for Pricing American Options

Michael Kohler

Abstract In this article we give a review of regression-based Monte Carlo methods for pricing American options. The methods require in a first step that the generally in continuous time formulated pricing problem is approximated by a problem in discrete time, i.e., the number of exercising times of the considered option is assumed to be finite. Then the problem can be formulated as an optimal stopping problem in discrete time, where the optimal stopping time can be expressed by the aid of so-called continuation values. These continuation values represent the price of the option given that the option is exercised after time t conditioned on the value of the price process at time t. The continuation values can be expressed as regression functions, and regression-based Monte Carlo methods apply regression estimates to data generated by the aid of artificial generated paths of the price process in order to approximate these conditional expectations. In this article we describe various methods and corresponding results for estimation of these regression functions.

1 Pricing of American Options as Optimal Stopping Problem

In many financial contracts it is allowed to exercise the contract early before expiry. E.g., many exchange traded options are of American type and allow the holder any exercise date before expiry, mortgages have often embedded prepayment options such that the mortgage can be amortized or repayed, or life insurance contracts allow often for early surrender. In this article we are interested in pricing of options with early exercise features.

It is well-known that in complete and arbitrage free markets the price of a derivative security can be represented as an expected value with respect to the so called martingale measure, see for instance Karatzas and Shreve (1998). Furthermore, the

---

Michael Kohler, Department of Mathematics, Technische Universität Darmstadt, Schloßgartenstraße 7, 64289 Darmstadt, Germany
e-mail: kohler@mathematik.tu-darmstadt.de

L. Devroye et al. (eds.), *Recent Developments in Applied Probability and Statistics*,
DOI 10.1007/978-3-7908-2598-5_2,

price of an American option with maturity T is given by the value of the optimal stopping problem

$$V_0 = \sup_{\tau \in \mathcal{T}([0,T])} \mathbf{E}\left\{d_{0,\tau} g_\tau(X_\tau)\right\}, \tag{1}$$

where g_t is a nonnegative payoff function, $(X_t)_{0 \le t \le T}$ is a stochastic process, which models the relevant risk factors, $\mathcal{T}([0, T])$ is the class of all stopping times with values in $[0, T]$, and $d_{s,t}$ are nonnegative $\mathcal{F}((X_u)_{s \le u \le t})$-measurable discount factors satisfying $d_{0,t} = d_{0,s} \cdot d_{s,t}$ for $s < t$. Here, a stopping time $\tau \in \mathcal{T}([0, T])$ is a measurable function of $(X_t)_{0 \le t \le T}$ with values in $[0, T]$ with the property that for any $r \in [0, T]$ the event $\{\tau \le r\}$ is contained in the sigma algebra $\mathcal{F}_r = \mathcal{F}((X_s)_{0 \le s \le r})$ generated by $(X_s)_{0 \le s \le r}$.

There are various possibilities for the choice of the process $(X_t)_{0 \le t \le T}$. The most simple examples are geometric Brownian motions, as for instance in the celebrated Black-Scholes setting. More general models include stochastic volatility models, jump-diffusion processes or general Levy processes. The model parameters are usually calibrated to observed time series data.

The first step in addressing the numerical solution of (1) is to pass from continuous time to discrete time, which means in financial terms to approximate the American option by a so-called Bermudan option. The convergence of the discrete time approximations to the continuous time optimal stopping problem is considered in Lamberton and Pagès (1990) for the Markovian case but also in the abstract setting of general stochastic processes.

For simplicity we restrict ourselves directly to a discrete time scale and consider exclusively Bermudan options. In analogy to (1), the price of a Bermudan option is the value of the discrete time optimal stopping problem

$$V_0 = \sup_{\tau \in \mathcal{T}(0,\ldots,T)} \mathbf{E}\left\{f_\tau(X_\tau)\right\}, \tag{2}$$

where $X_0, X_1, \ldots, X_T$ is now a discrete time stochastic process, f_t is the discounted payoff function, i.e., $f_t(x) = d_{0,t} g_t(x)$, and $\mathcal{T}(0, \ldots, T)$ is the class of all $\{0, \ldots, T\}$-valued stopping times. Here a stopping time $\tau \in \mathcal{T}(0, \ldots, T)$ is a measurable function of $X_0, \ldots, X_T$ with the property that for any $k \in \{0, \ldots, T\}$ the event $\{\tau = k\}$ is contained in the sigma algebra $\mathcal{F}(X_0, \ldots, X_k)$ generated by $X_0, \ldots, X_k$.

2 The Optimal Stopping Time

In the sequel we assume that $X_0, X_1, \ldots, X_T$ is a $\mathbb{R}^d$-valued Markov process recording all necessary information about financial variables including prices of the underlying assets as well as additional risk factors driving stochastic volatility or stochastic interest rates. Neither the Markov property nor the form of the payoff as a function of the state X_t are very restrictive and can often be achieved by including supplementary variables.

The computation of (2) can be done by determination of an optimal stopping time $\tau^* \in \mathcal{T}(0, \ldots, T)$ satisfying

$$V_0 = \sup_{\tau \in \mathcal{T}(0,\ldots,T)} \mathbf{E}\{f_\tau(X_\tau)\} = \mathbf{E}\{f_{\tau^*}(X_{\tau^*})\}. \tag{3}$$

For $0 \leq t < T$ let

$$q_t(x) = \sup_{\tau \in \mathcal{T}(t+1,\ldots,T)} \mathbf{E}\{f_\tau(X_\tau)|X_t = x\} \tag{4}$$

be the so-called continuation value describing the value of the option at time t given $X_t = x$ and subject to the constraint of holding the option at time t rather than exercising it. For $t = T$ we define the corresponding continuation value by

$$q_T(x) = 0 \quad (x \in \mathbb{R}^d), \tag{5}$$

because the option expires at time T and hence we do not get any money if we sell it after time T.

In the sequel we will use techniques from the general theory of optimal stopping (cf., e.g., Chow et al. 1971 or Shiryayev 1978) in order to show that the optimal stopping time τ^* is given by

$$\tau^* = \inf\{s \in \{0, 1, \ldots, T\} : q_s(X_s) \leq f_s(X_s)\}. \tag{6}$$

Since $q_T(x) = 0$ and $f_T(x) \geq 0$ there exists always some index where $q_s(X_s) \leq f_s(X_s)$, so the right-hand side above is indeed well defined. The above form of τ^* allows a very nice interpretation: in order to sell the option in an optimal way, we have to sell it as soon as the value we get if we sell it immediately is at least as large as the value we get in the mean in the future, if we sell it in the future in an optimal way.

In order to prove (6) we need the following notations: Let $\mathcal{T}(t, t+1, \ldots, T)$ be the subset of $\mathcal{T}(0, \ldots, T)$ consisting of all stopping times which take on values only in $\{t, t+1, \ldots, T\}$ and let

$$V_t(x) = \sup_{\tau \in \mathcal{T}(t,t+1,\ldots,T)} \mathbf{E}\{f_\tau(X_\tau)\big|X_t = x\} \tag{7}$$

be the so-called value function which describes the value we get in the mean if we sell the option in an optimal way after time $t-1$ given $X_t = x$. For $t \in \{-1, 0, \ldots, T-1\}$ set

$$\tau_t^* = \inf\{s \geq t+1 : q_s(X_s) \leq f_s(X_s)\}, \tag{8}$$

hence $\tau^* = \tau_{-1}^*$. Then the following result holds:

Theorem 1. *Under the above assumptions we have for any $t \in \{-1, 0, \ldots, T\}$ and $\mathbf{P}_{X_t}$-almost all $x \in \mathbb{R}^d$:*

$$V_t(x) = \mathbf{E}\left\{ f_{\tau_{t-1}^*}(X_{\tau_{t-1}^*}) \middle| X_t = x \right\}. \tag{9}$$

Furthermore we have

$$V_0 = \mathbf{E}\{f_{\tau^*}(X_{\tau^*})\}. \tag{10}$$

The above theorem is well-known in literature (cf., e.g., Chap. 8 in Glasserman 2004), but usually not proven completely. For the sake of completeness we present a complete proof next.

Proof. We prove (9) by induction. For $t = T$ we have

$$\tau_{T-1}^* = T$$

and any $\tau \in \mathcal{T}(T)$ satisfies

$$\tau = T.$$

So in this case we have

$$\begin{aligned} V_T(x) &= \sup_{\tau \in \mathcal{T}(T)} \mathbf{E}\left\{ f_\tau(X_\tau) \middle| X_T = x \right\} = \mathbf{E}\left\{ f_T(X_T) \middle| X_T = x \right\} \\ &= \mathbf{E}\left\{ f_{\tau_{T-1}^*}(X_{\tau_{T-1}^*}) \middle| X_T = x \right\}. \end{aligned}$$

Let $t \in \{0, \ldots, T-1\}$ and assume that

$$V_s(x) = \mathbf{E}\left\{ f_{\tau_{s-1}^*}(X_{\tau_{s-1}^*}) \middle| X_s = x \right\}$$

holds for all $t < s \leq T$. In the sequel we prove (9). To do this, let $\tau \in \mathcal{T}(t, \ldots, T)$ be arbitrary. Then

$$\begin{aligned} f_\tau(X_\tau) &= f_\tau(X_\tau) \cdot 1_{\{\tau=t\}} + f_\tau(X_\tau) \cdot 1_{\{\tau>t\}} \\ &= f_t(X_t) \cdot 1_{\{\tau=t\}} + f_{\max\{\tau,t+1\}}(X_{\max\{\tau,t+1\}}) \cdot 1_{\{\tau>t\}}. \end{aligned}$$

Since $1_{\{\tau=t\}}$ and $1_{\{\tau>t\}} = 1 - 1_{\{\tau\leq t\}}$ are measurable with respect to $X_0, \ldots, X_t$ and since $(X_t)_{0\leq t\leq T}$ is a Markov process we have

$$\begin{aligned} &\mathbf{E}\{f_\tau(X_\tau)|X_t\} \\ &\quad = \mathbf{E}\{f_t(X_t) \cdot 1_{\{\tau=t\}}|X_0, \ldots, X_t\} \\ &\qquad + \mathbf{E}\{f_{\max\{\tau,t+1\}}(X_{\max\{\tau,t+1\}}) \cdot 1_{\{\tau>t\}}|X_0, \ldots, X_t\} \\ &\quad = f_t(X_t) \cdot 1_{\{\tau=t\}} + 1_{\{\tau>t\}} \cdot \mathbf{E}\{f_{\max\{\tau,t+1\}}(X_{\max\{\tau,t+1\}})|X_0, \ldots, X_t\} \\ &\quad = f_t(X_t) \cdot 1_{\{\tau=t\}} + 1_{\{\tau>t\}} \cdot \mathbf{E}\{f_{\max\{\tau,t+1\}}(X_{\max\{\tau,t+1\}})|X_t\}. \end{aligned}$$

Using the definition of V_{t+1} together with $\max\{\tau, t+1\} \in \mathcal{T}(t+1, \ldots, T)$ and the Markov property we get

$$\begin{aligned} \mathbf{E}\{f_{\max\{\tau,t+1\}}(X_{\max\{\tau,t+1\}})|X_t\} &= \mathbf{E}\{\mathbf{E}\{f_{\max\{\tau,t+1\}}(X_{\max\{\tau,t+1\}})|X_{t+1}\}|X_t\} \\ &\leq \mathbf{E}\{V_{t+1}(X_{t+1})|X_t\}, \end{aligned}$$

from which we can conclude

$$\begin{aligned}\mathbf{E}\{f_\tau(X_\tau)|X_t\} &\le f_t(X_t)\cdot 1_{\{\tau=t\}} + 1_{\{\tau>t\}}\cdot \mathbf{E}\{V_{t+1}(X_{t+1})|X_t\}\\ &\le \max\{f_t(X_t), \mathbf{E}\{V_{t+1}(X_{t+1})|X_t\}\}.\end{aligned}$$

Now we make the same calculations using $\tau = \tau^*_{t-1}$. We get

$$\begin{aligned}&\mathbf{E}\{f_{\tau^*_{t-1}}(X_{\tau^*_{t-1}})|X_t\}\\ &= f_t(X_t)\cdot 1_{\{\tau^*_{t-1}=t\}} + 1_{\{\tau^*_{t-1}>t\}}\cdot \mathbf{E}\{f_{\max\{\tau^*_{t-1},t+1\}}(X_{\max\{\tau^*_{t-1},t+1\}})|X_t\}.\end{aligned}$$

By definition of τ^*_t we have on $\{\tau^*_{t-1} > t\}$

$$\max\{\tau^*_{t-1}, t+1\} = \tau^*_t.$$

Using this, the Markov property and the induction hypothesis we can conclude

$$\begin{aligned}\mathbf{E}\{f_{\tau^*_{t-1}}(X_{\tau^*_{t-1}})|X_t\} &= f_t(X_t)\cdot 1_{\{\tau^*_{t-1}=t\}} + 1_{\{\tau^*_{t-1}>t\}}\cdot \mathbf{E}\{\mathbf{E}\{f_{\tau^*_t}(X_{\tau^*_t})|X_{t+1}\}|X_t\}\\ &= f_t(X_t)\cdot 1_{\{\tau^*_{t-1}=t\}} + 1_{\{\tau^*_{t-1}>t\}}\cdot \mathbf{E}\{V_{t+1}(X_{t+1})|X_t\}.\end{aligned}$$

Next we show

$$\mathbf{E}\{V_{t+1}(X_{t+1})|X_t\} = q_t(X_t). \tag{11}$$

To see this, we observe that by induction hypothesis, Markov property and because of $\tau^*_t \in \mathcal{T}(t+1,\ldots,T)$ we have

$$\begin{aligned}\mathbf{E}\{V_{t+1}(X_{t+1})|X_t\} &= \mathbf{E}\{\mathbf{E}\{f_{\tau^*_t}(X_{\tau^*_t})|X_{t+1}\}|X_t\} = \mathbf{E}\{f_{\tau^*_t}(X_{\tau^*_t})|X_t\}\\ &\le \sup_{\tau\in\mathcal{T}(t+1,\ldots,T)} \mathbf{E}\{f_\tau(X_\tau)|X_t\} = q_t(X_t).\end{aligned}$$

Furthermore the definition of V_{t+1} implies

$$\begin{aligned}\mathbf{E}\{V_{t+1}(X_{t+1})|X_t\} &= \mathbf{E}\left\{\sup_{\tau\in\mathcal{T}(t+1,\ldots,T)} \mathbf{E}\{f_\tau(X_\tau)|X_{t+1}\}|X_t\right\}\\ &\ge \sup_{\tau\in\mathcal{T}(t+1,\ldots,T)} \mathbf{E}\{\mathbf{E}\{f_\tau(X_\tau)|X_{t+1}\}|X_t\} = q_t(X_t).\end{aligned}$$

Using the definition of τ^*_{t-1} we conclude

$$\begin{aligned}&f_t(X_t)\cdot 1_{\{\tau^*_{t-1}=t\}} + 1_{\{\tau^*_{t-1}>t\}}\cdot \mathbf{E}\{V_{t+1}(X_{t+1})|X_t\}\\ &= f_t(X_t)\cdot 1_{\{\tau^*_{t-1}=t\}} + 1_{\{\tau^*_{t-1}>t\}}\cdot q_t(X_t)\\ &= \max\{f_t(X_t), q_t(X_t)\}.\end{aligned}$$

Summarizing the above results we have

$$V_t(x) = \sup_{\tau \in \mathcal{T}(t,t+1,\dots,T)} \mathbf{E}\left\{ f_\tau(X_\tau) \middle| X_t = x \right\} \le \max\{f_t(x), \mathbf{E}\{V_{t+1}(X_{t+1})|X_t = x\}\}$$
$$= \max\{f_t(x), q_t(x)\} = \mathbf{E}\{f_{\tau^*_{t-1}}(X_{\tau^*_{t-1}})|X_t = x\},$$

which proves

$$V_t(x) = \max\{f_t(x), q_t(x)\} = \mathbf{E}\{f_{\tau^*_{t-1}}(X_{\tau^*_{t-1}})|X_t = x\}. \tag{12}$$

In order to prove (10) we observe that by arguing as above we get

$$\begin{aligned}
V_0 &= \sup_{\tau \in \mathcal{T}(0,\dots,T)} \mathbf{E}\{f_\tau(X_\tau)\} \\
&= \sup_{\tau \in \mathcal{T}(0,\dots,T)} \mathbf{E}\left\{ f_0(X_0) \cdot 1_{\{\tau=0\}} + f_{\max\{\tau,1\}}(X_{\max\{\tau,1\}}) \cdot 1_{\{\tau>0\}} \right\} \\
&= \mathbf{E}\left\{ f_0(X_0) \cdot 1_{\{f_0(X_0) \ge q_0(X_0)\}} + f_{\tau_0^*}(X_{\tau_0^*}) \cdot 1_{\{f_0(X_0) < q_0(X_0)\}} \right\} \\
&= \mathbf{E}\left\{ f_0(X_0) \cdot 1_{\{f_0(X_0) \ge q_0(X_0)\}} + \mathbf{E}\{V_1(X_1)|X_0\} \cdot 1_{\{f_0(X_0) < q_0(X_0)\}} \right\} \\
&= \mathbf{E}\left\{ f_0(X_0) \cdot 1_{\{f_0(X_0) \ge q_0(X_0)\}} + q_0(X_0) \cdot 1_{\{f_0(X_0) < q_0(X_0)\}} \right\} \\
&= \mathbf{E}\{\max\{f_0(X_0), q_0(X_0)\}\} \\
&= \mathbf{E}\{f_{\tau^*}(X_{\tau^*})\}.
\end{aligned}$$
□

Remark 1. The continuation values and the value function are closely related. As we have seen already in the proof of Theorem 1 (cf., (11) and (12)) we have

$$q_t(x) = \mathbf{E}\{V_{t+1}(X_{t+1})|X_t = x\}$$

and

$$V_t(x) = \max\{f_t(x), q_t(x)\}.$$

Remark 2. Remark 1 shows that $q_s(X_s) \le f_s(X_s)$ is equivalent to $V_s(X_s) \le f_s(X_s)$. Hence the optimal stopping time can be also expressed via

$$\tau^* = \inf\{s \in \{0, \dots, T\} : V_s(X_s) \le f_s(X_s)\}. \tag{13}$$

3 Regression Representations for Continuation Values

The previous section shows that it suffices to determine the continuation values $q_0, \dots, q_{T-1}$ in order to determine the optimal stopping time. We show in our next theorem three different regression representations for q_t, which have been introduced in Longstaff and Schwartz (2001), Tsitsiklis and Van Roy (1999) and Egloff (2005), resp. In principle they allow a direct (and sometimes recursive) computation of the continuation values by computing conditional expectations.

Theorem 2. *Under the above assumptions for any $t \in \{0, \dots, T-1\}$ and $\mathbf{P}_{X_t}$-almost all $x \in \mathbb{R}^d$ the following relations hold:*

(a)

$$q_t(x) = \mathbf{E}\left\{ f_{\tau_t^*}(X_{\tau_t^*}) \middle| X_t = x \right\}, \tag{14}$$

(b)

$$q_t(x) = \mathbf{E}\left\{ \max\left\{ f_{t+1}(X_{t+1}), q_{t+1}(X_{t+1}) \right\} \middle| X_t = x \right\} \tag{15}$$

(c)

$$q_t(x) = \mathbf{E}\left\{ \Theta_{t+1,t+w+1}^{(w)} \middle| X_t = x \right\} \tag{16}$$

for any $w \in \{0, 1, \dots, T-t-1\}$, *where*

$$\begin{aligned}
&\Theta_{t+1,t+w+1}^{(w)} \\
&= \sum_{s=t+1}^{t+w+1} f_s(X_s) \cdot 1_{\{f_{t+1}(X_{t+1}) < q_{t+1}(X_{t+1}), \dots, f_{s-1}(X_{s-1}) < q_{s-1}(X_{s-1}), f_s(X_s) \geq q_s(X_s)\}} \\
&\quad + q_{t+w+1}(X_{t+w+1}) \cdot 1_{\{f_{t+1}(X_{t+1}) < q_{t+1}(X_{t+1}), \dots, f_{t+w+1}(X_{t+w+1}) < q_{t+w+1}(X_{t+w+1})\}}.
\end{aligned}$$

Proof. (a) By (11), Theorem 1 and Markov property we get

$$\begin{aligned}
q_t(X_t) &= \mathbf{E}\left\{ V_{t+1}(X_{t+1}) \middle| X_t \right\} \\
&= \mathbf{E}\left\{ \mathbf{E}\left\{ f_{\tau_t^*}(X_{\tau_t^*}) \middle| X_{t+1} \right\} \middle| X_t \right\} \\
&= \mathbf{E}\left\{ \mathbf{E}\left\{ f_{\tau_t^*}(X_{\tau_t^*}) \middle| X_0, \dots, X_{t+1} \right\} \middle| X_0, \dots, X_t \right\} \\
&= \mathbf{E}\left\{ f_{\tau_t^*}(X_{\tau_t^*}) \middle| X_0, \dots, X_t \right\} \\
&= \mathbf{E}\left\{ f_{\tau_t^*}(X_{\tau_t^*}) \middle| X_t \right\}.
\end{aligned}$$

(b) Because of

$$\begin{aligned}
f_{\tau_t^*}(X_{\tau_t^*}) &= f_{t+1}(X_{t+1}) \cdot 1_{\{\tau_t^* = t+1\}} + f_{\tau_t^*}(X_{\tau_t^*}) \cdot 1_{\{\tau_t^* > t+1\}} \\
&= f_{t+1}(X_{t+1}) \cdot 1_{\{f_{t+1}(X_{t+1}) \geq q_{t+1}(X_{t+1})\}} \\
&\quad + f_{\tau_{t+1}^*}(X_{\tau_{t+1}^*}) \cdot 1_{\{f_{t+1}(X_{t+1}) < q_{t+1}(X_{t+1})\}}
\end{aligned}$$

we can conclude from (a) and Markov property

$$\begin{aligned}
q_t(X_t) &= \mathbf{E}\Big\{ f_{t+1}(X_{t+1}) \cdot 1_{\{f_{t+1}(X_{t+1}) \geq q_{t+1}(X_{t+1})\}} \\
&\qquad + f_{\tau_{t+1}^*}(X_{\tau_{t+1}^*}) \cdot 1_{\{f_{t+1}(X_{t+1}) < q_{t+1}(X_{t+1})\}} \Big| X_t \Big\} \\
&= \mathbf{E}\left\{ \mathbf{E}\left\{ \dots \middle| X_0, \dots, X_{t+1} \right\} \middle| X_0, \dots, X_t \right\} \\
&= \mathbf{E}\Big\{ f_{t+1}(X_{t+1}) \cdot 1_{\{f_{t+1}(X_{t+1}) \geq q_{t+1}(X_{t+1})\}} \\
&\qquad + \mathbf{E}\left\{ f_{\tau_{t+1}^*}(X_{\tau_{t+1}^*}) \middle| X_{t+1} \right\} \cdot 1_{\{f_{t+1}(X_{t+1}) < q_{t+1}(X_{t+1})\}} \Big| X_t \Big\} \\
&= \mathbf{E}\Big\{ f_{t+1}(X_{t+1}) \cdot 1_{\{f_{t+1}(X_{t+1}) \geq q_{t+1}(X_{t+1})\}} \\
&\qquad + q_{t+1}(X_{t+1}) \cdot 1_{\{f_{t+1}(X_{t+1}) < q_{t+1}(X_{t+1})\}} \Big| X_t \Big\} \\
&= \mathbf{E}\Big\{ \max\big\{ f_{t+1}(X_{t+1}), q_{t+1}(X_{t+1}) \big\} \Big| X_t \Big\}.
\end{aligned}$$

(c) For any $w \in \{0, 1, \dots, T-t-1\}$ we have

$$
\begin{aligned}
&f_{\tau_t^*}(X_{\tau_t^*})\\
&= \sum_{s=0}^{w} f_{t+s+1}(X_{t+s+1}) \cdot 1_{\{\tau_t^*=t+s+1\}} + f_{\tau_t^*}(X_{\tau_t^*}) \cdot 1_{\{\tau_t^*>t+w+1\}}\\
&= \sum_{s=0}^{w} f_{t+s+1}(X_{t+s+1})\\
&\quad \cdot 1_{\{f_{t+1}(X_{t+1})<q_{t+1}(X_{t+1}),\dots,f_{t+s}(X_{t+s})<q_{t+s}(X_{t+s}),f_{t+s+1}(X_{t+s+1})\geq q_{t+s+1}(X_{t+s+1})\}}\\
&\quad + f_{\tau_{t+w}^*}(X_{\tau_{t+w}^*}) \cdot 1_{\{f_{t+1}(X_{t+1})<q_{t+1}(X_{t+1}),\dots,f_{t+w+1}(X_{t+w+1})<q_{t+w+1}(X_{t+w+1})\}}.
\end{aligned}
$$

Using a) and Markov property we conclude

$$
\begin{aligned}
q_t(X_t) &= \mathbf{E}\left\{f_{\tau_t^*}(X_{\tau_t^*})\middle|X_t\right\}\\
&\quad \times \mathbf{E}\left\{\sum_{s=0}^{w} f_{t+s+1}(X_{t+s+1})\right.\\
&\quad \cdot 1_{\{f_{t+1}(X_{t+1})<q_{t+1}(X_{t+1}),\dots,f_{t+s}(X_{t+s})<q_{t+s}(X_{t+s}),f_{t+s+1}(X_{t+s+1})\geq q_{t+s+1}(X_{t+s+1})\}}\\
&\quad + f_{\tau_{t+w}^*}(X_{\tau_{t+w}^*})\\
&\quad \left. \cdot 1_{\{f_{t+1}(X_{t+1})<q_{t+1}(X_{t+1}),\dots,f_{t+w+1}(X_{t+w+1})<q_{t+w+1}(X_{t+w+1})\}}\middle|X_t\right\}\\
&= \mathbf{E}\{\mathbf{E}\{\dots|X_0,\dots,X_{t+w+1}\}|X_0,\dots,X_t\}\\
&= \mathbf{E}\left\{\sum_{s=0}^{w} f_{t+s+1}(X_{t+s+1})\right.\\
&\quad \cdot 1_{\{f_{t+1}(X_{t+1})<q_{t+1}(X_{t+1}),\dots,f_{t+s}(X_{t+s})<q_{t+s}(X_{t+s}),f_{t+s+1}(X_{t+s+1})\geq q_{t+s+1}(X_{t+s+1})\}}\\
&\quad + \mathbf{E}\{f_{\tau_{t+w}^*}(X_{\tau_{t+w}^*})|X_{t+w+1}\}\\
&\quad \left. \cdot 1_{\{f_{t+1}(X_{t+1})<q_{t+1}(X_{t+1}),\dots,f_{t+w+1}(X_{t+w+1})<q_{t+w+1}(X_{t+w+1})\}}\middle|X_t\right\}\\
&= \mathbf{E}\left\{\sum_{s=0}^{w} f_{t+s+1}(X_{t+s+1})\right.\\
&\quad \cdot 1_{\{f_{t+1}(X_{t+1})<q_{t+1}(X_{t+1}),\dots,f_{t+s}(X_{t+s})<q_{t+s}(X_{t+s}),f_{t+s+1}(X_{t+s+1})\geq q_{t+s+1}(X_{t+s+1})\}}\\
&\quad + q_{t+w+1}(X_{t+w+1})\\
&\quad \left. \cdot 1_{\{f_{t+1}(X_{t+1})<q_{t+1}(X_{t+1}),\dots,f_{t+w+1}(X_{t+w+1})<q_{t+w+1}(X_{t+w+1})\}}\middle|X_t\right\},
\end{aligned}
$$

which implies the assertion. □

Remark 3. Because of

$$\Theta_{t+1,t+1}^{(0)} = \max\{f_{t+1}(X_{t+1}), q_{t+1}(X_{t+1})\}$$

and

$$\Theta_{t+1,T}^{(T-t-1)} = f_{\tau_t^*}(X_{\tau_t^*})$$

the regression representation (16) includes (14) (for $t = T - t - 1$) and (15) (for $t = 0$) as special cases.

Remark 4. There exists also regression representations for the value functions. E.g., as we have seen already in Theorem 1 and its proof we have

$$V_t(x) = \mathbf{E}\{f_{\tau^*_{t-1}}(X_{\tau^*_{t-1}})|X_t = x\}$$

and

$$V_t(x) = \max\{f_t(x), \mathbf{E}\{V_{t+1}(X_{t+1})|X_t = x\}\}.$$

Furthermore, similarly to Theorem 2 it can be shown

$$V_t(x) = \mathbf{E}\{\Theta^{(w+1)}_{t,t+w+1}|X_t = x\}.$$

Using Theorem 2 or Remark 4 we can compute the continuation values and the value functions by (recursive) evaluation of conditional expectations. However, in applications the underlying distributions will be rather complicated and therefore it is not clear how to compute these conditional expectations in practice.

4 Outline of Regression-based Monte Carlo Methods

The basic idea of regression-based Monte Carlo methods is to use regression estimates as numerical procedures to compute the above conditional estimations approximately. To do this artificial samples of the price process are generated which are used to construct data for the regression estimates. The algorithms either construct estimates $\hat{q}_{n,t}$ of the continuation values q_t or estimates $\hat{V}_{n,t}$ of the value functions. Comparing the regression representations for the continuation values like

$$q_t(x) = \mathbf{E}\left\{\max\left\{f_{t+1}(X_{t+1}), q_{t+1}(X_{t+1})\right\}\big|X_t = x\right\}$$

with the regression representation for the value function like

$$V_t(x) = \max\{f_t(x), \mathbf{E}\{V_{t+1}(X_{t+1})|X_t = x\}\},$$

we see that in the later relation the maximum occurs outside of the expectation and as a consequence the value function will be in generally not differentiable. In contrast in the first relation the maximum will be smoothed by taking its conditional expectation. Since it is always easier to estimate smooth regression functions there is some reason to focus on continuation values, which we will do in the sequel.

Let $X_0, X_1, \ldots, X_T$ be a $\mathbb{R}^d$-valued Markov process and let f_t be the discounted payoff function. We assume that the data generating process is completely known, i.e., that all parameters of this process are already estimated from historical data. In order to estimate the continuation values q_t recursively, we generate in a first step artificial independent Markov processes $\{X_{i,t}\}_{t=0,\ldots,T}$ $(i = 1, 2, \ldots, n)$ which are identically distributed as $\{X_t\}_{t=0,\dots,T}$. Then we use these so-called Monte Carlo samples in a second step to generate recursively data to estimate q_t by using one of the regression representation given in Theorem 2.

We start with

$$\hat{q}_{n,T}(x) = 0 \quad (x \in \mathbb{R}^d).$$

Given an estimate $\hat{q}_{n,t+1}$ of q_{t+1}, we estimate

$$\begin{aligned} q_t(x) &= \mathbf{E}\left\{ f_{\tau_t^*}(X_{\tau_t^*}) \middle| X_t = x \right\}, \\ &= \mathbf{E}\left\{ \max\{f_{t+1}(X_{t+1}), q_{t+1}(X_{t+1})\} \middle| X_t = x \right\} \\ &= \mathbf{E}\left\{ \Theta^{(w)}_{t+1,t+w+1} \middle| X_t = x \right\} \end{aligned}$$

by applying a regression estimate to an "approximative" sample of (X_t, Y_t) where

$$Y_t = Y_t(X_{t+1}, \dots, X_T, q_{t+1}, \dots, q_T)$$

is either given by

$$Y_t = Y_t(X_{t+1}, \dots, X_t, q_{t+1}, \dots, q_T) = f_{\tau_t^*}(X_{\tau_t^*}),$$
$$Y_t = Y_t(X_{t+1}, q_{t+1}) = \max\{f_{t+1}(X_{t+1}), q_{t+1}(X_{t+1})\}$$

or

$$Y_t = Y_t(X_{t+1}, \dots, X_{t+w+1}, q_{t+1}, \dots, q_{t+w+1}) = \Theta^{(w)}_{t+1,t+w+1}.$$

With the notation

$$\hat{Y}_{i,t} = Y_t(X_{i,t+1}, \dots, X_{i,T}, \hat{q}_{n,t+1}, \dots, \hat{q}_{n,T})$$

(where we have suppressed the dependency of $\hat{Y}_{i,t}$ on n) this "approximative" sample is given by

$$\left\{ \left(X_{i,t}, \hat{Y}_{i,t} \right) : i = 1, \dots, n \right\}. \tag{17}$$

After having computed the estimates $\hat{q}_{0,n}, \dots, \hat{q}_{n,T}$ we can use them in two different ways to produce estimates of V_0. Firstly we can estimate

$$V_0 = \mathbf{E}\{\max\{f_0(X_0), q_0(X_0)\}\}$$

(cf. proof of Theorem 1) by just replacing q_0 by its estimate, i.e., by a Monte Carlo estimate of

$$\mathbf{E}\left\{ \max\{f_0(X_0), \hat{q}_{0,n}(X_0)\} \right\}. \tag{18}$$

Secondly, we can use our estimates to construct a plug-in estimate

$$\hat{\tau} = \inf\{s \in \{0, 1, \dots, T\} \geq 0 : \hat{q}_{n,s}(X_s) \leq f_s(X_s)\} \tag{19}$$

of the optimal stopping rule τ^* and estimate V_0 by a Monte Carlo estimate of

$$\mathbf{E}\{f_{\hat{\tau}}(X_{\hat{\tau}})\}. \tag{20}$$

Here in (19) and in (20) the expectation is taken only with respect to $X_0, \ldots, X_T$ and not with respect to the random variables used in the definition of the estimates $\hat{q}_{n,s}$.

This kind of recursive estimation scheme was firstly proposed by Carriér (1996) for the estimation of value functions. In Tsitsiklis and Van Roy (1999) and Longstaff and Schwartz (2001) it was used to construct estimates of continuation values.

In view of a theoretical analysis of the estimates it usually helps if new variables of the price process are used for each recursive estimation step. In this way the error propagation (i.e., the influence of the error of $\hat{q}_{n,t+1}, \ldots, \hat{q}_{n,T}$) can be analyzed much easier, cf. Kohler et al. (2010) or Kohler (2008a).

5 Algorithms Based on Linear Regression

In most applications the algorithm of the previous section is applied in connection with linear regression. Here basis functions

$$B_1, \ldots, B_K : \mathbb{R}^d \to \mathbb{R}$$

are chosen and the estimate $\hat{q}_{n,t}$ is defined by

$$\hat{q}_{n,t} = \sum_{k=1}^{K} \hat{a}_k \cdot B_k, \tag{21}$$

where $\hat{a}_1, \ldots, \hat{a}_K \in \mathbb{R}$ are chosen such that

$$\frac{1}{n}\sum_{i=1}^{n}\left|\hat{Y}_{i,t} - \sum_{k=1}^{K} \hat{a}_k \cdot B_k(X_{i,t})\right|^2 = \min_{a_1,\ldots,a_K \in \mathbb{R}} \frac{1}{n}\sum_{i=1}^{n}\left|\hat{Y}_{i,t} - \sum_{k=1}^{K} a_k \cdot B_k(X_{i,t})\right|^2. \tag{22}$$

Here $\hat{Y}_{i,t}$ are defined either by

$$\hat{Y}_{i,t} = \max\{f_{t+1}(X_{i,t+1}), \hat{q}_{n,t+1}(X_{i,t+1})\}$$

in case of the Tsitsiklis-Van-Roy algorithm, or by

$$\hat{Y}_{i,t} = f_{\hat{\tau}_{i,t}}(X_{i,\hat{\tau}_{i,t}})$$

where

$$\hat{\tau}_{i,t} = \inf\{s \in \{t+1, \ldots, T\} : f_s(X_{i,s}) \geq \hat{q}_{n,s}(X_{i,s})\}$$

in case of the Longstaff-Schwartz algorithm.

The estimate can be computed easily by solving a linear equation system. Indeed, it is well-known from numerical analysis (cf., e.g., Stoer 1993, Chap. 4.8.1) that (22) is equivalent to

$$\mathbf{B}^T \mathbf{B} \hat{\mathbf{a}} = \mathbf{B}^T \mathbf{Y} \tag{23}$$

where

$$\mathbf{Y} = (\hat{Y}_{1,t}, \ldots, \hat{Y}_{n,t})^T, \qquad \mathbf{B} = (B_k(X_{i,t}))_{i=1,\ldots,n,k=1,\ldots,K}$$

and

$$\hat{\mathbf{a}} = (\hat{a}_1, \ldots, \hat{a}_K)^T.$$

It was observed e.g. in Longstaff and Schwartz (2001) that the above estimate combined with the corresponding plug-in estimate (19) of the optimal stopping rule is rather robust with respect to the choice of the basis functions. The most simplest possibility are monomials, i.e.,

$$B_k(u_1, \ldots, u_d) = u_1^{s_{1,k}} \cdot u_2^{s_{2,k}} \cdots u_d^{s_{d,k}}$$

for some nonnegative integers $s_{1,k}, \ldots s_{d,k}$. For $d = 1$ this reduce to fitting a polynomial of a fixed degree (e.g., $K-1$) to the data. For d large the degree of the multinomial polynomial (e.g. defined by $s_{1,k} + \cdots + s_{d,k}$ or by $\max_{j=1,\ldots,d,k=1,\ldots,K} s_{j,k}$ has chosen to be small in order to avoid that there are too many basis functions. It is well-known in practice that the estimate gets much better if the payoff function is chosen as one of the basis functions.

The Longstaff-Schwartz algorithm was proposed in Longstaff and Schwartz (2001). It was further theoretical examined in Clément et al. (2002). The Tsitsiklis-Van-Roy algorithm was introduced and theoretical examined in Tsitsiklis and Van Roy (1999, 2001).

6 Algorithms Based on Nonparametric Regression

Already in Carriér (1996) it was proposed to use nonparametric regression to estimate value functions. In the sequel we describe various nonparametric regression estimates of continuation values.

According to Györfi et al. (2002) there are four (related) paradigms for defining nonparametric regression estimates. The first is local averaging, where the estimate is defined by

$$\hat{q}_{n,t}(x) = \sum_{i=1}^{n} W_{n,i}(x, X_{1,t}, \ldots, X_{n,t}) \cdot \hat{Y}_{i,t} \tag{24}$$

with weights $W_{n,i}(x, X_{1,t}, \ldots, X_{n,t}) \in \mathbb{R}$ depending on the x-values of the sample. The most popular example of local averaging estimates is the Nadaraya-Watson kernel estimate, where a kernel function

$$K : \mathbb{R}^d \to \mathbb{R}$$

(e.g., the so-called naive kernel $K(u) = 1_{\{\|u\| \leq 1\}}$ or the Gaussian kernel $K(u) = \exp(-\|u\|^2/2)$) and a so-called bandwidth $h_n > 0$ are chosen and the weights are defined by

$$W_{n,i}(x, X_{1,t}, \ldots, X_{n,t}) = \frac{K(\frac{x-X_{i,t}}{h_n})}{\sum_{j=1}^{n} K(\frac{x-X_{j,t}}{h_n})}.$$

Here the estimate is given by

$$\hat{q}_{n,t}(x) = \frac{\sum_{i=1}^{n} K(\frac{x-X_{i,t}}{h_n}) \cdot \hat{Y}_{i,t}}{\sum_{j=1}^{n} K(\frac{x-X_{j,t}}{h_n})}.$$

The second paradigm is global modeling (or least squares estimation), where a function space $\mathfrak{F}_n$ consisting of functions $f : \mathbb{R}^d \to \mathbb{R}$ is chosen and the estimate is defined by

$$\hat{q}_{n,t} \in \mathfrak{F}_n \quad \text{and} \quad \frac{1}{n}\sum_{i=1}^{n} |\hat{Y}_{i,t} - \hat{q}_{n,t}(X_{i,t})|^2 = \min_{f \in \mathfrak{F}_n} \frac{1}{n}\sum_{i=1}^{n} |\hat{Y}_{i,t} - f(X_{i,t})|^2. \quad (25)$$

In case that $\mathfrak{F}_n$ is a linear vector space (with dimension depending on the sample size) this estimate can be computed by solving a linear equation system corresponding to (23). Such linear function spaces occur e.g. in the definition of least squares spline estimates with fixed knot sequences, where the set $\mathfrak{F}_n$ is chosen as a set of piecewise polynomials satisfying some global smoothness conditions (like differentiability).

Especially for large d it is also useful to consider nonlinear function spaces. The most popular example are neural networks, where for the most simple model $\mathfrak{F}_n$ is defined by

$$\mathfrak{F}_n = \left\{ \sum_{i=1}^{k_n} c_i \cdot \sigma(a_i^T x + b_i) + c_0 : a_i \in \mathbb{R}^d,\ b_i \in \mathbb{R} \right\} \quad (26)$$

for some sigmoid function $\sigma : \mathbb{R} \to [0, 1]$. Here it is assumed that the sigmoid function σ is monotonically increasing and satisfies

$$\sigma(x) \to 0 \quad (x \to -\infty) \quad \text{and} \quad \sigma(x) \to 1 \quad (x \to \infty).$$

An example of such a sigmoid function is the logistic squasher defined by

$$\sigma(x) = \frac{1}{1+e^{-x}} \quad (x \in \mathbb{R}).$$

There exists a deepest decent algorithm (so-called backfitting) which computes the corresponding least squares estimate approximately (cf., e.g., Rumelhart and McClelland 1986).

The third paradigm is penalized modeling. Instead of restricting the set of functions over which the so called empirical L_2 risk

$$\frac{1}{n}\sum_{i=1}^{n}|\hat{Y}_{i,t} - f(X_{i,t})|^2$$

is minimized, in this case a penalty term penalizing the roughness of the function is added to the empirical L_2 risk and this penalized empirical L_2 risk is basically minimized with respect to all functions.The most popular example of this kind of estimates are smoothing spline estimates. Here the estimate is defined by

$$\hat{q}_{n,t}(\cdot) = \arg \min_{f\in W^k(\mathbb{R}^d)} \left(\frac{1}{n}\sum_{i=1}^{n}|f(X_{i,t}) - \hat{Y}_{i,t}|^2 + \lambda_n \cdot J_k^2(f) \right), \tag{27}$$

where $k \in \mathbb{N}$ with $2k > d$, $W^k(\mathbb{R}^d)$ denotes the Sobolev space

$$\left\{ f : \frac{\partial^k f}{\partial x_1^{\alpha_1} \cdots \partial x_d^{\alpha_d}} \in L_2(\mathbb{R}^d) \text{ for all } \alpha_1, \ldots, \alpha_d \in \mathbb{N} \text{ with } \alpha_1 + \cdots + \alpha_d = k \right\},$$

and

$$J_k^2(f) = \sum_{\alpha_1,\ldots,\alpha_d\in\mathbb{N},\alpha_1+\cdots+\alpha_d=k} \frac{k!}{\alpha_1! \cdot \ldots \cdot \alpha_d!} \int_{\mathbb{R}^d} \left| \frac{\partial^k f}{\partial x_1^{\alpha_1} \cdots \partial x_d^{\alpha_d}}(x) \right|^2 dx.$$

Here $\lambda_n > 0$ is the smoothing parameter of the estimate.

The fourth (and last) paradigm is local modeling. It is similar to global modeling, but this time the function is fitted only locally to the data and a new function is used for each point in $\mathbb{R}^d$. The most popular example of this kind of estimate are local polynomial kernel estimates. Here the estimate, which depends on a nonnegative integer M and a kernel function $K : \mathbb{R}^d \to \mathbb{R}$, is given by

$$\hat{q}_{n,t}(x) = \hat{p}_x(x) \tag{28}$$

where

$$\hat{p}_x(\cdot) \in \mathcal{F}_M = \left\{ \sum_{0\le j_1,\ldots,j_d\le M} a_{j_1,\ldots,j_d} \cdot \ldots \cdot (x^{(1)})^{j_1} \cdots (x^{(d)})^{j_d} : a_{j_1,\ldots,j_d} \in \mathbb{R} \right\} \tag{29}$$

satisfies

$$\frac{1}{n}\sum_{i=1}^{n}|\hat{p}_x(X_{i,t}) - \hat{Y}_{i,t}|^2 K\left(\frac{x - X_i}{h_n}\right) = \min_{p\in\mathcal{F}_M} \frac{1}{n}\sum_{i=1}^{n}|p(X_{i,t}) - \hat{Y}_{i,t}|^2 K\left(\frac{x - X_i}{h_n}\right). \tag{30}$$

The estimate can be computed again by solving a linear equation system, but this time of size n times n (instead K_n times K_n as for least squares estimates).

Each estimate above contains a smoothing parameter which determines how smooth the estimate should be. E.g., for the Nadaraya-Watson kernel estimate it is the bandwidth $h_n > 0$, where a small bandwidth leads to a very rough estimate. For the smoothing spline estimate it is the parameter $\lambda_n > 0$, and for the least squares neural network estimate the smoothing parameter is the number k_n of neurons. For a successful application of the estimates these parameters need to be chosen data-dependent. The most simple way of doing this is splitting of the sample (cf., e.g., Chap. 7 in Györfi et al. 2002): Here the sample is divided into two parts, the first part is used to compute the estimate for different values of the parameter, and the second part is used to compute the empirical error of each of these estimates and that estimate is chosen where this empirical error is minimal. Splitting of the sample is in case of regression-based Monte Carlo methods the best method to choose the smoothing parameter, because there the data is chosen artificially with arbitrary sample size so it does not hurt at all if the estimate depends primary on the first part of the sample (since this first part can be as large as possible in view of computation of the estimate).

The first article where the use of nonparametric regression for the estimation of continuation values was examined theoretically was Egloff (2005). There nonparametric least squares estimates have been used, where the parameters where chosen by complexity regularization (cf., e.g., Chap. 12 in Györfi et al. 2002) and the consistency for general continuation values and the rate of convergence of the estimate in case of smooth continuation values has been investigated. For smooth continuation values Egloff (2005) showed the usual optimal rate of convergence for estimation of smooth regression functions. However, due to problems with the error propagation the estimate was defined such that it was very hard to compute it in practice, and it was not possible to check with simulated data whether nonparametric regression is not only useful asymptotically (i.e., for sample size tending to infinity, as was shown in the theoretical results), but also for finite sample size.

In Egloff et al. (2007) the error propagation was simplified by generating new data for each time point which was (conditioned on the data corresponding to time t) independent of all previously used data. In addition, a truncation of the estimate was introduced which allowed to choose linear vector spaces as function spaces for the least squares spline estimates, so that they can be computed by solving a linear equation system. The parameter (here the vector space dimension of the function space) of the least squares estimates were chosen by splitting of the sample. As regression representation the general formula of Egloff (2005) (cf. Theorem 2(c)) has been used. Consistency and rate of convergence results for these estimates have been derived, where as a consequence of the truncation of the estimate the rates contained an additional logarithmic factor. But the main advantage of these estimates is that they are easy to compute, so it was possible to analyze the finite sample size behavior of the estimates.

In Kohler et al. (2010), Kohler (2008a) and Kohler and Krzyżak (2009) the error propagation was further simplified by generation of new paths of the price process for each recursive estimation step and by using only the simple regression representation of Tsitsiklis and Van Roy (1999) (cf. Theorem 2(b)). As a consequence it was possible to analyze the estimates by using results derived in Kohler (2006) for regression estimation in case of additional measurement errors in the dependent variable. Kohler et al. (2010) investigated least squares neural network estimates, which are very promising in case of large d, and Kohler (2008a) considered smoothing spline estimates. In both papers results concerning consistency and rate of convergence of the estimates have been derived. Kohler and Krzyżak (2009) presents a unifying theory which contains the results of the previous papers as well as results concerning new estimates (e.g., orthogonal series estimates).

The above papers focus on properties of the estimates of the continuation values, i.e., they consider the error between the continuation values and its estimates. As was pointed out by Belomestny (2009), sometimes much better rate of convergence results can be derived for the Monte Carlo estimate of (20) considered as estimate of the price V_0 of the option. Because in view of a good performance of the stopping time it is not important that the estimate of the continuation values are close to the continuation values, instead it is important that they lead to the same decision as the optimal stopping rule. And for this it is only important that

$$f_t(X_t) \geq \hat{q}_{n,t}(X_t)$$

is equivalent to

$$f_t(X_t) \geq q_t(X_t)$$

and not that $\hat{q}_{n,t}(X_t)$ and $q_t(X_t)$ are close. Belomestny (2009) introduces a kind of margin condition (similar to margin conditions in pattern recognition) measuring how quickly $q_t(X_t)$ approaches $f_t(X_t)$, and shows under this margin condition much better rate of convergence for the estimate (20) than previous results on the rates of convergence of the continuation values imply for the estimate (19).

7 Dual Methods

The above estimates yield estimates

$$\hat{\tau} = \inf\left\{s \in \{0,\ldots,T\} : \hat{q}_s(X_{n,s}) \leq f_s(X_s)\right\}$$

of the optimal stopping time τ^*. By Monte Carlo these estimates yields estimates of V_0, such that expectation

$$\mathbf{E}\{f_{\hat{\tau}}(X_{\hat{\tau}})\}$$

of the estimate is less than or equal to the true price V_0. It was proposed independently by Rogers (2001) and Haugh and Kogan (2004) that by using a dual method Monte Carlo estimates can be constructed such that the expectation of the estimate is greater than or equal to V_0. The key idea is the next theorem, which is already well-known in literature (cf., e.g., Sect. 8.7 in Glasserman 2004).

Theorem 3. *Let $\mathcal{M}$ be the set of all martingales $M_0,\ldots,M_T$ with $M_0 = 0$. Then*

$$V_0 = \inf_{M \in \mathcal{M}} \mathbf{E}\left\{\max_{t=0,\ldots,T} (f_t(X_t) - M_t)\right\} = \mathbf{E}\left\{\max_{t=0,\ldots,T} \left(f_t(X_t) - M_t^*\right)\right\}, \quad (31)$$

where

$$M_t^* = \sum_{s=1}^{t} \left(\max\{f_s(X_s), q_s(X_s)\} - \mathbf{E}\left\{\max\{f_s(X_s), q_s(X_s)\}|X_{s-1}\right\}\right). \qquad (32)$$

For the sake of completeness we present next a complete proof of Theorem 3.

Proof. We first prove

$$\max_{t=0,\dots,T}\left(f_t(X_t) - \sum_{s=1}^{t}\left(\max\{f_s(X_s), q_s(X_s)\} - \mathbf{E}\left\{\max\{f_s(X_s), q_s(X_s)\}|X_{s-1}\right\}\right)\right)$$
$$= \max\{f_0(X_0), q_0(X_0)\}. \qquad (33)$$

To do this, we observe that we have by Theorem 2(b)

$$\max_{t=0,\dots,T}\left(f_t(X_t) - \sum_{s=1}^{t}\left(\max\{f_s(X_s), q_s(X_s)\} - \mathbf{E}\left\{\max\{f_s(X_s), q_s(X_s)\}|X_{s-1}\right\}\right)\right)$$
$$= \max_{t=0,\dots,T}\left(f_t(X_t) - \sum_{s=1}^{t}\left(\max\{f_s(X_s), q_s(X_s)\} - q_{s-1}(X_{s-1})\right)\right).$$

For any $t \in \{1, \dots, T\}$ we have

$$f_t(X_t) - \sum_{s=1}^{t}\left(\max\{f_s(X_s), q_s(X_s)\} - q_{s-1}(X_{s-1})\right)$$
$$\leq f_t(X_t) - \sum_{s=1}^{t-1}\left(q_s(X_s) - q_{s-1}(X_{s-1})\right) - \left(f_t(X_t) - q_{t-1}(X_{t-1})\right)$$
$$= q_0(X_0),$$

furthermore in case $t = 0$ we get

$$f_t(X_t) - \sum_{s=1}^{t}\left(\max\{f_s(X_s), q_s(X_s)\} - q_{s-1}(X_{s-1})\right) = f_0(X_0),$$

which shows

$$\max_{t=0,\dots,T}\left(f_t(X_t) - \sum_{s=1}^{t}\left(\max\{f_s(X_s), q_s(X_s)\} - q_{s-1}(X_{s-1})\right)\right)$$
$$\leq \max\{f_0(X_0), q_0(X_0)\}.$$

But for $t = \tau^*$ we get in case of $q_0(X_0) > f_0(X_0)$ by definition of τ^*

$$
\begin{aligned}
& f_{\tau^*}(X_{\tau^*}) - \sum_{s=1}^{\tau^*} (\max\{f_s(X_s), q_s(X_s)\} - q_{s-1}(X_{s-1})) \\
&= f_{\tau^*}(X_{\tau^*}) - \sum_{s=1}^{\tau^*-1} (q_s(X_s) - q_{s-1}(X_{s-1})) - (f_{\tau^*}(X_{\tau^*}) - q_{\tau^*-1}(X_{\tau^*-1})) \\
&= q_0(X_0),
\end{aligned}
$$

and in case of $q_0(X_0) \leq f_0(X_0)$ (which implies $\tau^* = 0$) we have

$$
f_{\tau^*}(X_{\tau^*}) - \sum_{s=1}^{\tau^*} (\max\{f_s(X_s), q_s(X_s)\} - q_{s-1}(X_{s-1})) = f_0(X_0). .
$$

This completes the proof of (33).

As shown at the end of the proof of Theorem 1 we have

$$
V_0 = \mathbf{E}\{\max\{f_0(X_0), q_0(X_0)\}\}.
$$

Using this together with (33) we get

$$
\mathbf{E}\left\{\max_{t=0,\dots,T} \left(f_t(X_t) - M_t^*\right)\right\} = \mathbf{E}\{\max\{f_0(X_0), q_0(X_0)\}\} = V_0.
$$

Thus it suffices to show: For any martingale $M_0, \dots, M_T$ with $M_0 = 0$ we have

$$
\mathbf{E}\left\{\max_{t=0,\dots,T} (f_t(X_t) - M_t)\right\} \geq \sup_{\tau \in \mathcal{T}(0,\dots,T)} \mathbf{E}\{f_\tau(X_\tau)\} = V_0.
$$

But this follows from the optional sampling theorem, because if $M_0, \dots, M_T$ is a martingale with $M_0 = 0$ and τ is a stopping time we know

$$
\mathbf{E}M_\tau = \mathbf{E}M_0 = 0
$$

and hence

$$
\mathbf{E}\{f_\tau(X_\tau)\} = \mathbf{E}\{f_\tau(X_\tau) - M_\tau\} \leq \mathbf{E}\left\{\max_{t=0,\dots,T} (f_t(X_t) - M_t)\right\}.
$$

This completes the proof. □

Given estimates $\hat{q}_{n,s}$ ($s \in \{0, 1, \dots, T\}$) of the continuation values, we can estimate the martingale (32) by

$$
\hat{M}_t = \sum_{s=1}^{t} \left(\max\{f_s(X_s), \hat{q}_{n,s}(X_s)\} - \mathbf{E}^*\left\{\max\{f_s(X_s), \hat{q}_{n,s}(X_s)\} | X_{s-1}\right\}\right). \quad (34)
$$

Provided we use unbiased and $\mathcal{F}(X_0, \ldots, X_t)$-measurable estimates $\mathbf{E}^*$ of the inner expectation in (32) (which can be constructed, e.g., by nested Monte Carlo) this leads to a martingale, too. This in turn can be used to construct Monte Carlo estimates of V_0, for which the expectation

$$\mathbf{E}\left\{\max_{t=0,\ldots,T}\left(f_t(X_t) - \hat{M}_t\right)\right\}$$

is greater than or equal to V_0. As a consequence we get two kind of estimates with expectation lower and higher than V_0, resp., so we have available an interval in which our true price should be contained.

In connection with linear regression these kind of estimates have been studied in Rogers (2001) and Haugh and Kogan (2004). Jamshidian (2007) studies multiplicative versions of this method. A comparative study of multiplicative and additive duals is contained in Chen and Glasserman (2007). Andersen and Broadie (2004) derive upper and lower bounds for American options based on duality. Belomestny et al. (2009) propose in a Brownian motion setting estimates with expectation greater than or equal to the true price, which can be computed without nested Monte Carlo (and hence are quite easy to compute).

In Kohler (2008b) dual methods have been combined with nonparametric smoothing spline estimates of the continuation values and consistency of the resulting estimates was shown for all bounded Markov processes. In Kohler et al. (2008) it is shown how these estimates can be modified such that less nested Monte Carlo steps are needed in an application.

8 Application to Simulated Data

The PhD thesis Todorovic (2007) contains various comparisons of regression-based Monte Carlo methods on simulated data. Using the standard monomial basis for linear regression (without including the payoff function) it turns out that for linear regression the regression representation of Longstaff and Schwartz (2001) produces often better results than the regression representation of Tsitsiklis and Van Roy (1999) in view of the performance of the estimated stopping rule on new data. But for nonparametric regression it does not seem to make a difference whether the regression representation of Longstaff and Schwartz (2001), of Tsitsiklis and Van Roy (1999) or the more general form of Egloff (2005) is used. Furthermore Todorovic (2007) shows that nonparametric regression estimate lead sometimes to much better performance than the linear regression estimates (and in his simulations never really worse performance) as long as the payoff function is not included in the basis function.

It turns out that this is less obvious if the payoff function is used as one of the basis functions for linear regression. But as we show below, in this case a very high sample size for the Monte Carlo estimates leads again to better results for

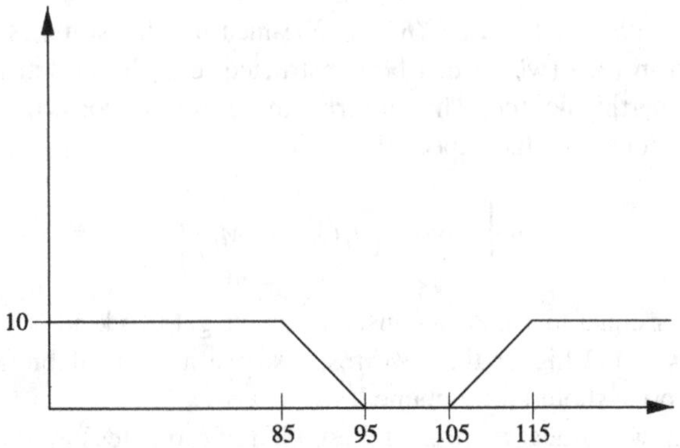

Fig. 1 Strangle spread payoff with strike prices 85, 95, 105 and 115

the nonparametric regression estimate. The reason for this is that the bias of the nonparametric regression estimates can be decreased by increasing the sample size, which is not true for linear regression.

In the sequel we consider an American option based on the average of three correlated stock prices. The stocks are ADECCO R, BALOISE R and CIBA. The stock prices were observed from Nov. 10, 2000 until Oct. 3, 2003 on weekdays when the stock market was open for the total of 756 days. We estimate the volatility from data observed in the past by the historical volatility

$$\sigma = (\sigma_{i,j})_{1 \le i,j \le 3} = \begin{pmatrix} 0.3024 & 0.1354 & 0.0722 \\ 0.1354 & 0.2270 & 0.0613 \\ 0.0722 & 0.0613 & 0.0717 \end{pmatrix}.$$

We simulate the paths of the underlying stocks with a Black-Scholes model by

$$X_{i,t} = x_0 \cdot e^{r \cdot t} \cdot e^{\sum_{j=1}^{3} (\sigma_{i,j} \cdot W_j(t) - \frac{1}{2} \cdot \sigma_{i,j}^2 t)} \quad (i = 1, \ldots, 3),$$

where $\{W_j(t) : t \in \mathbb{R}_+\}$ $(j = 1, \ldots, 3)$ are three independent Wiener processes and where the parameters are chosen as follows: $x_0 = 100$, $r = 0.05$ and components $\sigma_{i,j}$ of the volatility matrix as above. The time to maturity is assumed to be one year. To compute the payoff of the option we use a strangle spread function (cf. Fig. 1) with strikes 85, 95, 105 and 115 applied to the average of the three correlated stock prices.

We discretize the time interval $[0, 1]$ by dividing it into $m = 48$ equidistant time steps with $t_0 = 0 < t_1 < \cdots < t_m = 1$ and consider a Bermudan option with payoff function as above and exercise dates restricted to $\{t_0, t_1, \ldots, t_m\}$. We choose discount factors $e^{-r \cdot t_j}$ for $j = 0, \ldots, m$. For all three algorithms we use sample size $n = 40000$ for the regression estimates of the continuation values.

For the nonparametric regression estimate we use smoothing splines as implemented in the routine Tps from the library "fields" in the statistics package R, where the smoothing parameter is chosen by generalized cross-validation. For the

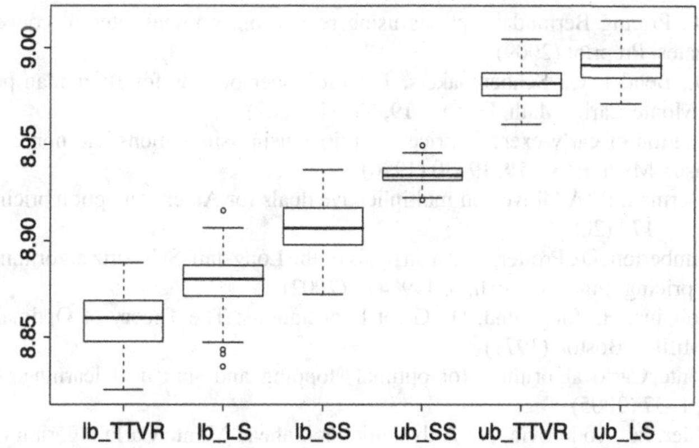

Fig. 2 Boxplots for 100 Monte Carlo estimates of lower bounds (lb) and upper bounds (ub) based on the estimates of the continuation values generated by the algorithm of Tsitsiklis and Van Roy (TTVR), Longstaff and Schwartz (LS) and nonparametric smoothing splines (SS)

Longstaff–Schwartz and Tsitsiklis–Van Roy algorithms we use linear regression as implemented in *R* with degree 1 and payoff function included in the basis.

For each of these algorithms we compute Monte Carlo estimates of lower bounds on the option price defined using the corresponding estimated stopping rule, and Monte Carlo estimates of upper bounds on the option price using the corresponding estimated optimal martingale. Here we use 100 nested Monte Carlo steps to approximate the conditional expectation occurring in the optimal martingale. The sample size of the Monte Carlo estimates is 10000 in case of estimation of upper bounds and 40000 in case of estimation of lower bounds.

We apply all six algorithms for computing lower or upper bounds to 100 independently generated sets of paths and we compare the algorithms using boxplots for the 100 lower or upper bounds computed for each algorithm. We would like to stress that for all three algorithms computing upper bounds the expectation of the values are upper bounds to the true option price, hence lower values indicates a better performance of the algorithms, and that for all three algorithms computing lower bounds the expectation of the values are lower bounds to the true option price, hence higher values indicates a better performance of the algorithms.

As we can see in Fig. 2, the algorithms based on nonparametric regression are superior to Longstaff–Schwartz and Tsitsiklis–Van Roy algorithms, since the lower boxplot of the upper bounds for this algorithm and the higher boxplot of the lower bounds for this algorithm indicate better performance.

References

Andersen, L., Broadie, M.: Primal-dual simulation algorithm for pricing multidimensional American options. Manag. Sci. **50**, 1222–1234 (2004)

Belomestny, D.: Pricing Bermudan options using regression: optimal rates of convergence for lower estimates. Preprint (2009)
Belomestny, D., Bender, C., Schoenmakers, J.: True upper bounds for Bermudan products via non-nested Monte Carlo. Math. Finance **19**, 53–71 (2009)
Carriér, J.: Valuation of early-exercise price of options using simulations and nonparametric regression. Insur. Math. Econ. **19**, 19–30 (1996)
Chen, N., Glasserman, P.: Additive and multiplicative duals for American option pricing. Finance Stoch. **11**, 153–179 (2007)
Clément, E., Lamberton, D., Protter, P.: An analysis of the Longstaff-Schwartz algorithm for American option pricing. Finance Stoch. **6**, 449–471 (2002)
Chow, Y.S., Robbins, H., Siegmund, D.: Great Expectations: The Theory of Optimal Stopping. Houghton Mifflin, Boston (1971)
Egloff, D.: Monte Carlo algorithms for optimal stopping and statistical learning. Ann. Appl. Probab. **15**, 1–37 (2005)
Egloff, D., Kohler, M., Todorovic, N.: A dynamic look-ahead Monte Carlo algorithm for pricing American options. Ann. Appl. Probab. **17**, 1138–1171 (2007)
Glasserman, P.: Monte Carlo Methods in Financial Engineering. Springer, Berlin (2004)
Györfi, L., Kohler, M., Krzyżak, A., Walk, H.: A Distribution-Free Theory of Nonparametric Regression. Springer Series in Statistics. Springer, Berlin (2002)
Haugh, M., Kogan, L.: Pricing American Options: A Duality Approach. Oper. Res. **52**, 258–270 (2004)
Jamshidian, F.: The duality of optimal exercise and domineering claims: a Doob-Meyer decomposition approach to the Snell envelope. Stochastics **79**, 27–60 (2007)
Karatzas, I., Shreve, S.E.: Methods of Mathematical Finance. Applications of Math., vol. 39. Springer, Berlin (1998)
Kohler, M.: Nonparametric regression with additional measurement errors in the dependent variable. J. Stat. Plan. Inference **136**, 3339–3361 (2006)
Kohler, M.: A regression based smoothing spline Monte Carlo algorithm for pricing American options. AStA Adv. Stat. Anal. **92**, 153–178 (2008a)
Kohler, M.: Universally consistent upper bounds for Bermudan options based on Monte Carlo and nonparametric regression. Preprint (2008b)
Kohler, M., Krzyżak, A.: Pricing of American options in discrete time using least squares estimates with complexity penalties. Preprint (2009)
Kohler, M., Krzyżak, A., Todorovic, N.: Pricing of high-dimensional American options by neural networks. Math. Finance (2010, to appear)
Kohler, M., Krzyżak, A., Walk, H.: Upper bounds for Bermudan options on Markovian data using nonparametric regression and a reduced number of nested Monte Carlo steps. Stat. Decis. **26**, 275–288 (2008)
Lamberton, D., Pagès, G.: Sur l'approximation des réduites. Ann. Inst. H. Poincaré **26**, 331–355 (1990)
Longstaff, F.A., Schwartz, E.S.: Valuing American options by simulation: a simple least-squares approach. Rev. Financ. Stud. **14**, 113–147 (2001)
Rogers, L.: Monte Carlo valuation of American options. Math. Finance **12**, 271–286 (2001)
Rumelhart, D.E., McClelland, J.L.: Parallel Distributed Processing: Explorations in the Microstructure of Cognition, vol. 1: Foundations. MIT Press, Cambridge (1986)
Shiryayev, A.N.: Optimal Stopping Rules. Applications of Mathematics. Springer, Berlin (1978)
Stoer, J.: Numerische Mathematik, vol. 1. Springer, Berlin (1993)
Todorovic, N.: Bewertung Amerikanischer Optionen mit Hilfe von regressionsbasierten Monte-Carlo-Verfahren. Shaker, Aachen (2007)
Tsitsiklis, J.N., Van Roy, B.: Optimal stopping of Markov processes: Hilbert space theory, approximation algorithms, and an application to pricing high-dimensional financial derivatives. IEEE Trans. Autom. Control **44**, 1840–1851 (1999)
Tsitsiklis, J.N., Van Roy, B.: Regression methods for pricing complex American-style options. IEEE Trans. Neural Netw. **12**, 694–730 (2001)

Binomial Trees in Option Pricing—History, Practical Applications and Recent Developments

Ralf Korn and Stefanie Müller

Abstract We survey the history and application of binomial tree methods in option pricing. Further, we highlight some recent developments and point out problems for future research.

1 Introduction

In many disciplines, there is the classical question on which came first, egg or hen; but not so in the history of binomial option valuation. There is no denying the fact that the diffusion model underlying the famous Black-Scholes formula (see Black and Scholes 1973) triggered the development of the binomial approach to option pricing. At first sight it seems surprising that the binomial approach originates from the Black-Scholes model although the mathematics behind diffusion models are clearly much more involved than that behind the discrete-time and finite state space binomial models (the reason why we interpret the Black-Scholes model as the *hen* and the simpler binomial model as the *egg*). To understand why the hen came first, we need to recognize option valuation as a discipline that brings together mathematical modeling skills and economic interpretations of real-world markets.

Ralf Korn, Center for Mathematical and Computational Modeling (CM)² and Department of Mathematics, University of Kaiserslautern, 67653 Kaiserslautern, Germany
e-mail: korn@mathematik.uni-kl.de

Stefanie Müller, Center for Mathematical and Computational Modeling (CM)² and Department of Mathematics, University of Kaiserslautern, 67653 Kaiserslautern, Germany
e-mail: stefanie@mathematik.uni-kl.de

Both authors thank the Rheinland-Pfalz Cluster of Excellence DASMOD and the Research Center (CM)² for support. Ralf Korn would like to dedicate this survey to the memory of Prof. Dr. Jürgen Lehn.

L. Devroye et al. (eds.), *Recent Developments in Applied Probability and Statistics*,
DOI 10.1007/978-3-7908-2598-5_3,

The Black-Scholes formula caused a shock amongst the economists at the time of its introduction. The economic ideas underlying the Black-Scholes approach, such as the principles of risk-neutrality and riskless portfolios, shook the theory of option pricing to its core. However, its involved mathematical background based on diffusion models might have appeared too academic or even awkward. This motivated various economists to search for a simpler modeling framework that preserves the economically relevant properties of the Black-Scholes framework but that is at the same time more easily accessible. The binomial approach to option pricing grew out of a discussion between M. Rubinstein and W.F. Sharpe at a conference in Ein Borek, Israel (see Rubinstein 1992 for the historical background). They realized that the economic idea behind the Black-Scholes model can be reduced to the following principle: If an economy incorporating three securities can only attain two future states, one such security will be redundant; i.e. each single security can be replicated by the other two, a fact later referred to as market completeness. With this insight at hand, it was obvious that one should introduce such a two-state model and verify that the economic properties of the Black-Scholes diffusion approach are preserved. This was the birth of binomial option pricing.

In this survey, we will first explain how to use binomial trees for option pricing in the corresponding discrete-time financial market. However, in practical applications, binomial trees are preferably used as numerical approximation tools for pricing options in more complex, continuous-time stock market models. We will present early approaches to binomial trees: the models suggested by Cox et al. (1979) and by Rendleman and Bartter (1979). In Black-Scholes settings, the application of the binomial approach to numerical option pricing can be justified by Donsker's Theorem on random walk approximations to a Brownian motion. Donsker's Theorem implies that as the period length tends to zero, the sequence of corresponding binomial models (appropriately scaled in time) converges weakly to a geometric Brownian motion, which underlies the Black-Scholes stock price model (provided that the first two moments of the one-period log-returns are matched). The application of binomial models as numerical pricing tools will be explained in detail. We will focus on aspects of practical relevance such as the convergence behavior of binomial estimates to Black-Scholes option prices, the speed of convergence and the algorithmic implementation. In particular, we discuss how to generalize the approximation by binomial models to option pricing in the multi-asset Black-Scholes setting.

2 Option Pricing and Binomial Tree Models: the Single Asset Case

An n-period binomial tree is a simple stochastic model for the dynamics of a stock price evolving over time. More precisely, it is a discrete-time stochastic process $\{S^{(n)}(i), i \in \{0, 1, \ldots, n\}\}$ such that

$$S^{(n)}(i+1) = \begin{cases} uS^{(n)}(i), & \text{if the price increases from period } i \text{ to } i+1, \\ dS^{(n)}(i), & \text{if the price decreases from period } i \text{ to } i+1, \end{cases}$$

where we require $u > d$ and $S^{(n)}(0) = s$. By our convention $u > d$, u is the favorable one-period return. The time spacing is assumed to be equidistant, so that each period has length $\Delta t = T/n$, where T is the time horizon. We also assume that at each state ("node") of the tree, we have the same probability $p \in (0, 1)$ to achieve the favorable one-period return u.

If we assume that in addition to trading this stock, the investor can also invest in a bank account with a continuously compounded interest rate r (i.e. investment grows by the factor $e^{r\Delta t}$ per period), we will require the no-arbitrage relation

$$u > e^{r\Delta t} > d. \tag{1}$$

If the above relation is violated, one can generate money without investing own funds by either selling the stock short (in the case $u \leq e^{r\Delta t}$) or financing a stock purchase by a credit (in the case of $d \geq e^{r\Delta t}$). In the following, we assume that the market is arbitrage-free (i.e. that (1) holds).

In the highly simplified financial market introduced above, an option is a functional $B = f(S^{(n)}(i), i = 1, \ldots, n)$ of the path of the stock price process. The owner of the option receives the payment B at the time horizon T (the maturity). As the final payment is a function of the stock price process, it is not known at the purchasing date. Consequently, trading the option can be identified with a bet on the evolution of the stock price $S^{(n)}$. Analogous to the Black-Scholes setting, the "fair" option price in the binomial model can be obtained via *the principle of replication*, i.e. one determines the costs required to set up a trading strategy in the stock and the riskless investment opportunity that will realize the same final payment B as received by holding the option (independently of the realized stock price movements!). We have the following basic result (see Bjoerk 2004):

Theorem 1 (Risk-Neutral Valuation and Replication). *Each option B in an n-period binomial model can be replicated by an investment strategy in the stock and the bond. The initial costs of this strategy determine the option price and are given by*

$$c_0 = E_{Q^{(n)}}\left(e^{-rT} B\right),$$

where the measure $Q^{(n)}$ is the product measure of the one-period transition measures $Q_i^{(n)}$ which are determined by

$$Q_i^{(n)}\left(\frac{S^{(n)}(i+1)}{S^{(n)}(i)} = u\right) = q = \frac{\exp(r\Delta t) - d}{u - d},$$

and for which we have

$$S^{(n)}(0) = E_{Q^{(n)}}\left(e^{-rT} S^{(n)}(n)\right). \tag{2}$$

Equation (2) shows that under $Q^{(n)}$ the expected relative return of the stock and the bond coincide. This motivates calling $Q^{(n)}$ the *risk-neutral measure* (note that it can easily be verified that $Q^{(n)}$ is the unique equivalent probability measure with this property). The risk-neutral probability q gives us the market view on the likelihood that the favorable one-period return u is attained. It can be different from the physical probability p. Then, if $E^{(n)}(e^{-rT}B)$ is computed with respect to p, we have $E^{(n)}(e^{-rT}B) \neq E_{Q^{(n)}}(e^{-rT}B)$, where $E_{Q^{(n)}}(e^{-rT}B)$ is the option price.

Note. The above result is identical to the result in the Black-Scholes setting: The underlying market is complete (i.e. every (suitably integrable) final payment can be replicated by appropriate trading in the underlying and the riskless investment) and the resulting option price is obtained as the net present value of the option payment under the risk-neutral measure. The risk-neutral measure is equivalent to the physical measure for the stock price evolution. As under the risk-neutral measure the corresponding discounted price processes of both assets are martingales, it is also called the equivalent martingale measure.

As seen above, the binomial approach leads to a modeling framework for option pricing that is technically easy and contains economically meaningful insights. However, the question remains whether the binomial model is in any reasonable way related to the Black-Scholes stock price model for which the stock price $\{S(t), t \in [0, T]\}$ is assumed to follow a geometric Brownian motion; that is

$$S(t) = s \cdot \exp\left(\left(r - \frac{1}{2}\sigma^2\right)t + \sigma W(t)\right)$$

with $W(t)$ a one-dimensional Brownian motion (under the risk-neutral measure Q associated with the continuous-time financial market) and $\sigma > 0$ a given constant describing the volatility of the stock price movements. The above question can be made more precise in two different ways: As the period length tends to zero,

- do we have (weak) convergence of the stock price paths in the sequence of increasing binomial models to the given geometric Brownian motion?
- does the sequence of binomial option prices $(E_{Q^{(n)}}(e^{-rT}B))_n$ converge to the corresponding option price in the Black-Scholes model?

The answers to these questions are intimately related to the concept of weak convergence of the corresponding stochastic processes. In particular, if we are only interested in the terminal value of the stock $S(T)$, the questions are answered by the classical Central Limit Theorem: Let X_n denote the (random) number of up-movements of the stock price in an n-period binomial model. We obviously have

$$X_n \sim B(n, p),$$

where $B(n, p)$ denotes the Binomial distribution with n trials and success probability p. Rewriting the stock price in the n-period binomial model as

$$S^{(n)}(n) = s \cdot u^{X_n} \cdot d^{n-X_n} = s \cdot e^{X_n \cdot \ln\left(\frac{u}{d}\right) + n \cdot \ln(d)},$$

using the choice $p = 1/2, b = r - \frac{1}{2}\sigma^2$ and

$$u = e^{b\Delta t + \sigma\sqrt{\Delta t}}, \qquad d = e^{b\Delta t - \sigma\sqrt{\Delta t}},$$

the Central Limit Theorem implies that

$$\begin{aligned} S^{(n)}(n) &= s \cdot \exp\left(bT + \sigma\sqrt{T}\left(\frac{2X_n - n}{\sqrt{n}}\right)\right) \\ &\xrightarrow{D} s \cdot \exp\left(bT + \sigma W(T)\right) = S(T). \end{aligned}$$

Hence, for the above parameter specifications, the terminal stock price in the binomial model $S^{(n)}(n)$ converges in distribution to the terminal value in the Black-Scholes model $S(T)$. Furthermore, provided the function $g : \mathbb{R}^+ \to \mathbb{R}^+$ satisfies suitable regularity conditions, the sequence $(E^{(n)}(e^{-rT}g(S^{(n)}(n))))_n$ obtained along the increasing binomial models converges to the quantity $E_Q(e^{-rT}g(S(T)))$ obtained in the Black-Scholes model; for instance, it clearly suffices that g is bounded and continuous. Yet $E_Q(e^{-rT}g(S(T)))$ is the Black-Scholes price for an option with payment $B = g(S(T))$. Consequently, for path-independent options (i.e. options that depend only on the terminal stock price), the above argument allows us to apply the binomial model specified above as a numerical valuation tool to approximate the option price in the Black-Scholes model. Of course, this is useful in practical applications if an explicit pricing formula is not known in the Black-Scholes setting. However, for the given parameter specifications, the sequence of binomial option prices $(E_{Q^{(n)}}(e^{-rT}g(S^{(n)}(n))))_n$ does in general not converge to the Black-Scholes option price (!). Hence, we observe that for numerical option pricing, it is only relevant whether the terminal distribution of the binomial model approximates the lognormal distribution specifying the terminal stock price $S(T)$. By contrast, it is irrelevant whether the corresponding probability p is determined according to the risk-neutral measure.

To introduce a general approximation technique that also works for option payments depending on the entire path of the stock price process, we have to invoke the concept of weak convergence of stochastic processes: Assume that the binomial model is such that the first two moments of the one-period log-returns of the stock price process S are matched. Then it follows from Donsker's Theorem (see e.g. Billingsley 1968) that (after linear interpolation) the binomial stock price process converges weakly to the geometric Brownian motion underlying the Black-Scholes stock price model. Hence, we have an affirmative answer to our first question whether the two stock price models can be related to one another. Furthermore, it follows from the definition of weak convergence that if the payoff function is bounded and continuous, the corresponding sequence $(E^{(n)}(e^{-rT}f(S^{(n)}(i), i = 1, \ldots, n)))_n$ converges to the option price in the Black-Scholes model. In fact, binomial option valuation can be justified for most common types of traded options; but this issue will not be addressed in this survey.

Note. According to the above arguments, the binomial method can be applied to numerical valuation of options in the Black-Scholes model. In this context, it is irrelevant whether the probability p coincides with the risk-neutral probability q (compare the above example where $p \neq q$). We only require that p be chosen such that the first two moments of the one-period log-returns are matched, so that the binomial stock price model approximates the Black-Scholes stock price model.

Of course, there are many possibilities to satisfy the moment matching conditions. The first suggestions were made by Cox, Ross and Rubinstein (CRR tree) and by Rendleman and Bartter (RB tree). The CRR tree is determined by the parameter specifications

$$u = e^{\sigma\sqrt{\Delta t}}, \qquad d = 1/u, \qquad p = \frac{1}{2}\left(1 + \left(r - \frac{1}{2}\sigma^2\right)\frac{1}{\sigma}\sqrt{\Delta t}\right).$$

Note that the probability p of an up-movement is only well-defined provided the grid size is sufficiently small; to be precise, we need that

$$n > \frac{(r - \frac{1}{2}\sigma^2)^2}{\sigma^2} T.$$

Note further that under the above specification of parameters, the second moment of the log-returns in the Black-Scholes model is only matched asymptotically; i.e. if grid size tends to zero. However, due to Slutsky's Theorem, it is clear that weak convergence is preserved. The CRR model is such that the log-tree (i.e. the tree containing the log-prices in the binomial model) is symmetric around the initial price $S^{(n)}(0)$. Upward or downward tendencies in the log-prices of the Black-Scholes model are incorporated into the binomial model via the above choice of the probability p.

The RB tree is given by the example considered above; that is,

$$p = \frac{1}{2}, \qquad u = e^{(r-\frac{1}{2}\sigma^2)\Delta t + \sigma\sqrt{\Delta t}}, \qquad d = e^{(r-\frac{1}{2}\sigma^2)\Delta t - \sigma\sqrt{\Delta t}}.$$

For this specification of parameters, the probabilities are automatically well-defined and symmetric. Upward or downward tendencies in the log-prices of the Black-Scholes model are incorporated into the discrete model via an appropriate form of the one-period returns u and d.

Both models ensure weak convergence to the Black-Scholes stock price model. Consequently, provided that the grid size is sufficiently small, the discounted expected value of the option payoff in the binomial models $E^{(n)}(e^{-rT} f(S^{(n)}(i); i = 1, \ldots, n))$ approximates the corresponding Black-Scholes option price. However, for both methods, the computed discounted expected option payoff does in general not coincide with the discrete-time option price because $p \neq q$. As a consequence, these models do not admit a simple economic interpretation. However, if the real-world market is modeled according to Black-Scholes, they can be applied to numerical option valuation.

3 Binomial Trees in Action—Implementation, Problems and Modifications

The binomial approach offers an attractive numerical pricing method because it can be implemented in form of an efficient backward algorithm. More precisely, for path-independent options with payment $B = f(S^{(n)}(i), i = 1, \ldots, n) = f(S^{(n)}(n))$, we have the following backward recursion:

Algorithm. Backward induction in the CRR tree

1. Set $V^{(n)}(T, S^{(n)}(n)) = f(S^{(n)}(n))$.
2. For $i = n-1, \ldots, 0$ do

$$\begin{aligned} V^{(n)} &\left(i \cdot \Delta t, S^{(n)}(i)\right) \\ &= \left[pV^{(n)}\left((i+1) \cdot \Delta t, uS^{(n)}(i)\right) \right. \\ &\quad \left. + (1-p)V^{(n)}\left((i+1) \cdot \Delta t, \frac{1}{u}S^{(n)}(i)\right)\right] \cdot e^{-r\Delta t}. \end{aligned}$$

3. Set $E^{(n)}(e^{-rT}B) = V^{(n)}(0, s)$ as the discrete-time approximation for the option price.

Algorithm. Backward induction in the RB tree

1. Set $V^{(n)}(T, S^{(n)}(n)) = f(S^{(n)}(n))$.
2. For $i = n-1, \ldots, 0$ do

$$\begin{aligned} &V^{(n)}\left(i \cdot \Delta t, S^{(n)}(i)\right) \\ &= \frac{1}{2}\left[V^{(n)}\left((i+1) \cdot \Delta t, uS^{(n)}(i)\right) + V^{(n)}\left((i+1) \cdot \Delta t, dS^{(n)}(i)\right)\right] \cdot e^{-r\Delta t}. \end{aligned}$$

3. Set $E^{(n)}(e^{-rT}B) = V^{(n)}(0, s)$ as the discrete-time approximation for the option price.

Apparently, due to the symmetry in probabilities, the RB model requires less operation counts for backward induction than the CRR model.

Note that for path-dependent options, it depends on the specific payoff functional whether there exist suitable modifications of the above algorithm. In particular, the algorithm can easily be adapted to the valuation of American options. Due to the widespread use of American options, this is an important advantage of the binomial method compared to alternative valuation techniques such as e.g. Monte Carlo methods. American options can be exercised at any time between the purchasing date and the expiration date T. In the Black-Scholes setting, this small conceptual difference causes a big difference in pricing because the optimal exercise date is not known on the date of purchase. Rather, it depends on the random evolution of the stock price process and is therefore itself random (mathematically, it is a stopping time with respect to the filtration generated by S). In contrast to the continuous-time American valuation problem, the American valuation problem can always (i.e. for any payoff function) be solved explicitly in the binomial model. Indeed, the main modification to the above backward induction algorithm is that for each node of the tree, the exercise value (i.e. the *intrinsic value* of the option) has to be compared to the value obtained by holding the option at least until the next time period and exercising it optimally afterwards. Let us illustrate binomial pricing of American options for the RB tree:

Algorithm. Backward induction for American options in the RB tree

1. Set $V^{(n)}(T, S^{(n)}(n)) = f(S^{(n)}(n))$.
2. For $i = n-1, \ldots, 0$ do

$$\tilde{V}^{(n)}\left(i \cdot \Delta t, S^{(n)}(i)\right)$$
$$= \frac{1}{2}\left[V^{(n)}\left((i+1)\cdot \Delta t, uS^{(n)}(i)\right) + V^{(n)}\left((i+1)\cdot \Delta t, dS^{(n)}(i)\right)\right]\cdot e^{-r\Delta t}$$

and set

$$V^{(n)}\left(i \cdot \Delta t, S^{(n)}(i)\right) = \max\left\{\tilde{V}^{(n)}\left(i \cdot \Delta t, S^{(n)}(i)\right), f\left(S^{(n)}(i)\right)\right\}.$$

3. Set $E^{(n)}(e^{-rT}B) = V^{(n)}(0, s)$ as the discrete-time approximation for the price of an American option with final payment f.

Note. Due to the simplified dynamics of binomial models (finite state space and discrete-time observations), binomial approximations to Black-Scholes option prices can be obtained by an easy and efficient backward induction algorithm. This is useful in practical applications if an analytic pricing formula is not known in the Black-Scholes setting. In particular, as seen above, the tree algorithm can easily be modified to the valuation of American options.

Although the binomial method is based on an efficient backward induction algorithm, it suffers from several drawbacks in practical applications. First, if the payoff function is discontinuous, the Berry-Esséen inequality on the rate of convergence of binomial price estimates is in general tight; i.e. convergence is no faster than $1/\sqrt{n}$. Second, for many types of options convergence is not smooth, but oscillatory: we observe low-frequency shrinking accompanied by high-frequency oscillations. Consequently, choosing a smaller grid size does not necessarily provide a better option price estimate, and extrapolation methods can typically not be applied. Well-known examples of irregular convergence behavior are the so-called *sawtooth effect* and the *even-odd problem*: Binomial price estimates obtained from the conventional methods described above often exhibit a sawtooth pattern. That is, if the grid size is increased ($n \to n+2$), the discretization error in the corresponding binomial option prices decreases to a negligible size. However, if the step size is further increased, the error rises abruptly. This is again followed by a period of decreasing errors. Figure 1 illustrates the sawtooth pattern. The sawtooth effect was first observed for

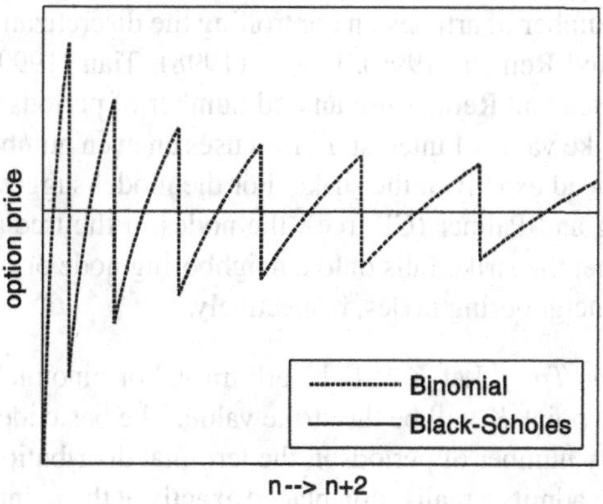

Fig. 1 The sawtooth effect

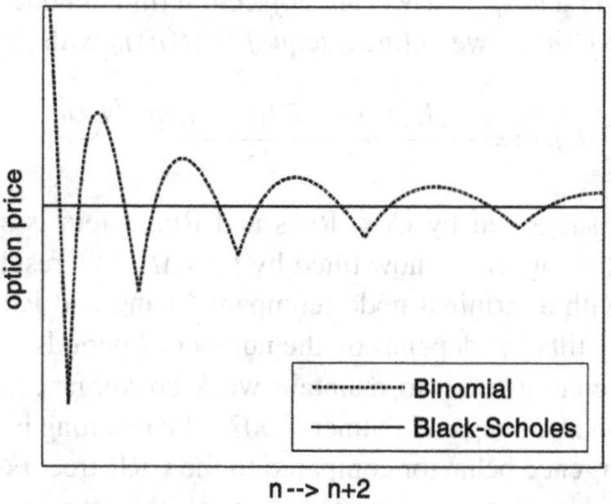

Fig. 2 Scallops

barrier options (i.e. options for which the right to exercise either originates or expires on certain regions of the path space of S) by Boyle and Lau (1994); yet it can also be present for other types of options. The price estimates obtained from conventional tree methods (for $n \rightarrow n+2$) also often converge in form of scallops. This is illustrated in Fig. 2. In this survey, we do not wish to explain where the different patterns come from, but let us stress that the convergence pattern observed depends both on the valuation problem under consideration and on the parameter specification of the tree method chosen. In addition to the irregular convergence behavior observed along $n \rightarrow n+2$, the binomial price estimates typically exhibit micro oscillations between even and odd values of n; the latter aspect is often referred to as the even-odd effect. The micro oscillations are superimposed on the macro oscillations considered previously; i.e. they are superimposed on the irregular convergence behavior along the even integers (the sawtooth pattern, scallops, etc.) and on the irregular convergence behavior along the odd integers.

There is a vast number of articles on controlling the discretization error, amongst which are Leisen and Reimer (1996), Leisen (1998), Tian (1999) and Chang and Palmer (2007). Leisen and Reimer use an odd number of periods with the tree centered around the strike value of interest. Leisen uses an even number of periods with the central node placed exactly at the strike. For the model suggested by Tian (Tian tree) and by Chang and Palmer (CP tree), the nodes in the tree are moved only a small distance so that the strike falls onto a neighboring node or onto the geometric average of the two neighboring nodes, respectively.

Example 1 (The Tian Tree). Let $K \in \mathbb{R}$ be arbitrary. For binomial valuation of call and put options, the point K will be the strike value. The basic idea behind the Tian model is that for any number of periods n, the terminal distribution of the CRR tree is modified so that it admits a realization placed exactly at the point K. To be precise, for each $n \in \mathbb{N}$, there is some integer $l(n)$ for which $K \in (s_n^{(n)}(l(n)-1), s_n^{(n)}(l(n))]$, where $s_n^{(N)}(l(n)-1)$ and $s_n^{(N)}(l(N))$ are adjacent terminal nodes in the CRR tree. Given the sequence $(l(n))_n$, we define a sequence $(t(n))_n$ with

$$t(n) := \frac{\ln(K/s_0) - (2l(n) - n)\sigma\sqrt{T/n}}{T}.$$

While the log-tree suggested by Cox, Ross and Rubinstein is symmetric around the starting value, the log-tree is now tilted by $t(n)\Delta t$. As a result, the strike value always coincides with a terminal node (compare Chang and Palmer 2007 for details[1]). Note that the tilt $t(n)$ depends on the number of periods n. It can be verified that the tilt is sufficiently small to maintain weak convergence to the stock price process S. As shown in Chang and Palmer (2007), the resulting binomial tree shows an improved convergence behavior compared to the CRR tree: For cash-or-nothing options (options with a piecewise constant payoff), the estimates still converge in order $1/\sqrt{n}$, but convergence is smooth, so extrapolation methods can be applied. For plain-vanilla options (options with a piecewise linear payoff), the estimates converge in order $1/n$. In contrast to the CRR tree, the coefficient of the leading error term is again constant.

Example 2 (The CP Tree). The CP model is such that for any number of periods n, the strike K is optimally located between two adjacent terminal nodes. The geometry of the CRR tree implies that the strike value is optimally placed if it is set at the geometric average of two adjacent nodes (compare Chang and Palmer 2007 for details). This can again be achieved by defining an appropriate sequence of tilt parameters: Let $(l(n))_n$ be defined as above. Then, the appropriate sequence of tilt parameters is given by $(\tilde{t}(n))_n$ with

$$\tilde{t}(n) = \frac{\ln(x/s_0) - (2l(n) - n - 1)\sigma\sqrt{T/n}}{T}.$$

[1] In the original article by Tian, the improved convergence behavior is illustrated by numerical examples. Theoretical results are given in Chang and Palmer (2007).

As for the Tian model, the resulting tree remains close enough to the CRR tree to ensure weak convergence. The convergence behavior of the CP model is further improved compared to the Tian model. For the latter, the probability to end up in the money (i.e. the likelihood that your bet on the stock price movement is correct) is consistently under- or overestimated. The CP model takes account of this problem. As a result, the rate of convergence for cash-or-nothing options is improved to $1/n$ (without extrapolation).

A conceptually different approach to improve the convergence behavior of the discretization error of binomial trees can be found in Rogers and Stapleton (1998). They fix some $\Delta x > 0$ and view the diffusion only at the discrete set of times at which it has moved Δx from where it was last observed. This approximation technique results in a random walk that approximates the diffusion uniformly closely, so that the convergence behavior can be improved without an explicit re-location of nodes in the tree. However, it leads to a pathwise binomial tree with a random number of periods.

Remark (Trinomial Trees). As the completeness of the binomial market is irrelevant for numerical valuation of Black-Scholes option prices, the approximation by tree methods is not limited to binomial models; that is, models that exhibit two-state movements. In fact, the Black-Scholes option price can be approximated by any k-nomial tree (i.e. a tree for which each node has the same number k of successor nodes) provided the tree model satisfies the moment matching conditions required to apply Donsker's Theorem. Furthermore, if the corresponding tree is *re-combining*, a backward induction algorithm similar to that described for binomial trees can be used to efficiently compute discounted expectations in the k-nomial model. In a recombining tree, paths with the same number of up- and down-movements end at the same node independently of the order in which the up- and down-movements have occurred. Of course, if the tree model allows for additional states, the computational effort required for backward induction increases compared to binomial trees. However, the application of multinomial models provides some additional free parameters because there are two moment matching conditions only—independently of the number of states in the discrete-time model. Consequently, recombining trinomial trees (i.e. trees for which each node has three successors) are sometimes used in practical applications to approximate the price of path-dependent options (such as barrier options) as they are more flexible than their binomial counterparts.

Note. Trinomial trees can be adapted to complex valuation problems—this can lead to an improved convergence behavior. However, trinomial option valuation is more costly than binomial pricing. Alternatively, the convergence behavior can be improved without increasing computational effort by applying advanced binomial models. The implementation of advanced binomial models can be more involved.

4 Multi-Asset Option Valuation and Binomial Trees

4.1 Standard Binomial Methods for Multi-Asset Options

Compared to the single-asset case, setting up multi-dimensional binomial trees is more complicated because the entire correlation structure between the underlying (log-)asset prices in the Black-Scholes setting have to be taken into account. More precisely, if we use a binomial approximation to the multi-asset Black-Scholes model given by

$$dS_i(t) = S_i(t)\,(rdt + \sigma_i dW_i)\,, \qquad \mathrm{Corr}\left(\frac{dS_i}{S_i}, \frac{dS_j}{S_j}\right) = \rho_{ij}dt, \quad i, j = 1, \ldots, m, \tag{3}$$

the correlations between the log-prices have to be matched in addition to the previous conditions on the expectation and the variance of the one-period log-returns. Consequently, the construction of multi-dimensional binomial trees becomes the more involved, the more assets are traded in the market. Furthermore, the additional moment matching conditions can lead to negative jump probabilities in the corresponding binomial model, so that the model is no longer well-defined. In particular, the above argument on weak convergence cannot be used anymore.

We define an m-dimensional n-period binomial tree via

$$S_i^{(n)}(k) = S_i^{(n)}((k-1))\, e^{\alpha_i \Delta t + \beta_i \sqrt{\Delta t} Z_{k,i}^{(n)}}, \quad k = 1, \ldots, n, \; i = 1, \ldots, m,$$

where as before each period has length $\Delta t = \frac{T}{n}$ and $Z_{k,i}^{(n)}$ are random variables taking values in $\{-1, 1\}$. We choose the constants α_i, β_i and the jump probabilities (defining the distribution of the random variables $Z_{k,i}^{(n)}$) such that

- the random vectors $(Z_{k,1}^{(n)}, \ldots, Z_{k,m}^{(n)})$, $k = 1, \ldots, n$, are i.i.d. for fixed n,
- the first two moments of the one-period log-returns in the Black-Scholes model coincide (at least) asymptotically with those in the tree; in particular, the covariances of the random variables $Z_{k,i}^{(n)}$ and $Z_{k,j}^{(n)}$ satisfy

$$\beta_i \beta_j \,\mathrm{Cov}\big(Z_{k,i}^{(n)}, Z_{k,j}^{(n)}\big) \overset{n\to\infty}{\longrightarrow} \rho_{ij}\sigma_i\sigma_j, \quad 1 \le i, j \le m.$$

Let us illustrate standard multi-asset binomial trees with multi-dimensional generalizations of the (one-dimensional) CRR tree and the (one-dimensional) RB tree:

Example 3 (The BEG Tree). The BEG tree as introduced in Boyle et al. (1989) is the m-dimensional generalization of the one-dimensional CRR tree; that is

$$S_i^{(n)}(k) = S_i^{(n)}((k-1))\, e^{\sigma_i \sqrt{\Delta t} Z_{k,i}^{(n)}}, \quad k = 1, \ldots, n, \; i = 1, \ldots, m,$$

where for each period k, the distribution of the random vector $(Z_{k,1}^{(n)}, \ldots, Z_{k,m}^{(n)})$ is defined via the set of jump probabilities

$$p_{BEG}^{(n)}(l) = \frac{1}{2^m}\left(1 + \sum_{j=1}^{m}\sum_{i=1}^{j-1}\delta_{ij}(l)\rho_{ij} + \sqrt{\Delta t}\sum_{i=1}^{m}\delta_i(l)\frac{r - \frac{1}{2}\sigma_i^2}{\sigma_i}\right), \quad 1 \le l \le 2^m,$$

with

$$\delta_i(l) = \begin{cases} 1 & \text{if } Z_{k,i}^{(n)} = 1, \\ -1 & \text{if } Z_{k,i}^{(n)} = -1, \end{cases} \qquad \delta_{ij}(l) = \begin{cases} 1 & \text{if } Z_{k,i}^{(n)} = Z_{k,j}^{(n)}, \\ -1 & \text{if } Z_{k,i}^{(n)} \neq Z_{k,j}^{(n)}. \end{cases} \tag{4}$$

As for the CRR tree, the log-prices in the BEG-tree are symmetric around the starting value. However, the probabilities depend in a complicated way on both the drift and the covariance structure of the continuous-time model.

Example 4 (The m-Dimensional RB Tree). The m-dimensional Rendleman-Barrter tree is described in Amin (1991) and in Korn and Müller (2009). It is given by

$$S_i^{(n)}(k) = S_i^{(n)}((k-1))\, e^{\left(r - \frac{1}{2}\sigma_i^2\right)\Delta t + \sigma_i\sqrt{\Delta t}Z_{k,i}^{(n)}}, \quad k = 1, \ldots, n, \; i = 1, \ldots, m,$$

where for each period k, the distribution of the random vector $(Z_{k,1}^{(n)}, \ldots, Z_{k,m}^{(n)})$ is defined via the set of jump probabilities

$$p_{RB}^{(n)}(l) = \frac{1}{2^m}\left(1 + \sum_{j=1}^{m}\sum_{i=1}^{j-1}\delta_{ij}(l)\rho_{ij}\right), \quad 1 \le l \le 2^m,$$

with $\delta_{ij}(.)$ given as in (4). While the log-prices in the RB-tree are non-symmetric, the probabilities depend only on the covariance structure of the continuous-time model. In particular, in contrast to the BEG tree, the jump probabilities depend neither on the number of periods n nor on the drift.

Let us emphasize that for both models, the jump probabilities are not necessarily non-negative! In contrast to the common belief, this problem cannot always be fixed by choosing a sufficiently large number of periods n (see Korn and Müller 2009 for an explicit example).

Remark (Incompleteness of Multi-Dimensional Binomial Trees). There is no analogue to Theorem 1 for multi-dimensional binomial trees. In particular, multi-dimensional binomial markets are incomplete for dimensions $m > 1$. This implies that there is in general no unique price for a given option payment. However, the incompleteness of the multi-dimensional binomial model is irrelevant for the application of the corresponding binomial model to numerical valuation of options in the corresponding (complete) multi-dimensional Black-Scholes market: Provided

- the model is well-defined, and
- the first two moments of the one-period log-returns are (asymptotically) matched,

the option price in the Black-Scholes model can be approximated by computing the expected discounted option payments in the multi-dimensional binomial models (with respect to the measure induced by the distribution of the random variables

$Z_{k,j}^{(n)}$). As for the one-dimensional setting, this procedure is justified by weak convergence arguments based on Donsker's Theorem. For an approximation of the multi-dimensional Black-Scholes model by a complete multinomial model, we refer to He (1990).

For the standard multi-dimensional trees considered above, the corresponding algorithm for binomial option pricing is conceptually the same as for the one-dimensional case. First, we assign possible payoff scenarios to the terminal nodes. Afterwards, we step backwards through the tree by computing the value at each node as the weighted average of the values assigned to its successor nodes. However, the number of terminal nodes grows exponentially in the number of traded assets (this effect is often referred to as the *curse of dimensionality*). Consequently, for high-dimensional options (i.e. options whose payments depend on a large number of assets), standard multi-dimensional trees are currently not practically useful. This is an inherent drawback of the binomial approach as a method based on the discretization of the underlying assets. However, up to dimension four, let us say, standard multi-dimensional trees can lead to results that are perfectly competitive and often superior to those obtained by Monte Carlo methods. Nevertheless, as inherited by their one-dimensional counterparts, standard multi-dimensional trees often exhibit an irregular convergence behavior.

Note. In principle, we can obtain multi-dimensional variants of the conventional one-dimensional binomial trees considered above. However, the corresponding trees are not well-defined for every parameter setting of the corresponding continuous-time model. Provided that the tree is well-defined and that it converges weakly to the multi-dimensional Black-Scholes stock price model, it can be applied to numerical valuation of options in the limiting Black-Scholes market. This yields an easy and efficient method to approximate the price of a multi-asset option, yet the corresponding tree model still suffers from several drawbacks with respect to practical applications.

In the following, we suggest orthogonal trees as an alternative to standard multi-dimensional trees.

4.2 Valuing Multi-Asset Options by Orthogonal Trees

The complications observed above motivate searching for alternative approaches to multi-dimensional trees that ensure well-defined jump probabilities and easy moment matching. In this context, Korn and Müller (2009) introduced orthogonal trees, which are based on a general decoupling method for multi-asset Black-Scholes settings. The model we suggest contains the two-dimensional example of Hull (2006) as a special case.

The construction of orthogonal trees consists of two steps: In contrast to standard multi-dimensional tree methods, we first decouple the components of the log-stock price process $(\ln(S_1), \ldots, \ln(S_m))$. Afterwards, we approximate the decoupled process by an m-dimensional tree defined as the product of appropriate one-dimensional trees. Of course, in order to apply the resulting tree to numerical option valuation, we have to apply a backtransformation to any time-layer of nodes that contributes to the option payment under consideration.

To explain the above procedure in detail, we consider the multi-asset Black-Scholes type market with m stocks following the price dynamics (3). The decoupling method is based on a decomposition of the volatility matrix Σ of the log-prices $(\ln(S_1(t)), \ldots, \ln(S_m(t)))$ via

$$\Sigma = GDG^T \tag{5}$$

with D a diagonal matrix. The above decomposition allows to introduce the transformed ("decoupled") log-price process Y defined by

$$Y(t) := G^{-1} \left(\ln\left(S_1(t)\right), \ldots, \ln\left(S_m(t)\right)\right)^T .$$

Note that the process $Y = (Y_1, \ldots, Y_m)$ follows the dynamics

$$\begin{aligned} dY_j(t) &= \mu_j dt + \sqrt{d_{jj}} d\bar{W}_j(t), \qquad \mu = G^{-1}\left(r\underline{1} - \frac{1}{2}\underline{\sigma}^2\right), \\ \underline{\sigma}^2 &= \left(\sigma_1^2, \ldots, \sigma_m^2\right)^T, \end{aligned} \tag{6}$$

where $\bar{W}(t)$ is an m-dimensional Brownian motion (see Korn and Müller 2009 for details). In particular, the component processes are independent. Consequently, the transformed process Y can be approximated by a product of (independent) one-dimensional trees; one for each component process Y_j. This implies that we have to match only the mean and variance of one-period log-returns for each component process. There is no need for correlation matching! In particular, it is easy to obtain well-defined transition probabilities.

The decoupling approach allows for the following choices:

- Which one-dimensional tree(s) should be used to approximate the components?
- Which matrix decomposition should we choose in (5)?

In fact, each component process Y_j can be approximated by an arbitrary one-dimensional (factorial) tree. As discussed previously, the one-dimensional RB tree leads to a backward induction algorithm that is cheaper than that obtained by the CRR tree. Consequently, we suggest to approximate each component Y_j by an appropriate one-dimensional RB tree. In particular, this choice implies that the resulting m-dimensional tree is always well-defined (independently of both the parameter setting and the number n of periods). To answer the second question, note that there is an infinite number of decompositions solving (5). In particular, the Cholesky decomposition and the spectral decomposition can be applied. However, as shown by numerical and theoretical considerations, it is typically more favorable to choose the spectral decomposition[2] (compare Korn and Müller 2009 for details). Applying the decoupling approach with the above choices results in the following basic steps for numerical valuation of path-independent options with payoff $B = f(S_1(T), \ldots, S_m(T))$:

Numerical Valuation of Path-Independent Options via Orthogonal RB Trees

1. Compute the spectral decomposition $\Sigma = GDG^T$ of the covariance matrix of the log-stock prices.
2. Introduce the process $Y(t) := G^T(\ln(S_1(t)), \ldots, \ln(S_m(t)))^T$ following the dynamics (6).
3. Approximate each component process Y_j by a one-dimensional RB tree which matches the mean and the variance of the one-period log-returns. This results in an m-dimensional discrete process $Y^{(n)}$ following the dynamics

$$Y^{(n)}(k+1) = \begin{pmatrix} Y_1^{(n)}(k) + \mu_1 \Delta t + Z_{k+1,1}^{(n)} \sqrt{d_{11}} \sqrt{\Delta t} \\ \vdots \\ Y_m^{(n)}(k) + \mu_m \Delta t + Z_{k+1,m}^{(n)} \sqrt{d_{mm}} \sqrt{\Delta t} \end{pmatrix},$$
$$Y^{(n)}(0) = Y(0),$$

   where $(Z_{k+1,1}^{(n)}, \ldots, Z_{k+1,m}^{(n)})$ is a random vector of independent components, each attaining the two values $+1$ and -1 with probability $\frac{1}{2}$. As above, we require that the random vectors $(Z_{k,1}^{(n)}, \ldots, Z_{k,m}^{(n)})$, $k = 1, \ldots, n$, are i.i.d. Further, d_{ii} are the eigenvalues of the covariance matrix Σ. By moment matching the drift vector $(\mu_1, \ldots, \mu_m)$ and the distribution of $(Z_{k+1,1}, \ldots, Z_{k+m,1})$ are determined in such a way that the process $Y^{(n)}$ approximates the decoupled process Y.
4. Apply the backtransformation of the decoupling rule to the terminal nodes in the tree associated with $Y^{(n)}$, i.e. with $h(\underline{x}) := (e^{G_1 \underline{x}}, \ldots, e^{G_m \underline{x}})$ (where G_i denotes the i th row of G) set

$$S^{(n)}(n) := h\left(Y^{(n)}(n)\right).$$

   Starting from the transformed terminal nodes (i.e. from the realizations of the random variable $S^{(n)}(n)$), we obtain an approximation to the Black-Scholes option price using the standard backward induction algorithm.

It remains to answer the question whether we can theoretically justify the above procedure to approximate Black-Scholes option prices via the decoupling approach. In fact, one can show by weak convergence arguments that this is ensured by continuity of the backtransformation map.

---

[2] The triangular structure of the Cholesky decomposition is favorable when additional assets enter the market. This issue will not be addressed in this survey.

Numerical Performance of Orthogonal RB Trees

For path-independent options, the above valuation algorithm is cheaper than the conventional multi-dimensional tree methods considered previously. This is due to the fact that under the above choice of the embedded one-dimensional trees, each path in the m-dimensional tree is equally likely. However, for path-dependent options, it does not suffice to apply the backtransformation to the terminal nodes only. Rather, the backtransformation has to be applied to every time layer in the tree that contributes to the option payments. In particular, for the valuation of American options, all time layers have to be transformed. This means that we have to transform the entire tree associated with $Y^{(n)}$ into a "valuation tree" associated with the discrete process $S^{(n)}$ defined by

$$S^{(n)}(k) := h\left(Y^{(n)}(k)\right), \quad k = 0, \ldots, n.$$

Due to the additional computational effort required for backtransformation, numerical valuation of path-dependent and American options is typically more expensive for orthogonal trees than for conventional multi-dimensional trees. However, in case that the latter are not well-defined, the decoupling approach at least justifies binomial option valuation. Furthermore, the decoupling approach often leads to a more regular convergence behavior than that observed for conventional multi-dimensional methods. This is a consequence of the fact that by applying the backtransformation to the nodes in the tree, the nodes are dislocated in an irregular (i.e. non-linear) way. As a result, the probability mass gets smeared in relation to fixed strike values or barrier levels. The benefits due to the more regular convergence behavior often overcompensate the additional computational effort required for backtransformation. In particular, the sawtooth effect can vanish completely, so that the order of convergence can be improved by applying extrapolation methods (for a detailed performance analysis see Korn and Müller 2009).

In addition to the above advantages, the decoupling approach is perfectly suited to cut down high-dimensional valuation problems to the "important dimensions". To explain this, note that the dynamics of stock markets can often be explained by a relatively small number of risk factors that is less than the number of traded assets. In such a market, it seems reasonable to value an option by a tree whose dimension is lower than the number of underlyings. The above algorithm is particularly suited to that purpose because it incorporates a principal component analysis in an implicit way. In particular, it considers the underlying risk factors (rather than the traded stocks) as the important ingredients. Consequently, if the number of relevant risk factors is small, the decoupling approach can give a fast first guess on high-dimensional valuation problems by considering the non-relevant risk factors as deterministic (compare Korn and Müller 2009).

Note. We suggest orthogonal trees as an alternative to standard multi-dimensional trees. On the one hand, the decoupling approach keeps the tree structure which results in an efficient backward induction algorithm; on the other hand, it is no longer based on a random walk approximation to the correlated asset price processes, which leads to advantages in practical applications. In particular, the above orthogonal tree procedure is always well-defined, and it can be combined with a principal component analysis, which leads to model reduction. This will allow for a fast first guess on the solution of high-dimensional valuation problems.

5 Conclusion and Outlook

Despite their conceptual simplicity, binomial trees can offer an efficient numerical method to approximate Black-Scholes option prices. This is in particular true for American options. However, multi-dimensional option valuation by binomial trees suffers from the inherent drawback that it is currently not of practical use for high-dimensional problems. Furthermore, as conventional binomial trees often lead to an irregular convergence behavior, controlling the discretization error is important in practical applications.

Binomial methods can also be applied to approximate option prices in continuous-time stock price models other than the Black-Scholes model. There is still ongoing research on the application of the binomial method to stock price models following non-Black-Scholes-type dynamics.

Further, the irregular convergence behavior of binomial methods remains a field of intensive study. Previous research often concentrates on a particular type of options (such as barrier options) and thus leads to highly specialized approaches. An exception is the orthogonal tree method of Korn and Müller (2009) which seems to exhibit a more smooth convergence behavior for the popular types of exotic options.

As shown above, the orthogonal tree procedure can be cut down to important risk factors. We suggest to analyse this issue for options on prominent indices. Further, one could think of alternative approaches to deal with the curse of dimensionality.

References

Amin, K.I.: On the computation of continuous time option prices using discrete approximations. J. Financ. Quant. Anal. **26**, 477–495 (1991)

Billingsley, P.: Convergence of Probability Measures. Wiley, New York (1968)

Bjoerk, T.: Arbitrage Theory in Continuous Time, 2nd edn. Oxford University Press, Oxford (2004)

Black, F., Scholes, M.S.: The pricing of options and corporate liabilities. J. Polit. Econ. **81**, 637–654 (1973)

Boyle, P.P., Lau, S.H.: Bumping up against the barrier with the binomial method. J. Deriv. **1**, 6–14 (1994)

Boyle, P.P., Evnine, J., Gibbs, S.: Numerical evaluation of multivariate contingent claims. Rev. Financ. Stud. **2**, 241–250 (1989)

Chang, L.-B., Palmer, K.: Smooth convergence in the binomial model. Finance Stoch. **11**, 91–105 (2007)

Cox, J.C., Ross, S.A., Rubinstein, M.: Option pricing: a simplified approach. J. Financ. Econ. **7**, 229–263 (1979)

He, H.: Convergence from discrete- to continuous-time contingent claim prices. Rev. Financ. Stud. **3**, 523–546 (1990)

Hull, J.C.: Options, Futures, and other Derivatives, 6th edn. Pearson/Prentice Hall, New Jersey (2006)

Korn, R., Müller, S.: The decoupling approach to binomial pricing of multi-asset options. J. Comput. Finance **12**, 1–30 (2009)

Leisen, D.P.J.: Pricing the American put option: a detailed convergence analysis for binomial models. J. Econ. Dyn. Control **22**, 1419–1444 (1998)

Leisen, D.P.J., Reimer, M.: Binomial models for option valuation—examining and improving convergence. Appl. Math. Finance **3**, 319–346 (1996)

Rendleman, R.J., Bartter, B.J.: Two-state option pricing. J. Finance **34**, 1093–1110 (1979)

Rogers, L.C.G., Stapleton, E.J.: Fast accurate binomial pricing. Finance Stoch. **2**, 3–17 (1998)

Rubinstein, M.: Guiding force. In: From Black-Scholes to Black Holes: New Frontiers in Options, Risk Magazine, November (1992)

Tian, Y.S.: A Flexible Binomial option pricing model. J. Futures Mark. **19**, 817–843 (1999)

Uncertainty in Gaussian Process Interpolation

Hilke Kracker, Björn Bornkamp, Sonja Kuhnt, Ursula Gather, and Katja Ickstadt

Abstract In this article, we review a probabilistic method for multivariate interpolation based on Gaussian processes. This method is currently a standard approach for approximating complex computer models in statistics, and one of its advantages is the fact that it accompanies the predicted values with uncertainty statements. We focus on investigating the reliability of the method's uncertainty statements in a simulation study. For this purpose we evaluate the effect of different objective priors and different computational approaches. We illustrate the interpolation method and the practical importance of uncertainty quantification in interpolation in a sequential design application in sheet metal forming. Here design points are added sequentially based on uncertainty statements.

1 Introduction

Interpolation is an ubiquitous problem in the natural sciences. The situation usually arises when a (typically continuous and smooth) function $y(\boldsymbol{x})$, $y : \mathcal{X} \mapsto \mathbb{R}$ with $\boldsymbol{x} \in \mathcal{X} \subset \mathbb{R}^d$, is difficult (*i.e.*, time consuming) to calculate, so that it is impracti-

Hilke Kracker, Fakultät Statistik, TU Dortmund, Dortmund, Germany
e-mail: kracker@statistik.tu-dortmund.de

Björn Bornkamp, Fakultät Statistik, TU Dortmund, Dortmund, Germany
e-mail: bornkamp@statistik.tu-dortmund.de

Sonja Kuhnt, Fakultät Statistik, TU Dortmund, Dortmund, Germany
e-mail: kuhnt@statistik.tu-dortmund.de

Ursula Gather, Fakultät Statistik, TU Dortmund, Dortmund, Germany
e-mail: gather@statistik.tu-dortmund.de

Katja Ickstadt, Fakultät Statistik, TU Dortmund, Dortmund, Germany
e-mail: ickstadt@statistik.tu-dortmund.de

L. Devroye et al. (eds.), *Recent Developments in Applied Probability and Statistics*,
DOI 10.1007/978-3-7908-2598-5_4,

cal to further work with $y(.)$ for subsequent analyses (*e.g.*, evaluation, optimization, etc). In these cases a good approximation $\hat{y}(.)$ of $y(.)$ is desirable, which is cheap to calculate. For this purpose one typically evaluates $y(.)$ at a set of carefully chosen design points $\boldsymbol{x}_1, \ldots, \boldsymbol{x}_n$, with $\boldsymbol{x}_i \in \mathcal{X}$ and builds an approximation $\hat{y}(.)$ that includes all information available on $y(.)$. A basic requirement for $\hat{y}(.)$, for example, would be that $\hat{y}(\boldsymbol{x}_i) = y(\boldsymbol{x}_i)$ for $i \in \{1, \ldots, n\}$. If the true function $y(.)$ is assumed to be continuous and smooth, $\hat{y}(.)$ should also be continuous and smooth. In some cases one might have additional information available, for example, on the gradient or the shape properties of $y(.)$ such as positivity, monotonicity, convexity or unimodality. However, even if all available information can be incorporated, it remains *uncertain* how well $\hat{y}(\boldsymbol{x})$ predicts the value of $y(\boldsymbol{x})$ for points $\boldsymbol{x} \notin \{\boldsymbol{x}_1, \ldots, \boldsymbol{x}_n\}$. Most approaches to interpolation do not explicitly account for this uncertainty in the sense that they report $\hat{y}(\boldsymbol{x})$ without any information on the certainty of this prediction.

A way of incorporating uncertainty into general problems in numerical analysis is by employing a probabilistic approach, see *e.g.* Diaconis (1988) or O'Hagan (1992) for reviews. In this paper we will investigate a particular probabilistic method for multivariate interpolation, called "Bayesian kriging" or "Gaussian process interpolation", which allows for quantifying the uncertainty involved in the interpolation process. The methodology originated in the geostatistics literature and has become quite popular in other research areas in recent years. The main idea is to impose a particular probabilistic model onto the interpolation process that post hoc allows to evaluate the uncertainty in the predicted value $\hat{y}(\boldsymbol{x})$ in probabilistic terms. We note that measures of uncertainty, other than probability, may be employed to accompany the prediction $\hat{y}(.)$, for example, deterministic upper bounds on an approximation error.

The application that is largely responsible for the increased interest in probabilistic interpolation methods in statistics is the approximation of complex computer models (see Santner et al. 2003). Computer models simulate a real world physical process by a mathematical formalization of the process. Areas of application include climate research, meteorology, hydrology (fluid dynamics), forming (continuum mechanics) and pharmacokinetics (chemical kinetics). Often the modeled process depends on input parameters $\boldsymbol{x} = (x^{(1)}, \ldots, x^{(d)})$, and particularly in engineering applications one is interested in finding a set of values for the input parameters so that the process works optimally in a certain pre-specified way.

In principle the computer model may be evaluated at any input configuration of interest, however, evaluations of the computer model can be extremely time consuming so that in many situations only a limited number of computer code evaluations, so called computer experiments, are feasible. Although conducting computer experiments is usually much cheaper than performing real physical experiments, both time-wise and financially, there is still a great potential of further speeding up the computer modeling process by the use of interpolation methods.

Approximating computer models with Gaussian process interpolation goes back to Sacks et al. (1989) and Currin et al. (1991) and has since then been accepted as the current standard approach in this field. One of the reasons for the success of this methodology is the fact that it allows for uncertainty statements of the predicted value. We will illustrate why uncertainty is important for analyzing computer models in the following two paragraphs by two specific examples.

One question arising naturally in the context of computer modeling is: How well does the computer model describe the real world? In this case one has to perform both real and simulated experiments. To judge the adequacy of an interpolation model, it is now important to have an uncertainty statement available. If the computer model approximation at a point deviates strongly from the experiment, it depends on the uncertainty in the prediction of the interpolation model, whether we would dismiss the computer model or not. If the uncertainty is large, we would possibly not dismiss the computer model, if the uncertainty is negligible however, we are sufficiently sure that the computer model is biased. A similar problem arises, when the computer model contains input variables that can be adjusted in the computer model but not in the real world, for example, unknown physical constants or numerical tuning parameters needed to evaluate $y(.)$. How can we tune the computer model so that it describes the real world as adequately as possible? This task is typically known as *calibration*, see Kennedy and O'Hagan (2001) and Bayarri et al. (2007) for a Bayesian approach for computer model calibration based on Gaussian process interpolation.

Another important aspect of computer experiments is the choice of the design $D = (\boldsymbol{x}_1, \ldots, \boldsymbol{x}_n)$, hence at which points to observe the function $y(.)$. In practice often space-filling designs are used. They fill up the design region $\mathcal{X}$ as well as possible with as few points as possible (see Santner et al. 2003 or Fang et al. 2006 for details), and might be regarded as general purpose designs. When one is interested in a specific characteristic of the computer model $y(.)$, other designs might turn out to be more adequate. Suppose for example one wants to find the minimum $\boldsymbol{x}^*$ of $y(.)$, and the computer experiments are conducted sequentially. After evaluating the computer model at a start design (for example, a space filling design), we would like to add design points so that we learn most about the location of the minimum of $y(.)$. In general it is not a good strategy to sample the next point at the minimum of the interpolating function. If the computer model has already been evaluated at this point there is no gain of knowledge to evaluate the deterministic function again at this point. To learn most about the minimum, we would rather like to choose a design point, which might be a potential minimum, but where we are still sufficiently uncertain about the specific function value. Thus it is smarter to sample the function based on a balance between the uncertainty in the predicted function value and the original goal of function minimization. One possibility is to place the new design point, where the probability of being smaller than the current minimum is largest. This effectively results in an exploration of the function for potential minima, rather than a concentration on the currently found optimum.

Gaussian process interpolation differs from other approaches to multivariate interpolation, because it delivers uncertainty statements, and one relies upon on them in subsequent analyses, as illustrated above. However we are currently not aware of papers that investigate the reliability of the uncertainty statements of the method. The methodology imposes a certain probabilistic model onto the interpolation process, involving particular assumptions. To evaluate their impact on the uncertainty statements, we will investigate three different, so called, uninformative or default choices of the prior probability distribution and compare two different computational approaches to analyse the probabilistic model, one based on numerical optimization and the other based on Markov chain Monte Carlo methods.

The outline of this paper is as follows: In Sect. 2 we describe the Gaussian process approach to multivariate interpolation, with particular focus on three so-called objective or default prior distributions for Gaussian process interpolation, as well as on two different computational approaches to evaluate the probabilistic model in Sect. 2.2. In Sect. 3 we will investigate the methodology in a variety of simulated as well as real examples to evaluate the Gaussian process approach with respect to the reliability of its uncertainty statements. Section 4 finally contains a real computer modeling example, where the purpose of the experiment was to learn about a contour of the computer model, *i.e.*, one would like to learn about the input values that result in a certain value of the computer model. Here the computer experiments will be conducted sequentially (explicitly using the uncertainty statements of the methodology) to obtain most information in as few computer code evaluations as possible. Section 5 concludes.

2 Gaussian Processes for Multivariate Interpolation

After their introduction into the computer modeling world by Sacks et al. (1989) and Currin et al. (1991), Gaussian process interpolation is currently seen as the defacto standard method for analyzing computer experiments (see, e.g., Santner et al. 2003) and forms the basis for the solutions of a variety of problems in the statistical analysis of complex computer models such as sensitivity analysis (Schonlau and Welch 2006; Oakley 2009), calibration and validation (Kennedy and O'Hagan 2001; Bayarri et al. 2007) or sequential design (Jones et al. 1998; Lehman et al. 2004).

The Gaussian process interpolation method itself, however has an even longer history in geostatistics. It was proposed by mining engineer Daniel Krige (Krige 1951) and further developed by Georges Matheron (Matheron 1963); Gaussian Process interpolation is therefore also known under the name of Kriging especially in spatial statistics (Cressie 1993). The connection to the Bayesian view on statistics, where the probabilistic element of Gaussian process interpolation plays a more dominant role was realized only later, see for example O'Hagan (1978). There exist surprisingly many cross links with other interpolation (and nonparametric regression) methods. Fang et al. (2006), for example, illustrate that the functional form of the predictor of local polynomial modeling and radial basis function interpolation is identical to the posterior mean (5) in Gaussian process interpolation for a certain choice of the underlying mean and covariance function. Also, interpolating splines resulting from a minimization of a roughness penalty subject to the interpolation condition can be obtained as a special case of Gaussian process interpolation for a certain (non-stationary) covariance function (Wahba 1978;

Gu 2002). There are also close connections to machine learning regression methods, such as neural networks and the support vector machine (see Seeger 2004) for details.

We will describe Gaussian process interpolation from the Bayesian perspective in Sect. 2.1 and will focus particularly on the choice of objective prior distributions in Sect. 2.2.

2.1 Gaussian Process Interpolation

We are interested in approximating the deterministic function $y(\boldsymbol{x})$ based on a small number of evaluations at the design $D = (\boldsymbol{x}_1, \ldots, \boldsymbol{x}_n)$ and the computer model function $y(.)$ is known to be continuous and smooth. In the Bayesian probabilistic formulation one defines a so-called prior distribution $\tilde{y}(.)$ for the function $y(.)$. The name *prior* distribution comes from the fact that it should represent information available before any evaluation of $y(.)$ has been performed (which is usually quite scarce). When one evaluates the function at the design D one obtains observations $y^n = (y(\boldsymbol{x}_1), \ldots, y(\boldsymbol{x}_n))'$ and one can calculate the distribution of $\tilde{y}(.)$ conditional on $\tilde{y}(\boldsymbol{x}_1) = y(\boldsymbol{x}_1), \ldots, \tilde{y}(\boldsymbol{x}_n) = y(\boldsymbol{x}_n)$, the so-called posterior distribution. This posterior distribution then contains the information in the prior distribution and the information based on the evaluation of $(\boldsymbol{x}_i, y(\boldsymbol{x}_i))$, $i = 1, \ldots, n$.

The main idea of Gaussian process interpolation is to use a Gaussian process as a prior for the function $y(.)$. By definition a Gaussian process $\tilde{y}(.)$ is a stochastic process which has the property that for any set of input points $\boldsymbol{x}_1, \ldots, \boldsymbol{x}_t$, $t = 1, 2, 3, \ldots$ in $\mathcal{X}$ the joint distribution of $\tilde{y}(\boldsymbol{x}_1), \ldots, \tilde{y}(\boldsymbol{x}_t)$ is multivariate normal with mean $\boldsymbol{\mu} = (\mu(\boldsymbol{x}_1), \ldots, \mu(\boldsymbol{x}_t))'$ and covariance matrix $\boldsymbol{\Gamma} \in \mathbb{R}^{t \times t}$ with $\boldsymbol{\Gamma}_{(i,j)} = c(\boldsymbol{x}_i, \boldsymbol{x}_j)$, for a mean function $\mu(.)$ and a covariance function $c(.,.)$. Hence it is a generalization of a normal distribution to an infinite space and its realizations are functions on $\mathcal{X}$ rather than vectors.

A variety of choices can be used for $\mu(.)$ and $c(.,.)$ (see Santner et al. 2003). Here we will present the most common choice in this setting, which uses a constant process variance σ^2, *i.e.*, $c(\boldsymbol{x}, \boldsymbol{x}') = \sigma^2 r(\boldsymbol{x}, \boldsymbol{x}')$ for a correlation function $r(.,.)$. Another assumption commonly made, is covariance stationarity of the Gaussian process, which implies that the correlation between two points depends only on the points' difference, *i.e.*, $r(\boldsymbol{x}, \boldsymbol{x}+\boldsymbol{h}) = r(\boldsymbol{x}', \boldsymbol{x}'+\boldsymbol{h})$ for $\boldsymbol{x}, \boldsymbol{x}+\boldsymbol{h}, \boldsymbol{x}', \boldsymbol{x}'+\boldsymbol{h} \in \mathcal{X}$. This is a relatively strong assumption as it implies a spatial homogeneity of the Gaussian process, in the sense that the function's fluctuations are the same throughout $\mathcal{X}$.

The mean function is typically chosen as a parametric, for example, linear regression model

$$\mu(.) = f(.)^T \boldsymbol{\beta}, \tag{1}$$

for a p-dimensional hyperparameter $\boldsymbol{\beta}$ and a fixed regression function $f(.)$. For approximating computer models simple mean functions, for example $f(\boldsymbol{x}) = 1$ or $f(\boldsymbol{x}) = (1, x^{(1)}, \ldots, x^{(d)})$, have proven to be useful in many applications as long

as no additional information on the mean is available.
A widely used stationary correlation function for modeling computer code output is the power exponential correlation function

$$r(\boldsymbol{x}, \boldsymbol{x}') = \exp\left(-d(\boldsymbol{x}, \boldsymbol{x}')\right) \quad \text{with } d(\boldsymbol{x}, \boldsymbol{x}') = \sum_{j=1}^{d} \xi_j \left| x^{(j)} - x'^{(j)} \right|^{a_j} \tag{2}$$

with correlation parameter vector $\boldsymbol{\psi} = (\boldsymbol{\xi}^T, \boldsymbol{a}^T)^T$ where $\boldsymbol{\xi} = (\xi_1, \ldots, \xi_d)^T$ and $\boldsymbol{a} = (a_1, \ldots, a_d)^T$ with $\xi_j > 0$ and $0 < a_j \leq 2$. If $a_j = 2$ for all $j = 1, \ldots, d$ we obtain the important case of the Gaussian correlation function and if all $\boldsymbol{\xi}_j = \xi$ and $a_j = a$ for all j the correlation function is isotropic. Note that the correlation between two points in (2) is monotonically decreasing in $d(\boldsymbol{x}, \boldsymbol{x}')$. Thus the computer model is assumed to be more similar if the distance between the points in $\mathcal{X}$ is smaller. For the power exponential correlation function it can be shown that the realizations are (almost surely) continuous and for the Gaussian correlation function (with $a_j = 2$, $j = 1, \ldots, d$) they are, in addition, infinitely often continuously differentiable, while for $a_j < 2$ realizations are not differentiable.

Gaussian processes are very flexible for modeling functions as their realizations can take various forms. Tokdar and Ghosh (2007) even prove full support on the space of continuous functions under relatively weak assumptions on the covariance function. This flexibility is also illustrated in Fig. 1, where three two-dimensional realizations with mean function 0 and exponential correlation function (2) are displayed corresponding to three different correlation parameters. The parameter vector $\boldsymbol{\xi}$ controls the wiggliness of the function: small values yield smooth functions whereas the realizations are fluctuating more strongly with increasing values for $\boldsymbol{\xi}$. The exponent correlation parameter $\boldsymbol{a}$ controls the differentiability: A value smaller than 2 results in non-differentiable functions. In this work we will subsequently assume the power exponential correlation function. Of course there are other possibilities, with the Matérn correlation function probably being the best known alternative/generalization. There are also extensions to non-stationary covariance functions as proposed in Paciorek and Schervish (2004) or Xiong et al. (2007).

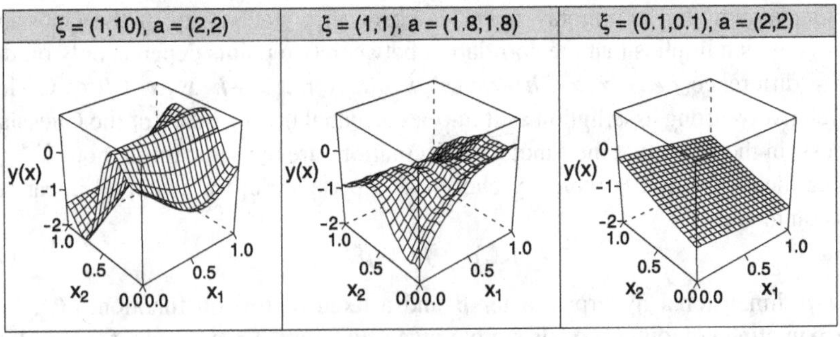

Fig. 1 Gaussian Process realizations with different correlation parameters ($\mu(\boldsymbol{x}) = 0$ and $\sigma^2 = 1$)

To sum up, the prior distribution $\tilde{y}^n = (\tilde{y}(\boldsymbol{x}_1), \ldots, \tilde{y}(\boldsymbol{x}_n))^T$ for the vector of computer model evaluations at a design $\boldsymbol{D}$ for given $\boldsymbol{\beta}, \sigma$ and $\boldsymbol{\psi}$, is a multivariate normal distribution

$$\tilde{y}^n \sim N\left(\boldsymbol{F}\boldsymbol{\beta}, \sigma^2 \boldsymbol{R}\right) \tag{3}$$

with regression matrix $\boldsymbol{F} = (f(\boldsymbol{x}_1)^T, \ldots, f(\boldsymbol{x}_n)^T)^T$ and correlation matrix $\boldsymbol{R}$ with entries $R_{i,i'} = r(\boldsymbol{x}_i, \boldsymbol{x}_{i'})$ for $i, i' \in 1, \ldots, n$. The other unknown parameters (so-called hyperparameters) in the mean function and the covariance function of the Gaussian process, $\boldsymbol{\beta}, \sigma^2$ and $\boldsymbol{\psi}$, also receive a prior distribution. Usually when modeling computer experiments there is no specific prior information for the hyperparameters available. Hence, so called non-informative or objective prior distributions are desirable. We factorize

$$p(\boldsymbol{\beta}, \sigma^2, \boldsymbol{\psi}) = p(\boldsymbol{\beta}, \sigma^2 | \boldsymbol{\psi}) \cdot p(\boldsymbol{\psi})$$

and use the standard flat prior $p(\boldsymbol{\beta}, \sigma^2 | \boldsymbol{\psi}) \propto \frac{1}{\sigma^2}$, where "$\propto$" means that the prior of $\boldsymbol{\beta}, \sigma^2$ given $\boldsymbol{\psi}$ is proportional to $\frac{1}{\sigma^2}$. It is the Jeffreys prior in this model (Paulo 2005) when treating $\boldsymbol{\psi}$ as a fixed value and assuming that $(\boldsymbol{\beta}, \sigma^2)'$ and $\boldsymbol{\psi}$ are independent a priori. We will first proceed with fixed $\boldsymbol{\psi}$ and discuss the choice of a prior for $\boldsymbol{\psi}$ in Sect. 2.2 in more detail.

In the Bayesian probabilistic framework interest centers on the distribution of $\tilde{y}(.)$ given the computer model evaluations, *i.e.* on the posterior distribution. Here an advantage of Gaussian process priors is that the prior to posterior analysis in large parts can be done analytically. This possibility of explicitly updating the posterior distribution is one of the main reasons for the success of Gaussian processes in practice. Let us hence look at the conditional posterior $p(\tilde{y}(.)|y^n, \boldsymbol{\psi})$. It follows from properties of the conditional distributions of the multivariate Gaussian distribution and standard Bayesian linear model theory (see O'Hagan and Forster 2004, pp. 393–398, for details) that the posterior process $\tilde{y}(.)|y^n$ is a so-called Student process

$$\tilde{y}(.)|y^n, \boldsymbol{\psi} \sim t_{n-p}\left(m^*(.), c^*(.,.)\right) \tag{4}$$

with mean function $m^*(.)$, scale function $c^*(.,.)$ and $n - p$ degrees of freedom. A Student process is the process generalization of the multivariate t-distribution, where the response at any set of inputs has a multivariate t-distribution. The posterior mean function $m^*(.)$ is given by

$$m^*(.) = E(\tilde{y}(.)|y^n, \boldsymbol{\psi}) = f(.)^T \hat{\boldsymbol{\beta}} + r(., \boldsymbol{D})^T \boldsymbol{R}^{-1}(y^n - \boldsymbol{F}\hat{\boldsymbol{\beta}}) \tag{5}$$

where $r(., \boldsymbol{D})$ is the vector of correlation function evaluations between a new point (.) and the design points and $\hat{\boldsymbol{\beta}} = \left(\boldsymbol{F}^T \boldsymbol{R}^{-1} \boldsymbol{F}\right)^{-1} \boldsymbol{F}^T \boldsymbol{R}^{-1} y^n$ equals the generalized least squares estimate for the regression parameter. The posterior covariance function can be calculated as

$$\mathrm{Cov}(\tilde{y}(\boldsymbol{x}), \tilde{y}(\boldsymbol{x}')|y^n, \boldsymbol{\psi}) = \frac{n-p}{n-p-2} c^*(\boldsymbol{x}, \boldsymbol{x}') \tag{6}$$

with

$$c^*(\boldsymbol{x}, \boldsymbol{x}') = \widehat{\sigma^2} \Big\{ r(\boldsymbol{x}, \boldsymbol{x}') - r(\boldsymbol{x}, \boldsymbol{D})^T \boldsymbol{R}^{-1} r(\boldsymbol{x}', \boldsymbol{D}) \\ - (f(\boldsymbol{x})^T - r(\boldsymbol{x}, \boldsymbol{D})^T \boldsymbol{R}^{-1} \boldsymbol{F})(\boldsymbol{F}\boldsymbol{R}^{-1}\boldsymbol{F})^{-1}(f(\boldsymbol{x})^T \\ - r(\boldsymbol{x}, \boldsymbol{D})^T \boldsymbol{R}^{-1} \boldsymbol{F})^T \Big\} \tag{7}$$

and $\widehat{\sigma^2} = \frac{1}{n-p}(\boldsymbol{y}^n - \boldsymbol{F}\hat{\boldsymbol{\beta}})^T \boldsymbol{R}^{-1}(\boldsymbol{y}^n - \boldsymbol{F}\hat{\boldsymbol{\beta}})$, the posterior process variance given $\boldsymbol{\psi}$.

The mean of the posterior distribution $m^*(.)$, given in (5), can now be used as a point estimate $\hat{y}(.)$ for $y(.)$. The first part in the formula for $m^*(.)$ is the estimated linear regression function whereas the second part weights the deviations $(\boldsymbol{y}^n - \boldsymbol{F}\hat{\boldsymbol{\beta}})$ from observations and estimated mean using the correlation structure so that the observations $\boldsymbol{y}^n$ at the design points are interpolated. Close to the design points the posterior mean thus depends mostly on the correlation structure and with increasing distance to the design points they are shrunken to the mean function. In the covariance of (6) the weighting using the correlation structure plays an important role as well: the variance is zero at the design points.

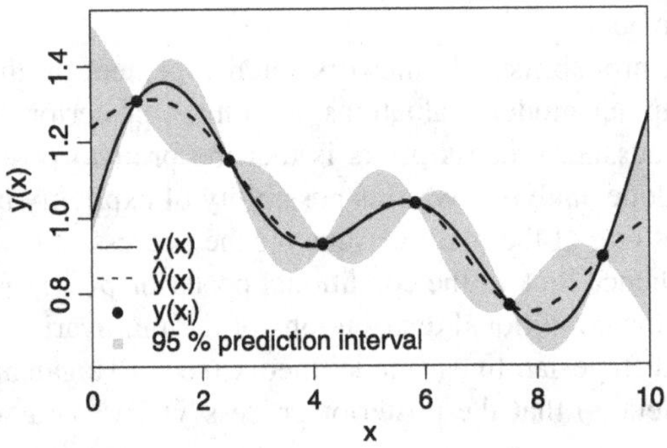

Fig. 2 Prediction for a one dimensional function $y(.)$ based on 6 function evaluations and corresponding 95% pointwise uncertainty intervals

To illustrate the theory described in this section we will employ a simple one dimensional example. Suppose we would like to interpolate the function $y(x) = (x-5)/10\sin(-x)+1$ on the interval [0, 10] based on a Gaussian process prior $\tilde{y}(.)$ for $y(.)$ with mean function $\mu(x) = \beta_0$ and Gaussian correlation function. For the hyperparameters $\boldsymbol{\beta}, \sigma$ we used the prior $p(\boldsymbol{\beta}, \sigma^2|\boldsymbol{\xi}) \propto \frac{1}{\sigma^2}$ and fixed ξ at the value 0.3.

Figure 2 shows a graphical representation of the conditional distribution $\tilde{y}(.)$ (that is the posterior process) given the function values at the design $D = (5/6, 15/6, 25/6, 35/6, 45/6, 55/6)^T$. The uncertainty at the design points is 0, so that all realizations from the posterior process interpolate the observations. The variance (and hence uncertainty) grows with the distance to the design points.

2.2 Priors for Correlation Structure

The parameters in the correlation function determine the smoothness of the posterior process (as illustrated in Fig. 1), and thus also considerably the length of the prediction intervals. So far we have treated the correlation parameter $\boldsymbol{\psi}$ as fixed, in this case the model allows for analytical prior to posterior updating. In this section we will discuss the prior to posterior analysis, when additionally $\boldsymbol{\psi}$ is treated as unknown.

The posterior distribution for the parameter $\boldsymbol{\psi}$ is obtained by integrating out all other parameters, which can be done analytically. The resulting marginal posterior for the correlation parameter is then proportional to (see also Handcock and Stein 1993)

$$p(\boldsymbol{\psi}|\boldsymbol{y}^n) \propto p(\boldsymbol{\psi}) \cdot \det(\boldsymbol{R})^{-1/2} \det(\boldsymbol{F}^T \boldsymbol{R}^{-1} \boldsymbol{F})^{-1/2} \widehat{\sigma^2}^{-(n-p)/2}, \tag{8}$$

which is not proportional to a known probability density.

The full posterior process is then formally given by

$$\int p(\tilde{y}(.)|\boldsymbol{y}^n, \boldsymbol{\psi}) p(\boldsymbol{\psi}|\boldsymbol{y}^n) d\boldsymbol{\psi}, \tag{9}$$

an integral which cannot be calculated analytically. One approach to resolve such a problem is to use Markov chain Monte Carlo methods to draw a large number of samples $\boldsymbol{\psi}^{(1)}, \ldots, \boldsymbol{\psi}^{(T)}$ from (8) and approximate (9) by $1/T \sum_{t=1}^{T} p(\tilde{y}(.)|\boldsymbol{y}^n, \boldsymbol{\psi}^{(t)})$. An alternative method of approximating (9) is by optimizing (8) and plugging the posterior mode $\boldsymbol{\psi}^*$ into the conditional predictive distribution in (4), so that the approximation of (9) is given by $p(\tilde{y}(.)|\boldsymbol{y}^n, \boldsymbol{\psi}^*)$. This alternative method is the current standard in the analysis of computer experiments, although it neglects the uncertainty about the correlation parameters and therefore will tend to underestimate the posterior variance. However only maximizing (8) is computationally much cheaper than Monte Carlo integration and we must keep in mind that the major goal of employing interpolation in computer models is to save computing time. In Sect. 3 we will investigate the differences of these two computational approaches and the impact they have on the uncertainty intervals in a simulation study.

As the prior for $p(\boldsymbol{\psi})$ might influence the posterior distribution, we will review the development of objective priors for $p(\boldsymbol{\psi})$ in the following based on the work Berger et al. (2001) and particularly Paulo (2005). In the following we assume that $a_j = 2$ for $j = 1, \ldots, d$ (*i.e.*, the Gaussian correlation function), because this is the most relevant case of smooth functions. We will compare three different priors for $\boldsymbol{\xi}$: (i) An improper uniform (flat) prior distribution on $[0, \infty)^d$, subsequently abbreviated $p^{(1)}(\boldsymbol{\xi})$, which is a classical prior distribution in the case if more specific information is lacking. Note that formally this is not a non-informative prior. Despite this fact it is often used, particularly when the chosen computational implementation is based on numerical optimization, since the results then coincide with the REML estimator, see Santner et al. (2003). (ii) A prior based on a product of exponential densities, subsequently abbreviated $p^{(2)}(\boldsymbol{\xi})$, with means equal to the ML estimators for the $\boldsymbol{\xi}_j$'s. This approach is hence an empirical Bayes approach, as the prior depends on the data. (iii) The third prior distribution is the marginal independence Jeffreys prior, $p^{(3)}(\boldsymbol{\xi})$, as described (among other objective priors) in Paulo (2005). In general the Jeffreys prior has some appealing properties: It is invariant with respect to re-parametrization of the model, it can be shown to minimize prior information relative to the posterior in a certain information theoretic sense and Bayesian credibility intervals resulting from Jeffreys prior often achieve their nominal frequentist coverage probability faster than credibility intervals from other arbitrary priors (see Kass and Wasserman 1996 for a detailed description of the properties of Jeffreys prior and variants based on it). In the multivariate Gaussian process situation there exist different variants of objective priors and Paulo (2005) describes several of them and compares the frequentist coverage probability of the credibility intervals for $\boldsymbol{\xi}$ in a simulation study. For this purpose Gaussian processes with fixed hyperparameters were simulated and the coverage probability of the credibility intervals for $\boldsymbol{\xi}$ were compared to their nominal level. The marginal independence Jeffreys prior is recommended over other approaches in Paulo (2005) since it lead to good performance in terms of frequentist coverage probability and was computationally the cheapest approach. This is the reason, why we employ it here. In summary we hence compare

$$p^{(1)}(\boldsymbol{\xi}) \propto \underline{1} \quad \text{for } \xi_j > 0,$$

$$p^{(2)}(\boldsymbol{\xi}) = \prod_{i=1}^{d} f_{Exp}\left(\boldsymbol{\xi}, \left(\hat{\xi}_{j,\mathrm{ML}}\right)^{-1}\right),$$

and

$$p^{(3)}(\boldsymbol{\xi}) \propto |I_J(\boldsymbol{\xi})|^{-1/2}$$

with matrix

$$I_J(\boldsymbol{\xi}) = \begin{pmatrix} n & \mathrm{tr}U_1 & \mathrm{tr}U_2 & \cdots & \mathrm{tr}U_d \\ & \mathrm{tr}U_1^2 & \mathrm{tr}U_1U_2 & \cdots & \mathrm{tr}U_1U_d \\ & & \ddots & \cdots & \vdots \\ & & & & \mathrm{tr}U_d^2 \end{pmatrix}$$

where f_{Exp} denotes the density of the exponential distribution and $U_j = (\frac{\partial}{\partial \xi_j}\boldsymbol{R})\boldsymbol{R}^{-1}$ for $j = 1, \ldots, d$ is the product of correlation matrix and the matrix's partial derivative with respect to ξ_j. Interestingly Jeffreys prior $p^{(3)}(\boldsymbol{\xi})$, depends on the design D through the correlation matrix R. In total the prior distributions for all hyperparameters are given by $p(\boldsymbol{\beta}, \sigma^2, \boldsymbol{\xi}) \propto p^{(m)}(\boldsymbol{\xi})(\sigma^2)^{-1}$, with $m = 1, 2, 3$ and assuming $\boldsymbol{\beta}$ and $(\sigma^2, \boldsymbol{\xi})$ are independent (Paulo 2005). Computationally Jeffreys prior is most complex, as it requires the calculation of a matrix inverse for evaluation. The product of exponentials prior is also relatively complex to calculate as it requires that the

ML estimate is calculated before it can be evaluated. Computationally the simplest prior is the flat prior.

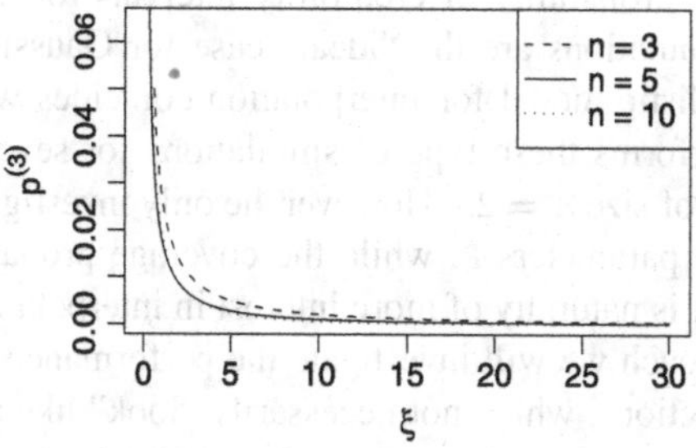

Fig. 3 Jeffreys prior in one dimension for equidistant designs on [0, 1] with different sample sizes ($n = 3, 5, 10$) and fixed $a_j = 2$

The shapes of the flat prior $p^{(1)}$ and the product of exponentials prior $p^{(2)}$ are well-known, while it is not obvious what the Jeffreys prior actually looks like. Figure 3 illustrates the Jeffreys prior for equidistant designs in one dimension for a Gaussian correlation function: smaller values for ξ have greater prior probability. The sample size determines how quickly the prior probability goes to zero with increasing ξ. This shape also generalizes to larger dimensions and is therefore in general more similar to the exponential than the flat prior.

When using improper priors (like $p^{(1)}$ and $p^{(3)}$ above) it is not guaranteed that the posterior is proper as the integral in (9) can be infinite. Berger et al. (2001) show in the one-dimensional case that an improper posterior distribution for ξ can arise in a certain quite common situation, while Paulo (2005) shows that the posterior is proper for the treated objective priors in the multivariate case for a factorial design and, for example, a constant mean function. In practice, however a proper posterior can be induced simply by bounding the parameter space at reasonable values for ξ and thus making the prior distributions proper. When only using the posterior mode for ξ and plugging this into (9), the issue is entirely bypassed.

3 Simulation Studies

In this section we will investigate the reliability of the uncertainty statements of Gaussian process interpolation for the three different priors $p^{(1)}$, $p^{(2)}$, $p^{(3)}$ and the two computational approaches described in the last section in a small simulation study. The computational approach based on plugging in the posterior mode ψ^* into (9) will subsequently be abbreviated as PM and the full Markov Chain Monte Carlo approach will be abbreviated as MCMC. To evaluate the performance we will use two approaches.

In the first approach we will simulate random functions from a Gaussian process and apply the Gaussian process interpolation methodology to each of the functions. We will then investigate the across function coverage probability for the function $y(.)$ and the coverage probability of credibility intervals for the hyperparameters $\boldsymbol{\xi}$. Note that these simulations are the "ideal" case for Gaussian process interpolation, as the probabilistic model for interpolation coincides with the true model. Paulo (2005) also performs these type of simulations for several objective priors and a factorial design of size $n = 25$. However, he only investigated the uncertainty intervals for the hyperparameters $\boldsymbol{\xi}$, while the coverage probability regarding the response function $y(.)$ is naturally of more interest in interpolation.

In the second approach we will investigate the performance for interpolation of three realistic test functions, which not necessarily "look" like a typical realization of a stationary Gaussian process. Here we will hence test the methodology, when the probabilistic assumptions underlying Gaussian process interpolation are not met.

In all cases we will use space filling designs for D, in particular maximin Latin hypercube designs, which are widely used for computer experiments. To generate the designs we employed the `lhs` package (Carnell 2009) for the `R` statistical computing language.

To calculate the posterior mode of the posterior density in (8) in the PM approach we used the `optim` function in `R` with the Nelder-Mead optimizer. The posterior density is usually multimodal so different starting values are used for the optimization. For the MCMC approach, sampling was accomplished employing the random walk Metropolis algorithm based on the posterior density for $\boldsymbol{\xi}$ given in (8). For this purpose we used the `R` package `mcmc` (Geyer 2009). To achieve a reasonably efficient Markov chain we tuned the algorithm in a starting phase. The step size of the algorithm was tuned to achieve an acceptance rate of around 0.3 and the covariance matrix of the proposal density was chosen proportional to the covariance matrix estimated from the iterations of this preliminary chain (see Roberts and Rosenthal 2001 for details on tuning the random walk Metropolis algorithm). Using this specification we then started the random walk Metropolis at the posterior mode and ran it for 15000 iterations. The first 100 iterations were discarded as burn in and only every 10th value is taken for the analysis.

To evaluate the uncertainty statements for the function $y(.)$ we used the across function coverage, *i.e.* the probability that the true function is contained in the credibility interval predicted by the method averaged over the input space. This measure seems reasonable from a practical viewpoint and a variant of it has also been used in the smoothing spline regression literature (see Chap. 3.3 of Gu 2002, where it is also investigated theoretically). For this purpose first pointwise prediction intervals $P_y(\boldsymbol{x}) = [L_y(\boldsymbol{x}), U_y(\boldsymbol{x})]$ are calculated, where $L_y(\boldsymbol{x})$ and $U_y(\boldsymbol{x})$ are the 0.025 and the 0.975 quantile of the posterior distribution at $\boldsymbol{x}$ (in the PM approach this is just a quantile of the t-distribution in (4), while in the MCMC approach empirical quantiles of samples from the predictive distribution are calculated). Then one checks whether the true value $y(\boldsymbol{x})$ is contained in $P_y(\boldsymbol{x})$ and denotes this value by $c_y(\boldsymbol{x})$, *i.e.* $c_y(\boldsymbol{x}) = \mathbb{1}_{[L_y(\boldsymbol{x}),U_y(\boldsymbol{x})]}(y(\boldsymbol{x}))$. To obtain a single value for the coverage over the entire function, we integrate over the design space to obtain one number:

$C_y = \int_{\mathcal{X}} c_y(\boldsymbol{x})d\boldsymbol{x}$. C_y is hence the average coverage over the design space. In addition to the coverage probability one is usually also interested in the length of the prediction intervals, hence we define by $H_y = \int_{\mathcal{X}} |U_y(\boldsymbol{x}) - L_y(\boldsymbol{x})|d\boldsymbol{x}$ the average length of the prediction intervals averaged over the design space. All integrals are approximated using a straightforward numerical integration on $\mathcal{X}$.

3.1 Simulations from a Gaussian Process

In a first step we simulated 100 realizations from a two dimensional Gaussian process with constant mean function $\mu(\boldsymbol{x}) = 1, \sigma^2 = 1.5, \boldsymbol{\xi} = (3, 0.5)^T$ and $a = (2, 2)$ on $\mathcal{X} = [0, 1]^2$. For the design D we used maximin Latin hypercube designs with sample sizes $n = 10, 15, 20, 25, 30$, to investigate the impact of sample size on the uncertainty statements.

For evaluation Gaussian process interpolation with an unknown constant mean function is used. The exponent $\boldsymbol{a}$ in the correlation function is fixed to $a_j = 2$, $j = 1, \ldots, d$, and the roughness parameter $\boldsymbol{\xi}$ is considered unknown. Additionally the three different priors introduced in the last section and the two different computational approaches were applied. In Table 1 one can observe the coverage probabilities C_y and expected prediction interval lengths H_y averaged over the 100 simulated functions. For the MCMC based analysis the coverage for the correlation parameters C_{ξ_j} is reported as well (coverage intervals for $\boldsymbol{\xi}_j$ were calculated based on the empirical quantiles from the MCMC simulations).

For all prior distributions the across function coverage for a small sample size $n = 10$ are below the target coverage with the best coverage values achieved using the flat prior or the data dependent product of exponentials prior and an MCMC based analysis. With increasing sample size they all approximately reach the nominal level 0.95 and results seem to be even a bit conservative for $n = 30$. One can observe the usual trade-off between length of the credibility interval and coverage probability: The posterior mode estimate for $\boldsymbol{\xi}$ yields shorter credibility intervals than MCMC, but (contrary to MCMC) does not achieve the nominal level in most scenarios considered. However, the differences between MCMC and PM get smaller for larger sample size. Overall none of the priors outperforms the other in terms of coverage probability and interval length.

For the correlation parameter the simulation results are quite similar: For small sample sizes the coverage is worse, but quickly approaches the nominal level. This can be expected for $\boldsymbol{\xi}_j$, as asymptotically Bayesian credibility intervals can also be interpreted as frequentist confidence intervals, particularly when objective priors are used.

In summary, the results are quite positive in the sense that the methodology works in the specific case it is designed for, provided the sample size is not too small.

Table 1 Coverage probabilities of 95% prediction/credibility intervals when sampling from a Gaussian process for the three different priors and the two computational approaches (MCMC and PM), averaged over 100 simulations (in round brackets the 0.1 and 0.9 quantile of the coverages observed in the simulation study)

sample size	$C_y^{(MCMC)}$	$C_y^{(PM)}$	$H_y^{(MCMC)}$	$H_y^{(PM)}$	$C_{\xi_1}^{(MCMC)}$	$C_{\xi_2}^{(MCMC)}$
Flat prior						
10	0.88 (0.64,1)	0.77 (0.34,1)	1.179	0.167	0.78	0.77
15	0.96 (0.84,1)	0.87 (0.64,1)	0.081	0.048	0.94	0.93
20	0.96 (0.88,1)	0.91 (0.72,1)	0.026	0.021	0.92	0.93
25	0.95 (0.84,1)	0.88 (0.69,1)	0.008	0.007	0.99	0.96
30	0.98 (0.95,1)	0.95 (0.88,1)	0.004	0.004	0.92	0.93
Exponential prior						
10	0.87 (0.66,1)	0.74 (0.41,1)	0.209	0.148	0.80	0.83
15	0.94 (0.78,1)	0.86 (0.61,1)	0.065	0.046	0.93	0.90
20	0.94 (0.76,1)	0.87 (0.60,1)	0.025	0.020	0.94	0.91
25	0.95 (0.84,1)	0.89 (0.64,1)	0.008	0.007	0.96	0.93
30	0.98 (0.95,1)	0.94 (0.87,1)	0.004	0.004	0.97	0.94
Jeffreys prior						
10	0.73 (0.24,1)	0.63 (0.26 ,0.98)	0.247	0.128	0.67	0.59
15	0.94 (0.76,1)	0.84 (0.51 ,1.00)	0.061	0.044	0.95	0.89
20	0.93 (0.74,1)	0.87 (0.60 ,1.00)	0.025	0.021	0.98	0.95
25	0.97 (0.89,1)	0.92 (0.76 ,1.00)	0.008	0.007	0.94	0.95
30	0.98 (0.92,1)	0.92 (0.80 ,1.00)	0.004	0.004	0.98	0.94

3.2 Evaluating Testfunctions

It is important to investigate how the method behaves when applied to test functions that are not necessarily realizations of a Gaussian process. For this purpose we consider three test functions of dimension two and seven with a complexity and sample size that is comparable to the application in sheet metal forming in Sect. 4.1.

The first test function f_1 is a two dimensional smooth convex combination of beta distribution functions $F_{\text{Beta}}(x, \alpha, \beta)$:

$$f_1(\boldsymbol{x}) = 1/3 \cdot F_{\text{Beta}}(0.5x_1 + 0.5x_2, 1, 1) + 1/3 \cdot F_{\text{Beta}}(0.7x_1 + 0.3x_2, 20, 10) \\ + 1/3 \cdot F_{\text{Beta}}(0.15x_1 + 0.85x_2, 15, 20).$$

This function (displayed in Fig. 4 left) increases moderately and should therefore be well described by a covariance stationary Gaussian process.

Additionally we look at a physical process, which can rapidly be simulated, because the function describing the process is available analytically. The piston simulator from Kenett and Zacks (1998) models the cycle time of a complete revolution of the piston's shaft depending on seven input factors: piston weight (M, 30–60 kg), piston surface area (S, 0.005–0.02 m^2), initial gas volume (V_0,

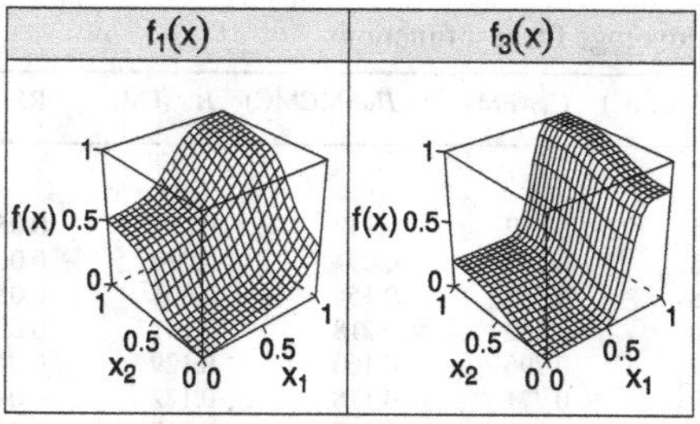

Fig. 4 Test functions $f_1(\boldsymbol{x})$ and $f_3(\boldsymbol{x})$

0.002–0.01 cm^3), spring coefficient (k, 1000–5000 N/m), atmospheric pressure (P_0, 9×10^4–11×10^4 N/m^2), ambient temperature (T, 290–396 K) and filling gas temperature (T_0, 340–3602 K) as

$$\text{cycletime} = 2\pi\sqrt{\frac{M}{k + S^2 \frac{P_0 V_0}{T_0} \frac{T}{V^2}}}$$

with $V = \frac{S}{2k}(\sqrt{A^2 + 4k\frac{P_0 V_0}{T_0} T} - A)$ and $A = P_0 S + 19.62M - \frac{k V_0}{S}$. This function is included as it is higher dimensional and based on a real example. For convenience we scale all inputs to [0, 1] and call the function describing the cycletime $f_2(\boldsymbol{x})$. As this example is higher dimensional here 20000 MCMC iterations are used, saving only every 10th value after a burn in of 1000.

For the third function we again consider a monotone combination of beta cdfs:

$$f_3(\boldsymbol{x}) = 0.8 \cdot F_{\text{Beta}}(x_1, 30, 15) + 0.2 \cdot F_{\text{Beta}}(x_2, 3, 4).$$

This function has one single steep increase but is almost constant apart from that (see Fig. 4 right). We included this comparably difficult example to investigate how the methodology works, when a function is evaluated, which is rather unlikely to be generated by a covariance stationary Gaussian process.

For each test function we consider three different sample sizes (5, 7 and 10 observations per dimension d), three different prior functions ($p^{(1)}$, $p^{(2)}$ and $p^{(3)}$) and two computational methods (MCMC and PM). The mean function is considered constant and the exponents a_j are fixed to 2 yielding smooth differentiable realizations of Gaussian processes. We report the coverage C_y of the test functions together with prediction interval lengths H_y in Table 2. Note that these values are results for a single function under consideration and not averaged over replications as in Sect. 3.1. In addition we evaluated the root mean square error $\text{RMSE} = \sqrt{\int_{\mathcal{X}} (\hat{y}(\boldsymbol{x}) - y(\boldsymbol{x}))^2 d\boldsymbol{x}}$, where $\hat{y}(\boldsymbol{x})$ is the posterior mean in the MCMC or the PM approach.

For all test functions the across function coverage using the MCMC approach is usually > 80% and for the posterior mode analysis in most cases > 70%. The difference between the MCMC and the PM approach is a bit larger here than in the

Table 2 Across function coverage for test functions

n / d	prior	C_y (MCMC)	C_y (PM)	H_y (MCMC)	H_y (PM)	RMSE (MCMC)	RMSE (PM)
$f_1(\boldsymbol{x})$ $(d = 2)$							
5	flat	0.875	0.512	0.149	0.082	0.056	0.059
5	exponential	0.578	0.397	0.094	0.070	0.064	0.064
5	Jeffreys	0.748	0.333	0.154	0.061	0.051	0.069
7	flat	1.000	0.862	0.218	0.141	0.037	0.044
7	exponential	0.891	0.796	0.160	0.129	0.045	0.046
7	Jeffreys	0.989	0.771	0.178	0.122	0.043	0.048
10	flat	0.891	0.760	0.087	0.067	0.022	0.024
10	exponential	0.844	0.703	0.078	0.063	0.024	0.026
10	Jeffreys	0.837	0.680	0.078	0.062	0.024	0.027
$f_2(\boldsymbol{x})$ $(d = 7)$							
5	flat	0.913	0.832	0.048	0.038	0.013	0.014
5	exponential	0.871	0.833	0.041	0.038	0.013	0.014
5	Jeffreys	1.000	0.811	2.608	0.061	0.059	0.025
7	flat	0.964	0.781	0.043	0.032	0.013	0.016
7	exponential	0.860	0.703	0.034	0.026	0.015	0.017
7	Jeffreys	0.863	0.771	0.050	0.039	0.020	0.020
10	flat	0.786	0.758	0.020	0.017	0.009	0.009
10	exponential	0.781	0.727	0.018	0.015	0.009	0.009
10	Jeffreys	0.791	0.789	0.020	0.018	0.009	0.009
$f_3(\boldsymbol{x})$ $(d = 2)$							
5	flat	1.000	0.900	1.011	0.210	0.123	0.067
5	exponential	0.961	0.816	0.237	0.191	0.077	0.076
5	Jeffreys	0.950	0.705	0.247	0.176	0.115	0.092
7	flat	0.986	0.812	0.170	0.115	0.041	0.039
7	exponential	0.902	0.730	0.148	0.128	0.037	0.035
7	Jeffreys	0.878	0.605	0.144	0.103	0.035	0.034
10	flat	0.506	0.385	0.073	0.061	0.054	0.056
10	exponential	0.426	0.327	0.066	0.056	0.058	0.059
10	Jeffreys	0.424	0.311	0.067	0.054	0.059	0.060

simulations in the last section. From the three different prior distributions the flat prior seems to be closest to the nominal level. Overall, however, the coverage does not reach a 95% coverage in many situations. As anticipated the performance for f_3 is worse, particularly for 10 observations per dimension. It is interesting to note that, contrary to the results in the last section, the coverage probability is not increasing with the sample size, but decreasing (particularly for f_2 and f_3). It seems that by increasing the sample size one introduces a "false" certainty: When the true function is deviating strongly from the assumptions underlying a stationary Gaussian process model, the uncertainty statements become less reliable. Overall however, particularly the MCMC approach seems to give reasonable results in the situations considered.

4 Application: Using Uncertainty in Interpolation for Sequential Designs

As discussed in the introduction one application of the uncertainty statements of Gaussian process interpolation are sequential designs for computer experiments. The basic idea is as follows: after evaluating the computer model at a start design, one uses the pointwise predictive distribution of the Gaussian process interpolation to determine sequentially the point where to evaluate the computer model next. This can reduce the number of computer model evaluations, especially, if the goal is to estimate a specific characteristic of $y(.)$ (such as the minimum, maximum or a contour) rather than to approximate the simulation for the entire region.

In choosing sequential designs for computer models the concept of expected improvement as proposed by Jones et al. (1998) is widely used and has been extended to various situations such as constraint optimization (Williams et al. 2000) or contour estimation (Ranjan and Bingham 2008). In this paper we will illustrate by means of an example in sheet metal forming how to sequentially design computer experiments for contour estimation.

4.1 Example: Contour Estimation for Springback in Deep Drawing

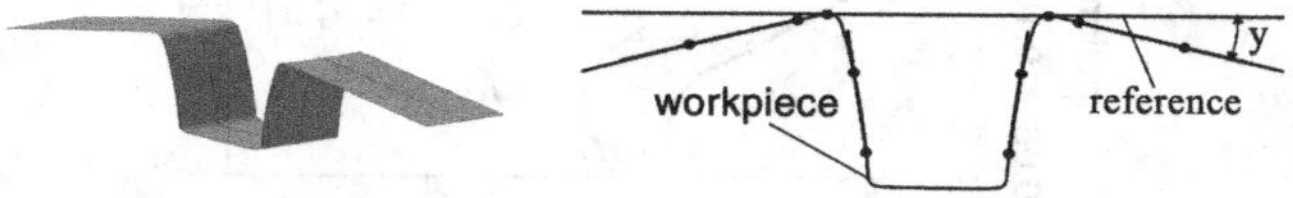

Fig. 5 Workpiece geometry with springback angle y

In deep drawing a sheet metal is drawn into a forming die by applying force with a punch. To simplify matters the shape of the formed workpiece is (most of all) determined by the shape of the die. However, when releasing the force from the punch the formed workpiece can spring back and the resulting shape deviates from the original one. Here we look at a simple workpiece geometry displayed in Fig. 5 which is very prone to springback (see Gösling et al. 2007). Overall springback is described here using the angle y and it is desired to deviate as little as possible from the target geometry (*i.e.* the angle should be as small as possible).

The springback angle depends on various factors, for example, on the die geometry and additional process parameters. Some of these parameters cannot be perfectly determined in the real production process such as the friction (quantified via the friction coefficient varying between 0.06 and 0.18) and blank thickness (varying between 0.8 and 1.2 mm). However, for the engineers it is of particular interest to find values for blank thickness and friction, that lead to an acceptable springback angle.

The entire process of forming and springback can be simulated using a finite element computer model. Here the commercial software PAMSTAMP 2G (2005) has been used. Now in principal the springback angle can be simulated for any combination of friction coefficient and blank thickness from the computer model. However, this computer code is time-consuming to run so that as few computer model evaluations as possible should be performed.

Now assume that the resulting workpiece is acceptable as long as the springback angle y is smaller than a specified angle $\alpha = 9°$. We are interested to identify values of the parameters friction and blank thickness such that the computer model yields an acceptable degree of springback. This corresponds to the estimation of a contour $S(\boldsymbol{x} : y(\boldsymbol{x}) = \alpha)$ which will divide the parameter space $\mathcal{X}$ in regions that have an acceptable or unacceptable degree of springback.

For the purpose of this paper we will use an approximation of the computer model based on previously performed computer model evaluations, rather than performing the actual finite element simulations. The corresponding function is displayed in Fig. 6. This approach has the advantage that the test function is close to a real computer model but still quick to evaluate, so that different sequential approaches can be compared for the purpose of this paper.

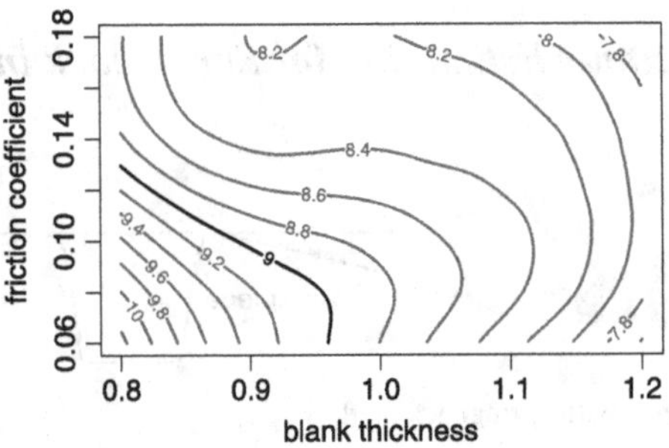

Fig. 6 Contours of the computer model function for springback (in black: contour $y(x) = 9$)

The sequential design algorithm adds design points sequentially using the expected improvement criterion and Gaussian process interpolation. The computer model is first evaluated at a space filling start design of size n_s, so that computer experiments $(\boldsymbol{x}_i, y(\boldsymbol{x}_i))_{i=1,\dots,n_s}$ are available. The Gaussian process interpolation yields a predictive process based on the current computer model evaluations. Design points are then added sequentially such that we can achieve an improvement in our goal of estimating the contour $S(\boldsymbol{x} : y(\boldsymbol{x}) = \alpha)$.

For example, it would be advantageous to sample in regions where $y(\boldsymbol{x})$ lies within a neighborhood $[\alpha - \epsilon(\boldsymbol{x}), \alpha + \epsilon(\boldsymbol{x})]$ of α. Therefore Ranjan and Bingham (2008) propose the following improvement function $I(\boldsymbol{x})$ for contour estimation

$$I(\boldsymbol{x}) = \begin{cases} \epsilon(\boldsymbol{x})^2 - (\tilde{y}(\boldsymbol{x}) - \alpha)^2 & \text{if } |\tilde{y}(\boldsymbol{x}) - \alpha| < \epsilon(\boldsymbol{x}), \\ 0 & \text{else,} \end{cases} \tag{10}$$

where the neighborhood is defined as $\epsilon(\boldsymbol{x}) = \gamma \cdot s(\boldsymbol{x})$ with a positive constant γ and $s(\boldsymbol{x}) = \sqrt{c^*(\boldsymbol{x},\boldsymbol{x})}$ from (7). The improvement is either zero when the deviation of $\tilde{y}(\boldsymbol{x})$ from α is too large or when the uncertainty in $\boldsymbol{x}$ is very small as $s^2(\boldsymbol{x})$ is proportional to the posterior variance (for an observed design point $\boldsymbol{x}_i$ it is always zero).

Now the uncertainty about $\tilde{y}(\boldsymbol{x})$ comes in to play: rather than maximizing the improvement function by plugging-in a fixed approximation $\hat{y}(\boldsymbol{x})$, we maximize the improvement that is expected under the predictive distribution of $\tilde{y}(\boldsymbol{x})|\boldsymbol{y}^n$. The expected improvement criterion is hence calculated as

$$E\left[I(\boldsymbol{x})\right] = \int_{\alpha-\epsilon(\boldsymbol{x})}^{\alpha+\epsilon(\boldsymbol{x})} \left(\epsilon(\boldsymbol{x})^2 - (\tilde{y}-\alpha)^2\right) p(\tilde{y}(\boldsymbol{x})|\boldsymbol{y}^n) d\tilde{y} \tag{11}$$

and for fixed $\boldsymbol{\psi}$ it yields the closed form expression

$$\begin{aligned} &E\left[I(\boldsymbol{x})|\boldsymbol{\psi}\right] \\ &= \left(\epsilon(\boldsymbol{x})^2 - \left(m^*(\boldsymbol{x}) - \alpha\right)^2 - s^2(\boldsymbol{x})\frac{n-p}{n-p-2}\right)\left(\mathcal{T}_{n-p}(c_2) - \mathcal{T}_{n-p}(c_1)\right) \\ &\quad - 2\left(m^*(\boldsymbol{x}) - \alpha\right)s(\boldsymbol{x})\left(\frac{n-p+c_1^2}{n-p-1}t_{n-p}(c_1) - \frac{n-p+c_2^2}{n-p-1}t_{n-p}(c_2)\right) \\ &\quad - s^2(\boldsymbol{x})\left(\frac{c_1(n-p)+c_1^3}{n-p-2}t_{n-p}(c_1) - \frac{c_1(n-p)+c_2^3}{n-p-2}t_{n-p}(c_2)\right) \end{aligned} \tag{12}$$

with $c_1 = \frac{\alpha-\epsilon(\boldsymbol{x})-m^*(\boldsymbol{x})}{s(\boldsymbol{x})}$, $c_2 = \frac{\alpha+\epsilon(\boldsymbol{x})-m^*(\boldsymbol{x})}{s(\boldsymbol{x})}$ and pdf $\mathcal{T}_\nu$ and density t_ν of the t-distribution with ν degrees of freedom. When $\boldsymbol{\psi}$ is treated as unknown the expected improvement is calculated by integrating (12) over the posterior of $\boldsymbol{\psi}$.

The next computer experiment is evaluated at the point $\tilde{\boldsymbol{x}} = \operatorname{argmax} E[I(\boldsymbol{x})]$, that maximizes the expected improvement. The entire sequential algorithm is as follows:

(i) Fit Gaussian process interpolation to the current data available.
(ii) Determine the point with highest expected improvement $\tilde{\boldsymbol{x}}$.
(iii) Evaluate the computer model at $\tilde{\boldsymbol{x}}$ and go back to step (i).

The procedure is stopped if either the budget of computer model evaluation is exhausted or some other stopping criterion is met.

An estimator for the contour is given by $\hat{S}(\boldsymbol{x} : \hat{y}(\boldsymbol{x}) = \alpha)$ and to illustrate the uncertainty about the contour we look at the region in $\mathcal{X}$ where α lies within pointwise 95 % uncertainty intervals for $\tilde{y}$.

Now we apply the sequential contour estimation to our example in sheet metal forming. We use a maximin Latin hypercube of size $n_s = 5$ as a start design and add 5 points sequentially based on the expected improvement criterion. In the sequential algorithm we chose $\gamma = 1$ for determining the contour neighborhood. For comparison a single step 10 point maximin Latin hypercube is considered as well.

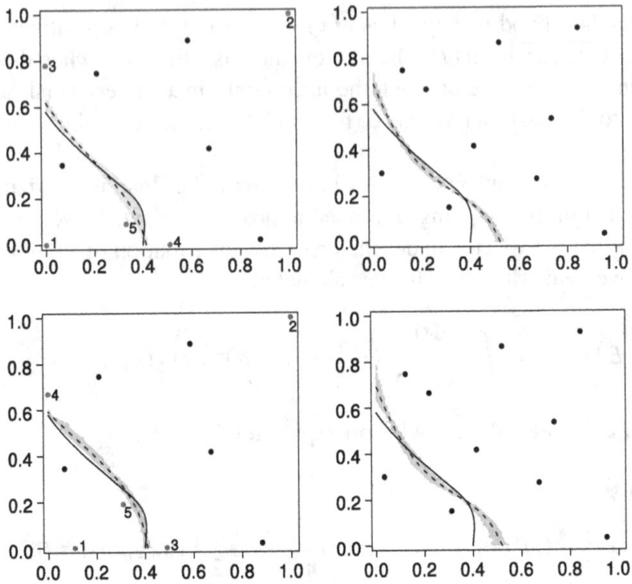

Fig. 7 Contour estimation using sequential design (left) and single step space filling design (right) in combination with posterior mode (top) and MCMC (bottom) computational approaches using Jeffreys prior. Legend: • design point, • sequential design point (with order number), – true contour, - - estimated contour using 10 design points, ■ contour uncertainty region

In the Gaussian process interpolation we use a constant mean, Gaussian correlation function and the seldom used Jeffreys prior $p^{(3)}(\boldsymbol{\xi})$ to gain further insight into its behavior in practice. Computations are accomplished using posterior mode and MCMC analysis based on a chain of length 10000 using every 30th value after a burnin of 1000 iterations.

The results using the posterior mode as well as the MCMC approach for analysis are displayed in Fig. 7. It can be seen that for both computational approaches the sequential design adds design points at quite similar locations (although in different order). The design points are added on the border of the design region (where uncertainty is usually largest) or close to the current estimate of the contour line as one might expect from the improvement criterion. Additionally, as hoped for, contour estimation based on the sequential design is closer to the truth than the single step design. Thus here a sequential design leads to better results as the goal is to estimate a specific characteristic, which can be exploited in the design criterion.

The computational approach seems to have only a minor influence in the sequential design and contour estimation for this example. Computationally however, the MCMC based approach is much more complex, as it requires a full MCMC run after each observation. To keep the computation time practical, especially for the expected improvement function, we chose a higher thinning rate for the chain. Recently Gramacy and Polson (2009) proposed a particle learning algorithm as an alternative to MCMC, that can cope more efficiently with data arriving sequentially.

Nevertheless in this example the posterior mode approach is a very attractive and computationally efficient alternative.

5 Conclusions

Gaussian process interpolation comes with a big promise: The possibility to accompany predictions from an interpolation model with an uncertainty statement. As illustrated, in several situations one subsequently relies on these uncertainty statements particularly in the analysis of computer experiments. However, these uncertainty statements are based on imposing a particular assumption (the prior probability measure) to the interpolation process, and it is often not clear how reliable they are for the practical situation under consideration.

In this article we studied the accuracy of the uncertainty statements for the methodology in situations, when the assumptions are exactly correct (by simulating functions to be interpolated from a Gaussian process in Sect. 3.1) and for realistic test functions (which might not necessarily look like a realization of a stationary Gaussian process). For this purpose we studied two computational approaches and three different choices of priors for the hyperparameters in the correlation function: The standard computational approach is the posterior mode approach described in Sect. 2.2. This plug-in approach ignores the uncertainty due to not knowing the correlation parameter and hence underestimates posterior uncertainty. As a second computational approach we studied a full Markov Chain Monte Carlo approach, to see to what extend it improves the uncertainty statements. The current standard approach for choosing a default prior distribution for hyperparameters in the correlation function consists of using an improper constant prior on the positive real numbers. Paulo (2005), however, proposed objective prior distributions based on general formal rules for deriving objective prior distributions. In this work we investigated, whether these formally objective prior distributions lead to an improvement in terms of representation of the uncertainty.

To study the reliability of uncertainty statements we evaluated the across-function coverage probability of the methodology, which we regard as the practically most relevant value. It is built by pointwisely observing whether the function lies in or outside a prediction interval resulting from the posterior and averaging this over the input space. When sampling functions from a Gaussian process the coverage probability did approximately reach the nominal level for all priors and computational approaches as long as the sample is not too small (with a slight advantage toward the MCMC based approach). In a second study, when evaluating test functions that are not necessarily a realization of a Gaussian process the coverage tends to be smaller than the nominal level. In this situation the MCMC approach gives more reliable results compared to the posterior mode approach. Note that for the extreme test function that does not look like a typical realization of a stationary Gaussian process the coverage does not approach the nominal level even with larger sample size.

Overall in our simulation study none of the objective priors turned out to be clearly superior over the other. For evaluating our test functions the flat prior seems to gain slightly higher coverages. The formally objective independence Jeffreys prior distribution overall seems to perform even slightly worse than the other two approaches, which might explain why this prior is seldom used in practice.

Regarding the computational implementation it seems that there are advantages for MCMC (particularly in the simulation study), but they certainly come at the cost of an increased computational effort, and it depends on time constraints, whether one would want to implement this approach. In the sequential design example in Sect. 4 the improvement due to an MCMC analysis are probably too small to justify the increase in computational time.

In summary it seems that in terms of uncertainty representation, there are no strong reasons to depart from the current standard procedures, *i.e.*, the PM approach and the flat prior distribution, although there are some advantages associated with an MCMC based evaluation.

The reduced coverage values observed for the extreme non-stationary test function in Sect. 3.2, are certainly discomforting, although it is not clear, whether or to what extent they affect for example the sequential design approaches. However, to achieve reasonable function coverage values also in more extreme situations, one could extend the prior distribution to non-stationarity, so that the model becomes more flexible. There are several approaches available for modeling non-stationarity, see for example Paciorek and Schervish (2004) or Xiong et al. (2007) for two recent examples. An alternative would be to use a more flexible mean function, for example, by using a spline basis for $f(\boldsymbol{x})$, but then one usually runs into numerical problems. Another way out are treed Gaussian processes that fit different covariance stationary Gaussian processes to data based partitions of the input space (see Gramacy and Lee 2008). One should keep in mind, however, that all these extensions require more parameters to be estimated from the data, which often requires a larger sample size in turn.

Acknowledgements

The authors would like to thank the Institute of Forming Technology and Ligthweight Construction at TU Dortmund, especially Marco Gösling, for the example in Sect. 4.1. Financial support of the Deutsche Forschungsgemeinschaft (SFB 708: '3D Surface Engineering' and Research Training Group 'Statistical Modelling') is gratefully acknowledged.

References

Bayarri, M.J., Berger, J.O., Paulo, R., Sacks, J., Cafeo, J.A., Cavendish, J., Lin, C.H., Tu, J.: A framework for validation of computer models. Technometrics **49**, 138–154 (2007)

Berger, J.O., de Oliveira, V., Sansó, B.: Objective Bayesian analysis of spatially correlated data. J. Am. Stat. Assoc. **96**, 1361–1374 (2001)

Carnell, R.: lhs: Latin Hypercube Samples. R Package Version 0.5 (2009)

Cressie, N.: Statistics for Spatial Data. Wiley, New York (1993)

Currin, C., Mitchell, T., Morris, M., Ylvisaker, D.: Bayesian prediction of deterministic functions with applications to the design and analysis of computer experiments. J. Am. Stat. Assoc. **86**, 953–963 (1991)

Diaconis, P.: Bayesian numerical analysis. In: Berger, J., Gupta, S. (eds.) Statistical Decision Theory and Related Topics IV, pp. 163–175. Springer, New York (1988)

Fang, K., Li, R., Sudjianto, A.: Design and Modeling for Computer Experiments. Computer Science & Data Analysis. Chapman & Hall/CRC, New York (2006)

Geyer, C.J.: mcmc: Markov Chain Monte Carlo. R Package Version 0.6 (2009)

Gramacy, R.B., Lee, H.K.H.: Bayesian treed Gaussian process models with an application to computer modeling. J. Am. Stat. Assoc. **103**, 1119–1130 (2008)

Gramacy, R.B., Polson, N.G.: Particle learning of Gaussian process models for sequential design and optimization. Available at http://www.arxiv.org/abs/0909.5262v1 (2009)

Gösling, M., Kracker, H., Brosius, A., Gather, U., Tekkaya, A.: Study of the influence of input parameters on a springback prediction by FEA. In: Proceedings of IDDRG 2007 International Conference, Gyor, Hungary, 21–23 May 2007, pp. 397–404 (2007)

Gu, C.: Smoothing Spline ANOVA Models. Springer, Berlin (2002)

Handcock, M.S., Stein, M.L.: A Bayesian analysis of kriging. Technometrics **35**, 403–410 (1993)

Jones, D., Schonlau, M., Welch, W.: Global optimization of expensive black-box functions. J. Glob. Optim. **13**, 455–492 (1998)

Kass, R.E., Wasserman, L.: The selection of prior distributions by formal rules. J. Am. Stat. Assoc. **91**, 1343–1370 (1996)

Kenett, R., Zacks, S.: Modern Industrial Statistics: Design and Control of Quality and Reliability. Duxbury, San Francisco (1998)

Kennedy, M., O'Hagan, A.: Bayesian calibration of computer models. J. R. Stat. Soc. B **63**, 425–464 (2001)

Krige, D.G.: A statistical approach to some basic mine valuation problems on the witwatersrand. J. Chem. Metall. Min. Soc. S. Afr. **52**, 119–139 (1951)

Lehman, J., Santner, T., Notz, W.: Designing computer experiments to determine robust control variables. Stat. Sin. **14**, 571–590 (2004)

Matheron, G.: Principles of geostatistics. Econ. Geol. **58**, 1246–1266 (1963)

Oakley, J.E.: Decision-theoretic sensitivity analysis for complex computer models. Technometrics **51**, 121–129 (2009)

O'Hagan, A.: Curve fitting and optimal design for prediction (with discussion). J. R. Stat. Soc. B **40**, 1–42 (1978)

O'Hagan, A.: Some Bayesian numerical analysis (with discussion). In: Bernardo, J.M., Berger, J.O., Dawid, A.P., Smith, A.F.M. (eds.) Bayesian Statistics 4, pp. 345–363. Oxford University Press, Oxford (1992)

O'Hagan, A., Forster, J.: Kendall's Advanced Theory of Statistics, vol. 2B: Bayesian Inference, 2nd edn. Arnold, London (2004)

Paciorek, C.J., Schervish, M.J.: Nonstationary covariance functions for Gaussian process regression. In: Thrun, S., Saul, L., Schölkopf, B. (eds.) Advances in Neural Information Processing Systems, vol. 16. MIT Press, Cambridge (2004)

Paulo, R.: Default priors for Gaussian processes. Ann. Stat. **33**, 556–582 (2005)

Ranjan, P., Bingham, D.: Sequential experiment design for contour estimation from complex computer codes. Technometrics **50**, 527–541 (2008)

Roberts, G.O., Rosenthal, J.S.: Optimal scaling for various Metropolis-Hastings algorithms. Stat. Sci. **16**, 351–367 (2001)

Sacks, J., Welch, W., Mitchell, T., Wynn, H.: Design and analysis of computer experiments. Stat. Sci. **4**, 409–435 (1989)

Santner, T., Williams, B., Notz, W.: Design & Analysis of Computer Experiments. Springer, New York (2003)

Schonlau, M., Welch, W.: Screening the input variables to a computer model via analysis of variance and visualization. In: Dean, A., Lewis, S. (eds.) Screening, pp. 308–327. Springer, New York (2006)

Seeger, M.: Gaussian processes for machine learning. Int. J. Neural Syst. **14**, 69–104 (2004)

Tokdar, S.T., Ghosh, J.K.: Posterior consistency of logistic Gaussian process priors in density estimation. J. Stat. Plan. Inference **137**, 34–42 (2007)

Wahba, G.: Improper priors, spline smoothing and the problem of guarding against model errors in regression. J. R. Stat. Soc. B **40**, 364–372 (1978)

Williams, B., Santner, T., Notz, W.: Sequential design of computer experiments to minimize integrated response functions. Stat. Sin. **10**, 1133–1152 (2000)

Xiong, Y., Chen, W., Apley, D., Ding, X.: A non-stationary covariance based kriging method for meta-modeling in engineering design. Int. J. Numer. Methods Eng. **71**, 733–756 (2007)

On the Inversive Pseudorandom Number Generator

Wilfried Meidl and Alev Topuzoğlu

Abstract The inversive generator was introduced by J. Eichenauer and J. Lehn in 1986. A large number of papers on this generator have appeared in the last three decades, some investigating its properties, some generalizing it. It has been shown that the generated sequence and its variants behave very favorably with respect to most measures of randomness.

In this survey article we present a comprehensive overview of results on the inversive generator, its generalizations and variants. As regards to recent work, our emphasis is on a particular generalization, focusing on the underlying permutation $P(x) = ax^{p-2} + b$ of $\mathbb{F}_p$.

1 Introduction

The celebrated inversive (congruential pseudorandom number) generator is defined as

$$x_n = \begin{cases} ax_{n-1}^{-1} + b & \text{for } x_{n-1} \neq 0, \\ b & \text{for } x_{n-1} = 0, \end{cases} \tag{1}$$

where x_0, a, b are in the finite field $\mathbb{F}_p$ with p elements. The sequence (x_n) having terms in $\mathbb{F}_p$, yields a sequence (y_n) of pseudorandom numbers in $[0,1)$ with $y_n = x_n/p$.

Since its introduction by Eichenauer and Lehn (1986), this generator gained a wide popularity. Early work on the inversive generator is due to the researchers in "Arbeitsgruppe 9 in TH Darmstadt" and their collaborators, see Eichenauer et al.

Wilfried Meidl, MDBF, Sabancı University, Orhanlı, 34956 Tuzla, İstanbul, Turkey
e-mail: wmeidl@sabanciuniv.edu

Alev Topuzoğlu, MDBF, Sabancı University, Orhanlı, 34956 Tuzla, İstanbul, Turkey
e-mail: alev@sabanciuniv.edu

L. Devroye et al. (eds.), *Recent Developments in Applied Probability and Statistics*,
DOI 10.1007/978-3-7908-2598-5_5,

(1987), Eichenauer and Niederreiter (1988), Eichenauer et al. (1989), Eichenauer-Herrmann (1991, 1992c), and Eichenauer-Herrmann and Ickstadt (1994).

It is not surprising though that the enchanting properties of the inversive generator managed to attract the attention of other researchers around the world rather quickly, leading to a vast amount of relevant work, that appeared since. Presenting a complete survey of results on the inversive generator therefore is an impossible task. The list of references we provide, though extensive, is also far from being complete. Interested readers may also check L'Ecuyer and Hellekalek (1998), Eichenauer-Herrmann (1992a, 1995), Eichenauer-Herrmann et al. (1998), Emmerich (1996), Hellekalek (1998), Niederreiter (1992, 1994), Shparlinski (2003), Topuzoğlu and Winterhof (2007).

Generalizations and modifications of the inversive generator include others using one inversion, like explicit inversive generators, generators with more than one inversions, and inversive generators over other structures, like Galois rings (see Sole and Zinoviev 2009). The underlying permutation

$$P(x) = ax^{p-2} + b \tag{2}$$

of $\mathbb{F}_p$ is, naturally, of interest also. Its cycle structure, for instance, yields the periods of the sequence (1) for different starting values.

A recent generalization of the inversive generator focuses on the permutation P in (2), and replaces P with a permutation involving more than one inversions. Results on the period lengths of sequences obtained by such permutations are presented in Sect. 1.5.1. This work clearly indicates that the analysis of these generators are more challenging. However one may expect that they also have favorable randomness properties. For instance similar bounds on the linear complexity profile, a randomness measure for sequences which will be recalled in Sect. 3, can be obtained with the same approach used for the classical inversive generator, when the number of inversions is small.

This recent work not only generalizes the inversive generator yielding an interesting class of sequences, but also provides a new approach to studying permutation polynomials over finite fields. These results are presented in detail in the last two sections.

2 Notation and Terminology

We start by recalling that a sequence is defined to be pseudorandom (PR) if it is generated by a deterministic algorithm, with the aim that it simulates a truly random sequence. A pseudorandom sequence in the unit interval [0, 1) is called a sequence of pseudorandom numbers (PRN). In particular, for a prime p, the finite field $\mathbb{F}_p$ is identified with the set $\{0, 1, \ldots, p-1\}$, and a sequence (s_n) in $\mathbb{F}_p$ gives rise to a sequence (u_n) of PRNs, satisfying $u_n = s_n/p$. The sequence (s_n) in this case is usually called a pseudorandom number generator.

Randomness of a given PR sequence is measured by various methods: In addition to many statistical tests that the sequence is expected to pass, a large number of theoretical tests have been developed. The significance of these theoretical tests were first established by the seminal work of Marsaglia (random numbers fall mainly in the planes, Marsaglia 1968), describing the weakness of the *linear generator*. Recall that the linear generator is defined as

$$s_{n+1} = as_n + b, \tag{3}$$

for $a, b \in \mathbb{F}_p, a \neq 0$. We shall focus on some of the theoretical quality measures, that are relevant for the analysis of the inversive generator and its generalizations. Comparison with the linear generator justifies the wide interest gained by the inversive generator. The reader is referred to L'Ecuyer and Hellekalek (1998), Hellekalek (1998), Niederreiter (1992), Shparlinski (2003), Topuzoğlu and Winterhof (2007) and references therein for detailed exposition of various other quality measures for randomness, and performance of popular PR sequences under them.

We shall restrict ourselves to the study of PR sequences over a finite field $\mathbb{F}_q$ of $q = p^r$ elements where r is a non-negative integer, and p is a prime. In order to obtain PRNs when $r \geq 2$, one chooses a fixed ordered basis $\{\beta_1, \ldots, \beta_r\}$ of $\mathbb{F}_q$ over $\mathbb{F}_p$. A sequence (y_n) of PRNs in the unit interval can then be obtained from (ξ_n) in $\mathbb{F}_q$ by $y_n = (k_r + k_{r-1}p + \cdots + k_1 p^{r-1})/q$ for $\xi_n = k_1\beta_1 + \cdots + k_r\beta_r$.

In particular the inversive generator (1) above can be expressed as a sequence over $\mathbb{F}_q$:

$$x_{n+1} = ax_n^{q-2} + b \tag{4}$$

where the starting value x_0, and a, b are in $\mathbb{F}_q$. For results on the inversive generator defined over the ring $\mathbb{Z}_m$, where m is a prime power, see for instance Eichenauer et al. (1988b), Eichenauer-Herrmann and Topuzoğlu (1990), Eichenauer-Herrmann and Grothe (1992), Eichenauer-Herrmann and Ickstadt (1994).

The sequences we consider are (purely) periodic, i.e., a sequence (ξ_n) satisfies $\xi_{n+t} = \xi_n$ for some positive integer t, for all $n \geq 0$. We recall that the smallest such t is defined to be the (least) period (or period length) of (ξ_n). For applications only parts of the sequences are used, therefore sequences with long periods are sought for. Hence knowledge of the period length of a given sequence is crucial.

It was shown in Eichenauer and Lehn (1986) that the inversive generator (1) in $\mathbb{F}_p$ has maximum possible period length p, if $x^2 - ax - b$ is a primitive polynomial over $\mathbb{F}_p$. See the work of Flahive and Niederreiter (1993) for a refinement, namely giving necessary conditions for x_n to have period p. Chou (1995a, 1995b) extended this result to the case of a prime power q. He also determined period lengths for all starting values. See Sect. 5 for a generalization of this work.

3 Analysis of the Inversive Generator and Its Variants

In this section we introduce some relevant randomness measures and describe the favorable performance of the inversive generator and its variants under these measures.

We start by focusing on the so-called *linear complexity*, a quality measure which is of particular importance for cryptographic applications. Let us first recall that a sequence $(s_n)_{n\geq 0}$ of elements of $\mathbb{F}_q$ is called a *(homogeneous) linear recurring sequence of order k* if there exist $a_0, a_1, \ldots, a_{k-1}$ in $\mathbb{F}_q$, satisfying the *linear recurrence of order k over* $\mathbb{F}_q$;

$$s_{n+k} = a_{k-1}s_{n+k-1} + a_{k-2}s_{n+k-2} + \cdots + a_0 s_n, \quad n = 0, 1, \ldots.$$

The *linear complexity profile* of a sequence (s_n) over $\mathbb{F}_q$ is the sequence $L(s_n, N)$, $N \geq 1$, where its Nth term is defined to be the smallest L such that a linear recurrence of order L over $\mathbb{F}_q$, can generate the first N terms of (s_n).

We put $L(s_n, N) = 0$ if the first N elements of (s_n) are all zero and $L(s_n, N) = N$ if the first $N-1$ elements of (s_n) are zero and $s_{N-1} \neq 0$.

The value

$$L(s_n) = \sup_{N\geq 1} L(s_n, N)$$

is called the linear complexity of the sequence (s_n). For the linear complexity of any periodic sequence of period t one can easily see that $L(s_n) = L(s_n, 2t) \leq t$.

Linear complexity and linear complexity profile of a given sequence can be determined by using the well-known Berlekamp-Massey algorithm (see e.g. Massey 1969). The algorithm is efficient for sequences with low linear complexity. Therefore such sequences can easily be predicted and hence should be avoided for use in cryptographical applications. The linear generator

$$s_{n+1} = as_n + b,$$

for $a, b \in \mathbb{F}_p, a \neq 0$ is a typical example having $L(s_n) \leq 2$. In fact the knowledge of three consecutive terms of this sequence is sufficient to retrieve the whole sequence. PR sequences with low linear complexity are shown to be unsuitable for some applications using quasi-Monte Carlo methods, also see Niederreiter (1992, 2003), Niederreiter and Shparlinski (2002a).

The expected values of linear complexity and linear complexity profile show that a "truly random" sequence (s_n), with least period t should have $L(s_n, N)$ close to $\min\{N/2, t\}$ for all $N \geq 1$.

As opposed to the linear generator failing this randomness test for any choice of parameters a, b, the inversive generator (with least period t) over $\mathbb{F}_p$ has almost optimal linear complexity profile, (see Shparlinski 2003):

$$L(x_n, N) \geq \min\{\lceil N/3 \rceil, \lceil t/2 \rceil\}. \quad (5)$$

The following variant of the inversive generator exhibits a particularly nice behavior with respect to the linear complexity profile. The *explicit inversive congruential generator* (z_n) was introduced by Eichenauer-Herrmann (1993) and is defined by the recursion

$$z_n = (an+b)^{p-2}, \quad n = 0, \dots, p-1, \qquad z_{n+p} = z_n, \quad n \geq 0, \tag{6}$$

with $a, b \in \mathbb{F}_p$, $a \neq 0$, and $p \geq 5$. It is shown by Meidl and Winterhof (2003) that

$$L(z_n, N) \geq \begin{cases} (N-1)/3, & 1 \leq N \leq (3p-7)/2, \\ N-p+2, & (3p-5)/2 \leq N \leq 2p-3, \\ p-1, & N \geq 2p-2. \end{cases} \tag{7}$$

Analogues of (7), when $q = p^r$, $r \geq 2$, are also given in Meidl and Winterhof (2003). In this case they are called *digital inversive generators*, see Eichenauer-Herrmann and Niederreiter (1994).

Note that the inversive generator (4) can be expressed as

$$x_{n+1} = f(x_n)$$

for $f(x) = ax^{q-2} + b \in \mathbb{F}_q[x]$. One can naturally consider any polynomial $f(x) \in \mathbb{F}_q[x]$ of degree $d \geq 2$, to obtain a *nonlinear congruential pseudorandom number generator* (u_n):

$$u_{n+1} = f(u_n), \quad n \geq 0, \tag{8}$$

with some initial value $u_0 \in \mathbb{F}_q$, see Eichenauer et al. (1988a), and Niederreiter (1988). Obviously, the sequence (u_n) is eventually periodic with some period $t \leq q$. Assuming it to be (purely) periodic, the following lower bound on the linear complexity profile holds, see Gutierrez et al. (2003).

$$L(u_n, N) \geq \min\left\{\log_d(N - \lfloor \log_d N \rfloor), \log_d t\right\}, \quad N \geq 1.$$

The linear complexity profile of pseudorandom number generators over $\mathbb{F}_p$, defined by a recurrence relation of order $m \geq 1$ is studied in Topuzoğlu and Winterhof (2005). Consider the recursively defined sequence

$$u_{n+1} = f(u_n, u_{n-1}, \dots, u_{n-m+1}), \quad n = m-1, m, \dots. \tag{9}$$

Here initial values $u_0, \dots, u_{m-1}$ are in $\mathbb{F}_p$ and $f \in \mathbb{F}_p(x_1, \dots, x_m)$ is a rational function in m variables over $\mathbb{F}_p$. The sequence (9) eventually becomes periodic with least period $t \leq p^m$. The fact that t can actually attain the value p^m gains nonlinear generators of higher orders a particular interest. In case of a polynomial f, lower bounds for the linear complexity and linear complexity profile of higher order generators are given in Topuzoğlu and Winterhof (2005).

A particular rational function f in (9) gives rise to a generalization of the inversive generator (4), as described below. Let (x_n) be the sequence over $\mathbb{F}_p$, defined by the linear recurring sequence of order $m+1$;

$$x_{n+1} = a_0 x_n + a_1 x_{n-1} + \cdots + a_m x_{n-m}, \quad n \geq m,$$

with $a_0, a_1, \ldots, a_m \in \mathbb{F}_p$ and initial values $x_0, \ldots, x_m \in \mathbb{F}_p$. An increasing function $N(n)$ is defined by

$$N(0) = \min\{n \geq 0 : x_n \neq 0\},$$
$$N(n) = \min\{l \geq N(n-1) + 1 : x_l \neq 0\},$$

and the nonlinear generator (z_n) is produced by

$$z_n = x_{N(n)+1} x_{N(n)}^{-1}, \quad n \geq 0$$

(see Eichenauer et al. 1987). It is easy to see that (z_n) satisfies

$$z_{n+1} = f(z_n, \ldots, z_{n-m+1}), \quad n \geq m-1,$$

whenever $z_n \cdots z_{n-m+1} \neq 0$, where the rational function f is given by

$$f(x_1, \ldots, x_m) = a_0 + a_1 x_1^{-1} + \cdots + a_m x_1^{-1} x_2^{-1} \cdots x_m^{-1}.$$

A sufficient condition for (z_n) to attain the maximal period length p^m is given in Eichenauer et al. (1987). It is shown in Topuzoğlu and Winterhof (2005) that the linear complexity profile $L(z_n, N)$ of (z_n) with the least period p^m satisfies

$$L(z_n, N) \geq \min\left(\left\lceil \frac{p-m}{m+1} \right\rceil p^{m-1} + 1, N - p^m + 1\right), \quad N \geq 1.$$

This result is in accordance with (5), i.e., the case $m = 1$.

Given a periodic sequence (s_n) over $\mathbb{F}_q$, the subspaces $L(s_n, s)$ of $\mathbb{F}_q^s$ for $s \geq 1$, spanned by the vectors $\mathbf{s}_n - \mathbf{s}_0$, $n = 1, 2, \ldots$ where

$$\mathbf{s}_n = (s_n, s_{n+1}, \ldots, s_{n+s-1}), \quad n = 0, 1, \ldots$$

have been considered in order to study its structural properties. Starting with the remarkable result of Eichenauer in 1991 (Inversive congruential pseudorandom numbers avoid the planes, Eichenauer-Herrmann 1991), lattice structure of the inversive generator has attracted particular attention: A sequence (s_n) is said to pass the s-*dimensional lattice test* for some $s \geq 1$, if $L(s_n, s) = \mathbb{F}_q^s$. It is obvious for example that the linear generator (3) can pass the s-dimensional lattice test at most for $s = 1$, whereas the inversive generator is well known to pass the test for all $s \leq (p+1)/2$ (see Eichenauer et al. 1988a).

On the other hand sequences, which pass the lattice test for large dimensions, can have bad statistical properties, see Eichenauer and Niederreiter (1988). Accordingly the notion of *lattice profile at* N is introduced by Dorfer and Winterhof (2003). For given $s \geq 1$ and $N \geq 2$, (s_n) passes the s-*dimensional* N-*lattice test* if the subspace spanned by the vectors $\mathbf{s}_n - \mathbf{s}_0$, $1 \leq n \leq N - s$, is $\mathbb{F}_q^s$. The largest s for which (s_n) passes the s-dimensional N-lattice test is called the *lattice profile at* N, and is denoted by $S(s_n, N)$.

It is shown in Dorfer and Winterhof (2003) that the lattice profile is closely related to the linear complexity profile. In fact, either

$$S(s_n, N) = \min\{L(s_n, N), N + 1 - L(s_n, N)\}$$

or

$$S(s_n, N) = \min\{L(s_n, N), N + 1 - L(s_n, N)\} - 1.$$

It therefore follows immediately that the inversive generators or its generalizations we described so far, behave favorably when the lattice profile is considered.

Now we turn our attention to another randomness measure. The *nonlinear complexity profile* $NL_m(s_n, N)$ of an infinite sequence (s_n) over $\mathbb{F}_q$ is the function, which is defined for every integer $N \geq 2$, as the smallest k such that a polynomial recurrence relation

$$s_{n+k} = g(s_{n+k-1}, \ldots, s_n), \quad 0 \leq n \leq N - k - 1,$$

with a polynomial $g(x_1, \ldots, x_k)$ over $\mathbb{F}_q$ of total degree at most m can generate the first N terms of (s_n). It is easy to see that

$$L(s_n, N) \geq NL_1(s_n, N) \geq NL_2(s_n, N) \geq \cdots.$$

Therefore lower bounds for the nonlinear complexity profile yield lower bounds for the linear complexity profile. See Gutierrez et al. (2003) for results on the nonlinear complexity profile for nonlinear and inversive generators.

One would expect that a periodic random sequence and a shift of it would have a low correlation. Autocorrelation measures the similarity between a sequence (s_n) of period t and its shifts by k positions, for $1 \leq k \leq t - 1$. See Eichenauer-Herrmann (1992b) for results concerning the inversive generator. Niederreiter and Rivat (2008) introduce two new types of inversive generators. Among other results, they obtain good correlation properties for the associated binary sequences, where they use a correlation measure introduced in Mauduit and Sarközi (1997).

So far we have considered randomness measures that are of particular importance for cryptological applications. For Monte Carlo applications, the uniformity of distribution is much more significant.

Let P be a point set (finite sequence) $\mathbf{y}_0, \mathbf{y}_1, \ldots, \mathbf{y}_{N-1}$ in $[0, 1)^s$ with $s \geq 1$. The *discrepancy* $D_N^{(s)}$ of P is

$$D_N^{(s)}(P) = D_N^{(s)}(\mathbf{y}_0, \mathbf{y}_1, \ldots, \mathbf{y}_{N-1}) = \sup_J \left| \frac{A_N(J)}{N} - V(J) \right|,$$

where the supremum is taken over all subboxes $J \subseteq [0, 1)^s$, $A_N(J)$ is the number of points $\mathbf{y}_0, \mathbf{y}_1, \ldots, \mathbf{y}_{N-1}$ in J and $V(J)$ is the volume of J. We put $D_N(P) = D_N^{(1)}(P)$.

The well-known Erdős-Turán inequality (see for instance Drmota and Tichy 1997), (10), enables one to estimate discrepancy by the use of bounds on exponential sums. Let P be the point set $y_0, y_1, \dots, y_{N-1}$ in $[0, 1)$. There exists an absolute constant C such that for any integer $H \geq 1$,

$$D_N(P) < C\left(\frac{1}{H} + \frac{1}{N}\sum_{h=1}^{H}\frac{1}{h}|S_N(h)|\right), \tag{10}$$

where $S_N(h) = \sum_{n=0}^{N-1}\exp(2\pi i h x_n)$. For the case $s \geq 2$ the generalized version of (10), the Erdős-Turán-Koksma inequality (see Drmota and Tichy 1997) is used.

It follows by the law of the iterated logarithm that the order of magnitude of discrepancy of N random points in $[0, 1)^s$ should be around $N^{-1/2}(\log\log N)^{1/2}$, see Niederreiter (1992). Accordingly, as a measure of randomness of a PRN sequence, one investigates the discrepancy of s-tuples of consecutive terms. It is possible to show that the distribution of PRNs obtained by the inversive generator is sufficiently irregular. To be precise, consider PRNs, produced by (1) over $\mathbb{F}_p$, having the least period p. Put

$$\mathbf{y}_n = (x_n/p, x_{n+1}/p, \dots, x_{n+s-1}/p) \in [0, 1)^s, \quad n = 0, \dots, p-1$$

where $s \geq 1$. Depending on the parameters $a, b \in \mathbb{F}_p$, and in particular on the average, $D_p^{(s)}(\mathbf{y}_0, \dots, \mathbf{y}_{p-1})$ has an order of magnitude between $p^{-1/2}$ and $p^{-1/2}(\log p)^s$ for every $s \geq 2$, see Niederreiter (1992). Compare with the result on the linear generator where the similar order of magnitude is $p^{-1}(\log p)^s \log\log p$ on the average (Niederreiter 1977).

We refer the reader to Eichenauer-Herrmann and Emmerich (1996), Emmerich (1996), Larcher et al. (1999) for other relevant results, and in particular some on another variant: the *compound inversive generator*. Further references are also listed especially in Emmerich (1996).

As we have remarked earlier, only parts of the period of a PR sequence are used in applications. Therefore bounds on the discrepancy of sequences in parts of the period are of great interest. First non-trivial bounds for parts of the period are obtained by Niederreiter and Shparlinski (1999). While bounds obtained for full period are often the best possible, as the corresponding lower bounds demonstrate (see Niederreiter 1992), bounds in Niederreiter and Shparlinski (1999) concerning nonlinear congruential generators are rather weak. Better results are obtained for the inversive congruential generators of period t. Niederreiter and Shparlinski (2001) showed that

$$D_N(y_0, y_1, \dots, y_{N-1}) = O(N^{-1/2}p^{1/4}\log p), \quad 1 \leq N \leq t.$$

For an average discrepancy bound over all initial values of a fixed inversive congruential generator see Niederreiter and Shparlinski (2002b).

For the distribution of (explicit) nonlinear generators see the series of papers Niederreiter and Winterhof (2000, 2005), and Winterhof (2006). In particular for the explicit inversive generator (6) one has

$$D_N(z_n/p) = O(\min\{N^{-1/2}p^{1/4}\log p, N^{-1}p^{1/2}(\log p)2\}), \quad 1 \leq N \leq p.$$

Now that the "randomness" of the inversive generator is well established, we turn our attention to a recent generalization of it, which is particularly interesting since it yields an alternative method to study permutations of a finite field.

4 Permuting the Elements of $\mathbb{F}_q$ by the Inversive Generator

Let $\wp \in \mathbb{F}_q[x]$ be a polynomial over the finite field $\mathbb{F}_q$. Then we can define a corresponding function from $\mathbb{F}_q$ into $\mathbb{F}_q$ by $x \to \wp(x)$ for all $x \in \mathbb{F}_q$. Conversely for every function f from $\mathbb{F}_q$ into $\mathbb{F}_q$ there exists a polynomial $\wp \in \mathbb{F}_q[x]$ such that $f(x) = \wp(x)$ for all $x \in \mathbb{F}_q$, i.e. every function on a finite field is a *polynomial function*. Given a function $f(x)$ on a finite field, the polynomial $\wp$ for which $f(x) = \wp(x)$ for all $x \in \mathbb{F}_q$ is not unique, but in fact the set of all functions on $\mathbb{F}_q$ equals the set of all polynomial functions on $\mathbb{F}_q$ of degree smaller or equal to $q-1$. More precisely, under the operation of composition and reduction modulo $x^q - x$, the set of polynomials in $\mathbb{F}_q[x]$ of degree $\leq q-1$ forms a semigroup that is isomorphic to the semigroup of functions on $\mathbb{F}_q$.

A polynomial $\wp \in \mathbb{F}_q[x]$ is called a *permutation polynomial* of $\mathbb{F}_q$ if the function on $\mathbb{F}_q$ induced by $\wp$ is a permutation of $\mathbb{F}_q$. As it is now obvious, the set of permutation polynomials of $\mathbb{F}_q$ of degree $\leq q-1$ forms a group which is isomorphic to S_q, the symmetric group on q elements.

A simple example of a permutation polynomial of $\mathbb{F}_q$ is the polynomial $P(x) = ax^{q-2}+b$, the polynomial underlying the inversive generator. For instance if $q = 11$, and we choose $a = 2$ and $b = 7$ we obtain the permutation

x:	0	1	2	3	4	5	6	7	8	9	10
$P(x)$:	7	9	8	4	2	3	0	1	10	6	5

Alternatively one can describe this permutation via its *cycle decomposition*: (0 7 1 9 6) (2 8 10 5 3 4), meaning that $P(0) = 7$, $P(7) = 1$, $P(1) = 9$, $P(9) = 6$ and $P(6) = 0$ (similarly for the second cycle starting with the element 2). As obvious, the cycle decomposition of $P(x) = 2x^9 + 7$ gives precisely the periods of the inversive generator $x_n = 2x_{n-1}^9 + 7$ over $\mathbb{F}_{11}$. If one chooses $u_0 = 0, 7, 1, 9$ or 6 then one obtains a 5-periodic sequence, for any other value for u_0 the period length will be 6.

Summarizing, the determination of the possible period lengths of the inversive generator

$$x_{n+1} = ax_n^{q-2} + b, \quad x_0 \in \mathbb{F}_q,$$

(see Chou 1995a, 1995b) is equivalent to analysing the cycle structure of the permutation given by $P(x) = ax^{q-2} + b$. We remark that the inversive generator has maximal possible period q if and only if the permutation $P(x) = ax^{q-2} + b$ is a single cycle (containing all elements of $\mathbb{F}_q$). In this case we will say that the permutation P is a *full cycle*. We recall that the inversive generator has maximal possible period q, i.e. the permutation $P(x) = ax^{q-2} + b$ is a full cycle, if $x^2 - ax - b$ is a primitive polynomial over $\mathbb{F}_q$.

In the following we present results on the period of the inversive generator over $\mathbb{F}_q$ in detail. We hereby define the inversive generator in a somewhat more general way:

$$x_n = \mathcal{P}(x_{n-1}), \quad x_0 \in \mathbb{F}_q \tag{11}$$

for a (permutation) polynomial of the form

$$\mathcal{P}(x) = (a_0 x + a_1)^{q-2} + a_2, \quad a_1, a_2 \in \mathbb{F}_q,\ a_0 \in \mathbb{F}_q^* = \mathbb{F}_q \setminus \{0\}. \tag{12}$$

Note that for $a_1 = 0$ this definition coincides with the classical inversive generator.

In a natural way we can correspond to the permutation $\mathcal{P}$ (12) the rational transformation $R(x) = (ax + b)/(cx + d)$ with $a = a_0 a_2, b = a_1 a_2 + 1, c = a_0$ and $d = a_1$. The permutation $\mathcal{P}$ can then be expressed as

$$\mathcal{P}(x) = \begin{cases} R(x) & \text{if } x \neq \frac{-d}{c}, \\ \frac{a}{c} & \text{if } x = \frac{-d}{c}. \end{cases} \tag{13}$$

Moreover as pointed out in Çeşmelioğlu et al. (2008a, 2008b), there is a one to one correspondence between the set of permutations $\mathcal{P}$ given as in (12) and the set of distinct permutations of the form (13) for a nonconstant rational transformation $R(x) = (ax + b)/(cx + d) \in \mathbb{F}_q(x), c \neq 0$. The cycle structure of permutations of the form (13) (or equivalently of the form $\mathcal{P}(x) = (a_0 x + a_1)^{q-2} + a_2$)—thus the possible period lengths of the sequence (11)—has been presented in Sect. 2 of Çeşmelioğlu et al. (2008b) (see also Chou 1995b): To a permutation given as in (13) we naturally associate the matrix

$$A = \begin{pmatrix} a & b \\ c & d \end{pmatrix}$$

in $GL(2, q)$, and its characteristic polynomial $f(x) = x^2 - \mathrm{tr}(A)x + \det(A)$ with $R(x)$ (or $\mathcal{P}(x)$). In what follows, $\mathrm{ord}(z)$ denotes the order of an element z in the multiplicative group of $\mathbb{F}_{q^2}$, and ϕ denotes the Euler ϕ-function. The following theorem is from Çeşmelioğlu et al. (2008a).

Theorem 1. *Let $\mathcal{P}$ be the permutation defined by* (13) *over $\mathbb{F}_q, q = p^r$. Suppose that $f(x)$ is the characteristic polynomial of the matrix A associated with $\mathcal{P}$ and $\alpha, \beta \in \mathbb{F}_{q^2}$ are the roots of $f(x)$.*

(1) *Suppose $f(x)$ is irreducible. If $k = \mathrm{ord}(\frac{\alpha}{\beta}) = \frac{q+1}{t}$, $1 \le t < \frac{q+1}{2}$, then $\mathcal{P}$ is a composition of $t-1$ cycles of length k and one cycle of length $k-1$. In particular $\mathcal{P}$ is a full cycle if $t = 1$.*
(2) *Suppose $\alpha, \beta \in \mathbb{F}_q$ and $\alpha \neq \beta$. If $k = \mathrm{ord}(\frac{\alpha}{\beta}) = \frac{q-1}{t}$, $t \ge 1$, then $\mathcal{P}$ is a composition of $t-1$ cycles of length k, one cycle of length $k-1$, and two cycles of length* 1.

(3) *Suppose* $f(x) = (x-\alpha)^2$, $\alpha \in \mathbb{F}_q^* = \mathbb{F}_q \setminus \{0\}$, *then* $\mathcal{P}$ *is a composition of one cycle of length* $p-1$, $p^{r-1}-1$ *cycles of length* p *and one cycle of length* 1.

For $q \geq 5$ *the number of distinct permutations* $\mathcal{P}$ *is equal to* $\phi(k)\frac{q-1}{2}q$ *in the cases* (1) *and* (2), *and is equal to* $(q-1)q$ *in the case* (3).

As explained earlier in detail, the inversive generator, and thus the underlying permutation $P(x) = ax^{q-2} + b$ of $\mathbb{F}_q$, has excellent behavior with respect to many quality measures. A further quality measure frequently used in the area of coding theory, is the dispersion of a permutation. In Çeşmelioğlu et al. (2009) the permutation $\mathcal{P}$ of the prime field $\mathbb{F}_p$

$$\mathcal{P}(x) = (a_0x + a_1)^{p-2} + a_2, \tag{14}$$

has been analysed with respect to its *(normalized) dispersion*, which is defined as follows: Let P be a permutation of the set $\{0, 1, \ldots, N-1\}$. The dispersion of P is the cardinality of the set

$$D(P) = \{(i_2 - i_1, P(i_2) - P(i_1)) \mid 0 \leq i_1 < i_2 \leq N-1\}.$$

The normalized dispersion is defined as

$$\gamma(P) = \frac{2\,|D(P)|}{N(N-1)}.$$

As obvious, the normalized dispersion can at most be 1. Permutations with normalized dispersion 1 are called Costas arrays. They were introduced by J.P. Costas in 1965 for applications in sonar design. For a recent exposition of Costas arrays including all currently known algebraic constructions we refer to Golomb and Gong (2007).

We may consider a permutation P to have good randomness properties with respect to dispersion if $\gamma(P)$ is close to the expected normalized dispersion of a random permutation (which is approximately 0.8), see Avenancio-Leon (2005), Corrada-Bravo and Rubio (2003), and Heegard and Wicker (1999).

In Çeşmelioğlu et al. (2009) the precise value for the dispersion of the permutation (14) is determined when $a_2 = 0$ and $p \equiv 5 \bmod 6$ (for $p \equiv 1 \bmod 6$ a lower bound is given). In order to state the result we have to introduce some notation: As usual we identify $\mathbb{F}_p$ with $\{0, 1, \ldots, p-1\}$ so that the element $-a_1/a_0 \in \mathbb{F}_p$ is identified with an integer between 0 and $p-1$. This integer shall be denoted by x_I. The next result is in Çeşmelioğlu et al. (2009).

Theorem 2. *Let* $p \equiv 5 \bmod 6$ *and let* $\underline{\mathcal{P}}(x) = (a_0x + a_1)^{q-2} \in \mathbb{F}_p[x]$, $a_0 \neq 0$. *The dispersion* $|D(\underline{\mathcal{P}})|$ *of* $\underline{\mathcal{P}}(x)$ *is then given by*

$$|D(\underline{\mathcal{P}})| = \begin{cases} \frac{(p+3)(p-1)+4x_I(p-2x_I-1)}{4}, & 0 \leq x_I \leq \frac{p-1}{2}, \\ \frac{(p+3)(p-1)+4(p-1-x_I)(2x_I-p+1)}{4}, & \frac{p-1}{2} < x_I \leq p-1. \end{cases}$$

Consequently by a judicious selection of a_0 and a_1, the dispersion of $\mathcal{P}$ can be pre-determined. The value for the normalized dispersion can be made close to the expected normalized dispersion of a random permutation (see Çeşmelioğlu et al. 2009):

Corollary 1. *Let* $p \equiv 5 \bmod 6$. *The maximum value for the normalized dispersion is* $\gamma(\mathcal{P}) = 0.75 + \frac{5}{4p}$. *This value is attained for* $x_I = \lfloor\frac{p-1}{4}\rfloor$ *and* $x_I = \lfloor\frac{3p-1}{4}\rfloor$.

One can also consider $D_p(\mathcal{P})$, obtained by calculating modulo p. Obviously $D_p(\mathcal{P}) \leq D(\mathcal{P})$. For $p \equiv 5 \bmod 6$ it is shown in Çeşmelioğlu et al. (2009) that any value for $D_p(\mathcal{P})$, which only depends on x_I and not on a_2, can be combined with any of the cycle decompositions given as in Theorem 1 by an appropriate choice of the parameters a_0, a_1 and a_2.

It is suggested that the cycle structure of a permutation P affects the performance of P, as well as its dispersion, when used as an interleaver (see Avenancio-Leon 2005; Corrada-Bravo and Rubio 2003, and Heegard and Wicker 1999 for background on interleavers in coding theory). Of particular interest seem to be permutations that decompose into cycles of the same length (see Rubio and Corrada-Bravo 2004; Rubio et al. 2008). Theorem 1 where the cycle decomposition of the permutations $\mathcal{P}(x)$ given as in (14) is presented, shows that $\mathcal{P}(x)$ has either only one nontrivial cycle (of length > 1) or all nontrivial cycles are of the same length except for one cycle having one element less. Therefore the fact that we can design inversive permutation polynomials with prescribed dispersion modulo p and cycle decomposition may suggest the use of this class of permutations as a toolkit for simulations, that may contribute to a better understanding of the affect of dispersion and cycle decomposition of an interleaver on its performance.

5 From the Inversive Generator to all Permutations of $\mathbb{F}_q$

A further generalization of the inversive generator over a finite field $\mathbb{F}_q$ is obtained by allowing more than one, say n inversions: For $a_1, a_{n+1} \in \mathbb{F}_q$, $a_i \in \mathbb{F}_q^*$ for $i = 0, 2, \ldots, n$ and a starting value $u_0 \in \mathbb{F}_q$ we can define a sequence (u_k) with terms in $\mathbb{F}_q$ by

$$u_k = (\cdots((a_0 u_{k-1} + a_1)^{q-2} + a_2)^{q-2} \cdots + a_n)^{q-2} + a_{n+1}. \tag{15}$$

A second motivation for considering constructions with more than one inversions comes from a classical result of Carlitz (1953): The symmetric group on q letters, which is isomorphic to the group of permutation polynomials of $\mathbb{F}_q$ of degree less than $q-1$ under the operation of composition and reduction modulo $x^q - x$, is generated by the linear polynomials $ax + b$, for $a, b \in \mathbb{F}_q$, $a \neq 0$, and x^{q-2}.

As a consequence any permutation of a finite field $\mathbb{F}_q$ can be represented by a polynomial of the form

$$\mathcal{P}_n(x) = (\cdots((a_0 x + a_1)^{q-2} + a_2)^{q-2} \cdots + a_n)^{q-2} + a_{n+1}, \quad n \geq 0, \tag{16}$$

where $a_1, a_{n+1} \in \mathbb{F}_q$, $a_i \in \mathbb{F}_q^*$ for $i = 0, 2, \ldots, n$. With this notation the sequence (15) above becomes $u_k = \mathcal{P}_n(u_{k-1})$ with a starting value $u_0 \in \mathbb{F}_q$.

In the series of papers Çeşmelioğlu et al. (2008a, 2008b) and Aksoy et al. (2009), permutations given in the form (16) have been analysed. In the following we summarize the results starting with the analysis of the cycle structure of $\mathcal{P}_n$ for small n. We hereby remark that similarly to the classical inversive generator, the period of the sequence (u_k), defined as in (15), depends on the cycle structure of the corresponding permutation (16). We first need some technical preliminaries.

For the polynomial $\mathcal{P}_n(x)$ we consider the rational function

$$r_n(x) = (\cdots((a_0x + a_1)^{-1} + a_2)^{-1} \cdots + a_n)^{-1} + a_{n+1}$$

and its continued fraction expansion,

$$a_{n+1} + 1/(a_n + 1/(\cdots + a_2 + 1/(a_0x + a_1)\cdots)),$$

so as to form the *nth convergent*

$$\mathcal{R}_n(x) = \frac{\alpha_{n+1}x + \beta_{n+1}}{\alpha_n x + \beta_n}, \tag{17}$$

where

$$\alpha_k = a_k\alpha_{k-1} + \alpha_{k-2} \quad \text{and} \quad \beta_k = a_k\beta_{k-1} + \beta_{k-2}, \tag{18}$$

for $k \geq 2$ and $\alpha_0 = 0, \alpha_1 = a_0, \beta_0 = 1, \beta_1 = a_1$. We remark that α_k and β_k cannot both be zero.

We define the set of *poles*, $\mathbf{O}_n$, as

$$\mathbf{O}_n = \left\{ \mathbf{x_i} : \mathbf{x_i} = \frac{-\beta_i}{\alpha_i}, \ i = 1, \ldots, n \right\} \subset \mathbb{P}^1(\mathbb{F}_q) = \mathbb{F}_q \cup \{\infty\}. \tag{19}$$

Obviously $\mathcal{P}_n(x) = \mathcal{R}_n(x)$ for $x \in \mathbb{F}_q \setminus \mathbf{O}_n$. Since the ordering and repetition of the poles will be crucial, we also consider the *string of poles* $\mathcal{O}_n$:

$$\mathcal{O}_n = \mathbf{x_1}, \mathbf{x_2}, \ldots, \mathbf{x_n}.$$

To every rational transformation $\mathcal{R}_n(x)$ of the form (17) we associate the permutation $\mathcal{F}_n(x)$ defined as in (13), i.e. $\mathcal{F}_n(x) = \mathcal{R}_n(x)$ for $x \neq \mathbf{x_n}$ and $\mathcal{F}_n(x_n) = \alpha_{n+1}/\alpha_n$ when $\mathbf{x_n} \in \mathbb{F}_q$.

5.1 On the Period Length of (u_k) for $\mathcal{P}_2$ and $\mathcal{P}_3$

We recall that the possible period lengths of the sequence $u_k = \mathcal{P}_n(u_{k-1})$ for a permutation $\mathcal{P}_n(x)$ of the form (16) and a starting value $u_0 \in \mathbb{F}_q$ are given by the cycle structure of the permutation $\mathcal{P}_n(x)$. In Çeşmelioğlu et al. (2008a) the cycle decomposition of the polynomial $\mathcal{P}_n(x)$ has been analysed for $n = 2$, and $n = 3$ and $a_4 = 0$. As already pointed out there is a one to one correspondence between the set of permutations $\mathcal{P}_1$ and the set of distinct permutations of the form (13). Therefore Theorem 1 on the cycle decomposition of permutations of the form (13) which turns out to be crucial for the analysis of the cycle decomposition of the permutations $\mathcal{P}_n$ covers the case $n = 1$.

As observed in Çeşmelioğlu et al. (2008a) the first three poles are distinct for any permutation $\mathcal{P}_n(x)$ and the first two are not poles at infinity. Furthermore for $\mathcal{P}_n(x)$ with $n = 2$, and $n = 3$ and $a_4 = 0$ we have

$$\mathcal{P}_2 = (\mathcal{F}_2(\mathbf{x_1})\mathcal{F}_2(\mathbf{x_2}))\mathcal{F}_2(x) \tag{20}$$

with $\mathcal{F}_2(x) = \mathcal{R}_2(x)$ if $x \neq \mathbf{x_2}$ and $\mathcal{F}_2(\mathbf{x_2}) = (a_2a_3 + 1)/a_2$, and

$$\mathcal{P}_3(x) = (\mathcal{F}_3(\mathbf{x_2})\mathcal{F}_3(\mathbf{x_1})\mathcal{F}_3(\mathbf{x_3}))\mathcal{F}_3(x) \tag{21}$$

with $\mathcal{F}_3(x) = \mathcal{R}_3(x)$ if $x \neq \mathbf{x_3}$ and $\mathcal{F}_3(\mathbf{x_3}) = a_2/(a_2a_3 + 1)$ if $\mathbf{x_3} \neq \infty$, i.e. $a_2a_3 + 1 \neq 0$. If $a_2a_3 + 1 = 0$ then $\mathbf{x_1} = -\frac{a_1}{a_0}$, $\mathbf{x_2} = -\frac{a_1a_2+1}{a_0a_2}$, $\mathbf{x_3} = \infty$ and

$$\mathcal{R}_3(x) = -a_2(a_0a_2x + a_1a_2 + 1) \tag{22}$$

is linear, thus $\mathcal{F}_3(x)$ and $\mathcal{R}_3(x)$ coincide, giving, (see Sect. 4 in Çeşmelioğlu et al. 2008a)

$$\mathcal{P}_3(x) = (\mathcal{F}_3(\mathbf{x_1})\mathcal{F}_3(\mathbf{x_2}))\mathcal{F}_3(x) = (-a_2 0)\mathcal{F}_3(x). \tag{23}$$

Clearly the cycle decomposition of $\mathcal{P}_2$ and $\mathcal{P}_3$ respectively, is determined by the cycle decomposition of $\mathcal{F}_2$ and $\mathcal{F}_3$ and the effect of the cycle $(\mathcal{F}_2(\mathbf{x_1})\mathcal{F}_2(\mathbf{x_2}))$ and $(\mathcal{F}_3(\mathbf{x_2})\mathcal{F}_3(\mathbf{x_1})\mathcal{F}_3(\mathbf{x_3}))$ (or $\mathcal{F}_3(\mathbf{x_1})\mathcal{F}_3(\mathbf{x_2})$) if $\mathbf{x_3} = \infty$) on the cycle decomposition of $\mathcal{F}_2$ and $\mathcal{F}_3$, respectively. Thus Theorem 1 on the cycle structure of permutations of the form (13) naturally plays a crucial role in order to understand the cycle structure of permutations $\mathcal{P}_2$, $\mathcal{P}_3$. In accordance with the notation introduced before stating Theorem 1, for

$$\mathcal{P}_2(x) = ((a_0x + a_1)^{q-2} + a_2)^{q-2} + a_3, \quad a_0a_2 \neq 0.$$

The characteristic polynomial associated to the rational transformation $\mathcal{R}_2(x)$ that corresponds to $\mathcal{P}_2(x)$ is of the form

$$f(x) = x^2 - (a_0(a_2a_3 + 1) + a_1a_2 + 1)x + a_0. \tag{24}$$

The characteristic polynomial associated to the rational transformation $\mathcal{R}_3(x)$ that corresponds to

$$\mathcal{P}_3(x) = (((a_0x + a_1)^{q-2} + a_2)^{q-2} + a_3)^{q-2}, \quad a_0a_2a_3 \neq 0 \tag{25}$$

is given by

$$f(x) = x^2 - (a_0a_2 + a_1(a_2a_3 + 1) + a_3)x - a_0. \tag{26}$$

We remark that in the case that $\mathbf{x_3} = \infty$, i.e. $a_2 a_3 + 1 = 0$, the polynomial (26) becomes the reducible polynomial

$$f(x) = x^2 - (a_0 a_2 - a_2^{-1})x - a_0 = (x - a_0 a_2)(x + a_2^{-1}).$$

With the above given preliminaries one can determine the cycle decomposition of permutations of the form $\mathcal{P}_2$ and $\mathcal{P}_3$ with $a_4 = 0$. As an example consider the case of a permutation $\mathcal{P}_2$ for which the associated characteristic polynomial $f(x)$ is irreducible and the roots $\alpha, \beta \in \mathbb{F}_{q^2}$ of $f(x)$ satisfy $\mathrm{ord}(\alpha/\beta) = (q+1)/2$. By Theorem 1 the permutation $\mathcal{F}_2$ that corresponds to $\mathcal{P}_2$ has then two cycles. If the poles $\mathbf{x_1}, \mathbf{x_2}$ are not in the same cycle, then the transposition $(\mathcal{F}_2(\mathbf{x_1})\mathcal{F}_2(\mathbf{x_2}))$ in (20) causes a merging of the two cycles, i.e the permutation $\mathcal{P}_2$ is then a full cycle. Whether $\mathbf{x_1}, \mathbf{x_2}$ are in the same cycle of $\mathcal{F}_2$ or not is determined by the parameter $\gamma_0 = (\beta - 1)/(\alpha - 1)$, more precisely $\mathbf{x_1}, \mathbf{x_2}$ are in the same cycle of $\mathcal{F}_2$ if and only if $\gamma_0^{(q+1)/2} = 1$. We refer to Çeşmelioğlu et al. (2008a) for the details.

Similarly the cycle decomposition can be found for all other possible cases. It turns out that one has to distinguish 7 cases for $\mathcal{P}_2$ (see Theorems 6 and 7 in Çeşmelioğlu et al. 2008a) and 20 cases for $\mathcal{P}_3$ with $a_4 = 0$ (see Theorems 11, 13 and 15 in Çeşmelioğlu et al. 2008a). We omit the presentation of all cases and refer to Çeşmelioğlu et al. (2008a). But we present all cases in which we obtain a full cycle since then the sequence (u_k) defined as in (15) has maximal possible period q. For the sake of completeness we include the permutations $\mathcal{P}_1$. Therefore we note that the characteristic polynomial associated to $\mathcal{P}_1(x) = (a_0 x + a_1)^{q-2} + a_2$ is given by

$$f(x) = x^2 - (a_0 a_2 + a_1)x - a_0. \tag{27}$$

Theorem 3. *The permutation $\mathcal{P}_1$ is a full cycle if and only if the polynomial $f(x)$ in (27) is irreducible and the roots $\alpha, \beta \in \mathbb{F}_{q^2}$ of $f(x)$ satisfy $\mathrm{ord}(\alpha/\beta) = (q+1)/2$.*
The permutation $\mathcal{P}_2$ is a full cycle if and only if

(1) (i) *the polynomial $f(x)$ in (24) is irreducible,*
(ii) *the roots $\alpha, \beta \in \mathbb{F}_{q^2}$ of $f(x)$ satisfy $\mathrm{ord}(\alpha/\beta) = (q+1)/2$,*
(iii) *$\gamma_0 = (\beta - 1)/(\alpha - 1)$ satisfies $\gamma_0^{(q+1)/2} \neq 1$, or*
(2) *$\mathbb{F}_q$ is a prime field and $f(x) = (x-1)^2$ (then $a_0 = 1$, $a_3 = -a_1$).*

The permutation $\mathcal{P}_3$ with $a_4 = 0$ is a full cycle if and only if one of the following conditions (1)–(4) is satisfied.

(1) (i) *The polynomial $f(x)$ in (26) is irreducible,*
(ii) *the roots $\alpha, \beta \in \mathbb{F}_{q^2}$ of $f(x)$ satisfy $\mathrm{ord}(\alpha/\beta) = q + 1$ so that $\mathcal{F}_3$ is a full cycle, and*
(iii) *the pole $\mathbf{x_1}$ lies between the poles $\mathbf{x_2}, \mathbf{x_3}$ in the cycle $\mathcal{F}_3$.*
(2) (i) *The polynomial $f(x)$ in (26) is irreducible,*
(ii) *3 divides $q + 1$, and the roots $\alpha, \beta \in \mathbb{F}_{q^2}$ of $f(x)$ satisfy $\mathrm{ord}(\alpha/\beta) = (q+1)/3$, i.e. $\mathcal{F}_3$ is composed of 2 cycles of length $(q+1)/3$ and 1 cycle of length $(q-2)/3$,*

(iii) *the elements* $\gamma_1 = (\beta - a_3)/(\alpha - a_3)$, $\gamma_2 = (a_2\beta + 1)/(a_2\alpha + 1)$, $\gamma_3 = (\beta - a_1)/(\alpha - a_1) \in \mathbb{F}_{q^2}$ *satisfy* $\gamma_1^{(q+1)/3}, \gamma_2^{(q+1)/3}, \gamma_3^{(q+1)/3} \neq 1$, *i.e. the poles* $\mathbf{x_1}, \mathbf{x_2}, \mathbf{x_3}$ *are in distinct cycles of* $\mathfrak{F}_3$.

(3) (i) *The polynomial* $f(x)$ *in* (26) *has two distinct roots* $\alpha, \beta \in \mathbb{F}_q$,
(ii) $ord(\alpha/\beta) = q - 1$, *i.e.* $\mathfrak{F}_3$ *is composed of one cycle of length* $q - 2$ *and two cycles of length* 1,
(iii) $a_2a_3 + 1 \neq 0$, *i.e. the pole* $\mathbf{x_3}$ *is in* $\mathbb{F}_q$,
(iv) $a_3 = -a_0/a_1$ *and* $a_2 = -1/a_1$, *i.e.* $\mathbf{x_1}, \mathbf{x_2}$ *are the fixed points of* $\mathfrak{F}_3$.

(4) (i) $a_2a_3 + 1 = 0$, *i.e.* $\mathfrak{F}_3(x)$ *is linear,*
(ii) $ord(-a_0a_2^2) = q - 1$, *and*
(iii) *either* $a_1 = a_0a_2$ *or* $a_2 = -1/a_1$, *i.e. either* $\mathbf{x_1}$ *or* $\mathbf{x_2}$ *is the fixed point of* $\mathfrak{F}_3$.

Remark 1. The parameters $\gamma_1, \gamma_2, \gamma_3$ are concerned with the distribution of the poles in the cycles of $\mathfrak{F}_3$. Indeed, $\mathbf{x_i}, \mathbf{x_3}$ are in the same cycle if and only if $\gamma_i^k = 1$ for $i = 1, 2$, and $\mathbf{x_1}, \mathbf{x_2}$ are in the same cycle if and only if $\gamma_3^k = 1$.

Based on the above results the number of distinct permutations of the types $\mathcal{P}_2$ and $\mathcal{P}_3$, $a_4 = 0$, with full cycle has been determined in Çeşmelioğlu et al. (2008a). Several of the cases in Theorem 3 had to be addressed with different approaches. In the following theorem we include permutations of the type $\mathcal{P}_1$ for the sake of completeness.

Theorem 4. *The number of distinct permutations of the form* $\mathcal{P}_1(x)$ *with full cycle is* $\phi(q-1)\frac{q-1}{2}q$.

The number of distinct permutations of the form $\mathcal{P}_2(x)$ *with full cycle is* $\phi(\frac{q+1}{2}) \times (q+1)q(q-1)/4$ *when* $q = p^r$ *for a prime* p *with* $r > 1$, *and* $\phi(\frac{p+1}{2})(p+1)p(p-1)/4 + p(p-1)$ *when* $q = p$ *is prime.*

The number of distinct permutations of the form $\mathcal{P}_3(x)$ *with* $a_4 = 0$ *and full cycle is*

$$\frac{\phi(q+1)(q-1)^2(q-2)}{4} + 3\phi(q-1)(q-1)$$

if $3 \mid (q+1)$, *and*

$$\frac{\phi(q+1)(q-1)^2(q-2)}{4} + 3\phi(q-1)(q-1) + \phi\left(\frac{q+1}{3}\right)(q-1)\frac{(q+1)^2}{9}$$

if $3 \mid (q+1)$.

The approach to permutations of a finite field via polynomials of the form $\mathcal{P}_n$ permits a description of all full cycle permutations of a prime field $\mathbb{F}_p$. As observed by Çeşmelioğlu (2009), a permutation $\wp(x)$ of the prime field $\mathbb{F}_p$ is a full cycle if and only if it has a representation of the form

$$\wp(x) = \mathcal{P}_{2k}(x) = (((\cdots(\cdots((x + c_1)^{p-2} + c_2)^{p-2} \cdots + c_{k+1})^{p-2} - c_k)^{p-2} - c_{k-1})^{p-2} - \cdots - c_2)^{p-2} - c_1 \quad (28)$$

for some integer k and $c_1 \in \mathbb{F}_p$, $c_i \in \mathbb{F}_p^*$, $2 \leq i \leq k+1$. Moreover the rth iterate of $\wp(x)$ is obtained by exchanging c_{k+1} in (28) with rc_{k+1}. As a consequence the set of permutations of the form (28) with fixed $c_1, \ldots, c_k$ and variable $c_{k+1} \in \mathbb{F}_p^*$ forms a Sylow p-subgroup of the group of permutations of $\mathbb{F}_p$, and vice versa all Sylow p-subgroups are obtained in this way.

A different approach to the construction of full cycle permutations of a finite field $\mathbb{F}_q$ expressed in the form $\mathcal{P}_n$ has been worked out in Sect. 5 of Çeşmelioğlu et al. (2008a). This approach uses results of Beck (1977) where a symmetric binary $k \times k$-matrix L, called *link relation matrix*, is assigned to a given sequence $\sigma_1, \sigma_2, \ldots, \sigma_k$ of transpositions with elements of a set T (e.g. $T = \mathbb{F}_q$). Let τ be a full cycle permutation of the set T, then the number of cycles of the permutation $\sigma_1\sigma_2 \cdots \sigma_k\tau$ can be determined from the rank of L.

The approach investigated in Çeşmelioğlu et al. (2008a) also permits the construction of permutations of $\mathbb{F}_q$ written as $\mathcal{P}_n$ with a given number of cycles (not necessarily one cycle). For details on the multiplication with transpositions, link relation matrices, and the construction of full cycle permutations of $\mathbb{F}_q$ written as $\mathcal{P}_n$ we refer the reader to Sect. 5 of Çeşmelioğlu et al. (2008a) and the references therein.

5.2 The Carlitz Rank of a Permutation—or the Smallest Number of Inversions Needed

We recall that following a classical result of Carlitz any permutation of a finite field $\mathbb{F}_q$ can be represented by a polynomial of the form (16). Such a representation of a permutation is of course not unique. For instance the polynomials

$$\mathcal{P}_6(x) = ((((((x+16)^{15}+8)^{15}+14)^{15}+5)^{15}+9)^{15}+2)^{15} \tag{29}$$

and

$$\mathcal{P}_4(x) = ((((16x+3)^{15}+3)^{15}+3)^{15}+13)^{15}+9 \tag{30}$$

describe the same permutation $\wp$ of $\mathbb{F}_{17}$. As one observes six "*inversions*" x^{q-2} are used in the description (29) whereas in description (30) of $\wp$ only four inversions are needed. In fact 4 is the minimum number of inversions needed in order to describe the permutation $\wp$ in the form (16).

In Aksoy et al. (2009) for a given permutation $\wp(x)$ of a finite field $\mathbb{F}_q$, the smallest n satisfying $\wp = \mathcal{P}_n$ for a permutation $\mathcal{P}_n$ of the form (16) was defined as the *Carlitz rank* of $\wp(x)$. In accordance with Aksoy et al. (2009) we denote the Carlitz rank of $\wp(x)$ by $\mathrm{Crk}(\wp)$ and emphasize that $\mathrm{Crk}(\wp) = n$ means that $\wp(x)$ is composed of at least n inversions x^{q-2} with n (or $n+1$) linear polynomials. Clearly linear polynomials are the permutation polynomials of Carlitz rank 0. The set of polynomials $\mathcal{P}_1$ of the form (12) that is in one-to-one correspondence with the set of permutations (13) defined by nonconstant rational linear transformations $(ax+b)/(cx+d)$ is precisely the set of permutation polynomials of Carlitz rank 1.

In the following we give an overview of the results on the Carlitz rank obtained in Aksoy et al. (2009).

Consider a permutation of the form

$$\mathcal{P}_s(x) = (\cdots((a_0x + a_1)^{q-2} + a_2)^{q-2} \cdots + a_s)^{q-2} + a_{s+1}$$

with the string of poles $\mathbf{x_1}, \mathbf{x_2}, \ldots, \mathbf{x_s}$ and the sth convergent $\mathcal{F}_s(x)$. As observed in Aksoy et al. (2009).

$$\mathcal{P}_s(x) = (\mathcal{F}_s(\mathbf{x_s})\mathcal{F}_s(\mathbf{x_{s-1}}))(\mathcal{F}_s(\mathbf{x_{s-1}})\mathcal{F}_s(\mathbf{x_{s-2}})) \cdots (\mathcal{F}_s(\mathbf{x_2})\mathcal{F}_s(\mathbf{x_1}))\mathcal{F}_s(x) \quad (31)$$

if $\mathbf{x_i} \neq \infty$, $i = 1, \ldots, s$. If $\mathbf{x_i} = \infty$ for some integer i, $3 \leq i \leq s-2$, then in equation (31) we substitute $(\mathcal{F}_s(\mathbf{x_{i+1}})\mathcal{F}_s(\mathbf{x_i}))(\mathcal{F}_s(\mathbf{x_i})\mathcal{F}_s(\mathbf{x_{i-1}}))$ by $(\mathcal{F}_s(\mathbf{x_{i+1}})\mathcal{F}_s(\mathbf{x_{i-1}}))$ (if $\mathbf{x_s} = \infty$ or $\mathbf{x_{s-1}} = \infty$, then the first transposition in (31) is $(\mathcal{F}_s(\mathbf{x_{s-1}})\mathcal{F}_s(\mathbf{x_{s-2}}))$ or $(\mathcal{F}_s(\mathbf{x_{s-2}})\mathcal{F}_s(\mathbf{x_{s-3}}))$, respectively). For the details we refer to Corollary 1 in Aksoy et al. (2009).

In all cases one then easily obtains a decomposition of $\mathcal{P}_s$ as

$$\mathcal{P}_s = \tau_1 \cdots \tau_m \mathcal{F}_s(x) \quad (32)$$

where $\tau_1, \ldots, \tau_m$ are disjoint cycles. Once a permutation in the above form is given, it is possible to determine the smallest integer n such that $\mathcal{P}_s = \mathcal{P}_n$ for a permutation $\mathcal{P}_n(x)$ and $\mathcal{F}_s = \mathcal{F}_n$, see Theorem 3 in Aksoy et al. (2009). If $n \leq (q-2)/2$ then $Crk(\mathcal{P}_s) = n$. Theorem 3 in Aksoy et al. (2009) is summarized as follows in the following theorem. We write $a \in \mathrm{supp}(\tau)$ if $a \in \mathbb{F}_q$ is not fixed by the cycle τ.

Theorem 5. *Suppose that $\mathcal{P}_s$ can be decomposed as*

$$\mathcal{P}_s(x) = \tau_1 \cdots \tau_m \mathcal{F}_s(x), \quad (33)$$

where $\tau_1, \ldots, \tau_m$ are disjoint cycles of length $l(\tau_j) = l_j \geq 2$, $1 \leq j \leq m$.

(a) *If $\mathcal{F}_s$ is not linear and $\mathcal{F}_s(\mathbf{x_s}) \in \mathrm{supp}(\tau_j)$ for some $1 \leq j \leq m$, then there exists a permutation $\mathcal{P}_n$ with $n = m + \sum_{j=1}^{m} l_j - 1$ such that $\mathcal{P}_s(x) = \mathcal{P}_n(x)$ for all $x \in \mathbb{F}_q$ (and additionally $\mathcal{F}_n(x) = \mathcal{F}_s(x)$).*

(b) *If $\mathcal{F}_s$ is not linear and $\mathcal{F}_s(\mathbf{x_s}) \notin \mathrm{supp}(\tau_j)$ for any $1 \leq j \leq m$, then there exists a permutation $\mathcal{P}_n$ with $n = m + \sum_{j=1}^{m} l_j + 1$ such that $\mathcal{P}_s(x) = \mathcal{P}_n(x)$ for all $x \in \mathbb{F}_q$ (and additionally $\mathcal{F}_n(x) = \mathcal{F}_s(x)$).*

(c) *If $\mathcal{F}_s$ is linear, then there exists a permutation $\mathcal{P}_n$ with $n = m + \sum_{j=1}^{m} l_j$ such that $\mathcal{P}_s(x) = \mathcal{P}_n(x)$ for all $x \in \mathbb{F}_q$ (and additionally $\mathcal{F}_n(x) = \mathcal{F}_s(x)$).*

In all three cases, $Crk(\mathcal{P}_s) = n$ if $n < (q-1)/2$.

Theorem 5 provides a method to determine the number $\mathcal{B}(n)$ of distinct permutations with given Carlitz rank n for $n < (q-1)/2$ by counting the number of distinct permutations of the form (32) yielding Carlitz rank n. For this purpose we recall the *associated Stirling numbers of the first kind*. Let $t, k \geq 1$ and $m \geq 0$ be integers. We denote the number of permutations of a set of k elements with cycle decomposition $\tau_1 \cdots \tau_m$, such that each cycle τ_j, $1 \leq j \leq m$, has length $l(\tau_j)$ at least t by

$\mathcal{S}(t, k, m)$. The numbers $\mathcal{S}(t, k, m)$ for variable t, k and m are called the associated Stirling numbers of the first kind (see for instance Comtet 1974 and id:A008306 in Sloane's On-line Encyclopdia at http://www.research.att.com/~njas/sequences). As the cycles in (32) have length at least 2 we need $\mathcal{S}(2, k, m)$, the number of permutations of a set of k elements having m cycles and no fixed points, for counting the permutations of $\mathbb{F}_q$ with fixed Carlitz rank. The associated Stirling numbers $\mathcal{S}(2, k, m)$ satisfy the recurrence relation

$$\mathcal{S}(2, k+1, m+1) = k\mathcal{S}(2, k, m+1) + k\mathcal{S}(2, k-1, m) \tag{34}$$

(with obvious starting values). The following theorem, where the formulas for $\mathcal{B}(n)$, $n < (q-1)/2$, that involve $\mathcal{S}(2, k, m)$ are presented, is the main result of Aksoy et al. (2009). The explicit calculation of $\mathcal{B}(n)$ requires the calculation of the numbers $\mathcal{S}(2, k, m)$ by the recursive formula (34).

Theorem 6. *The number $\mathcal{B}(n)$ of permutations of $\mathbb{F}_q$ with Carlitz rank n is given by*

$$\begin{aligned}\mathcal{B}(n) = (q^2-q) \sum_{m=1}^{\lfloor \frac{n+1}{3} \rfloor} \binom{q}{n+1-m} \mathcal{S}(2, n+1-m, m)(n+1-m) \\ + (q^2-q) \sum_{m=1}^{\lfloor \frac{n-1}{3} \rfloor} \binom{q}{n-1-m} \mathcal{S}(2, n-1-m, m)(q-(n-1-m)) \\ + (q^2-q) \sum_{m=1}^{\lfloor \frac{n}{3} \rfloor} \binom{q}{n-m} \mathcal{S}(2, n-m, m)\end{aligned}$$

for all $2 \le n < (q-1)/2$.

For small n one then easily obtains explicit expressions for $\mathcal{B}(n)$:

$$\mathcal{B}(0) = q(q-1), \qquad \mathcal{B}(1) = q^2(q-1), \qquad \mathcal{B}(2) = q^2(q-1)^2$$

$$\mathcal{B}(3) = \frac{1}{2}q^2(q-1)^2(2q-3), \qquad \mathcal{B}(4) = \frac{1}{6}q^2(q-1)^2(q-2)(6q-13),$$

$$\mathcal{B}(5) = \frac{1}{12}q^2(q-1)^2(q-2)(q-3)(12q-35),$$

$$\mathcal{B}(6) = \frac{1}{120}q^2(q-1)^2(q-2)(q-3)(120q^2-926q+1799),$$

$$\mathcal{B}(7) = \frac{1}{120}q^2(q-1)^2(q-2)(q-3)(q-4)(120q^2-1146q+2765),$$

$$\begin{aligned}\mathcal{B}(8) = \frac{1}{1260}q^2(q-1)^2(q-2)(q-3)(q-4)(q-5) \\ \times (1260q^2-14373q+41473),\end{aligned}$$

$$\mathcal{B}(9) = \frac{1}{5040}q^2(q-1)^2(q-2)(q-3)(q-4)(q-5)$$
$$\times (5040q^3 - 97182q^2 + 626590q - 1349523),$$
$$\mathcal{B}(10) = \frac{1}{10080}q^2(q-1)^2(q-2)(q-3)(q-4)(q-5)(q-6)$$
$$\times (10080q^3 - 223484q^2 + 1657175q - 4106319).$$

As easily seen, the vast majority of permutations of $\mathbb{F}_q$ have Carlitz rank larger than $(q-1)/2$, hence with Theorem 6 only a small proportion of the permutations of the finite field $\mathbb{F}_q$ can be counted.

An unsolved problem is the determination of the maximum value $Crk_M(q)$, of the Carlitz rank that a permutation of $\mathbb{F}_q$ can have. In other words we are interested in the smallest number $Crk_M(q)$ such that any permutation of $\mathbb{F}_q$ is a composition of linear polynomials and at most $Crk_M(q)$ "inversions" x^{q-2}. With a rough estimate in Aksoy et al. (2009) the upper bound $Crk_M(q) \le (3q-7)/2$ has been obtained.

The Carlitz rank and the "polynomial degree" can be regarded as two complexity measures for permutations of $\mathbb{F}_q$. It is well known that the vast majority of the permutations of $\mathbb{F}_q$ has large polynomial degree $q-2$, and as pointed out above the majority of the permutations of $\mathbb{F}_q$ does not have a small Carlitz rank. The following theorem shows that a permutation cannot be "simple" with respect to both complexity measures, polynomial degree and Carlitz rank, at the same time (see Theorem 4 in Aksoy et al. 2009).

Theorem 7. *Let $g(x)$ be a permutation polynomial in $\mathbb{F}_q[x]$ with $\deg(g) = d > 1$ and suppose that $Crk(g) = n$. Then*

$$n \ge q - 1 - d.$$

As observed in Aksoy et al. (2009) the lower bound $Crk_M(q) \ge q - 1 - \delta$ immediately follows from Theorem 7 and the fact that x^d is a permutation of $\mathbb{F}_q$ if $\gcd(d, q-1) = 1$, when $\delta > 1$ is the smallest integer satisfying $\gcd(\delta, q-1) = 1$.

We hope that we have been able to familiarize the reader with the basic properties and problems concerning the permutation polynomials $\mathcal{P}_n(x)$ (16). This viewpoint enables us to generate new types of PR sequences with advantages. We have already mentioned sequences obtained by (15). Alternatively one can choose the terms a_i in (16) suitably to obtain a sequence (u_n) defined by $u_n = \mathcal{P}_n(u_0)$. Depending on the period of the sequence (a_n), the period length of (u_n) can be increased. The predictability of (u_n) in this case is related to that of (a_n). Hence this generalization can yield sequences which are highly unpredictable—an advantage over inversive generators in the light of a recent result on the predictability of the inversive generator, see Blackburn et al. (2003). Naturally randomness properties of these new PR sequences are of interest. Work on this and related problems is under progress.

References

Aksoy, E., Çeşmelioğlu, A., Meidl, W., Topuzoğlu, A.: On the Carlitz rank of permutation polynomials. Finite Fields Appl. **15**, 418–440 (2009)

Avenancio-Leon, C.: Analysis of some properties of interleavers for Turbo codes. In: Proc. of NCUR, Lexington, USA (2005)

Beck, I.: Cycle decomposition of transpositions. J. Comb. Theory, Ser. A **23**, 198–207 (1977)

Blackburn, S., Gomez-Perez, D., Gutierrez, J., Shparlinski, I.: In: Predicting the inversive generator. Lecture Notes in Computer Science, vol. 2898, pp. 264–275. Springer, Berlin (2003)

Carlitz, L.: Permutations in a finite field. Proc. Am. Math. Soc. **4**, 538 (1953)

Çeşmelioğlu, A.: Personal communication (2009)

Çeşmelioğlu, A., Meidl, W., Topuzoğlu, A.: On the cycle structure of permutation polynomials. Finite Fields Appl. **14**, 593–614 (2008a)

Çeşmelioğlu, A., Meidl, W., Topuzoğlu, A.: Enumeration of a class of sequences generated by inversions. In: Li, Y.Q., et al. (eds.) Proceedings of the Int. Workshop on Coding and Cryptology, Fujian, China, June 2007, pp. 44–57 (2008b)

Çeşmelioğlu, A., Meidl, W., Topuzoğlu, A.: On a class of APN permutation polynomials. Preprint (2009)

Chou, W.-S.: On inversive maximal period polynomials over finite fields. Appl. Algebra Eng. Commun. Comput. **6**, 245–250 (1995a)

Chou, W.-S.: The period lengths of inversive pseudorandom vector generations. Finite Fields Appl. **1**, 126–132 (1995b)

Comtet, L.: Advanced Combinatorics, the Art of Finite and Infinite Expansions. Reidel, Dordrecht (1974)

Corrada-Bravo, C.J., Rubio, I.M.: Deterministic interleavers for Turbo codes with random-like performance and simple implementation. In: Proc. of the 3rd International Symposium on Turbo Codes and Related Topics, Brest, France, pp. 555–558 (2003)

Dorfer, G., Winterhof, A.: Lattice structure and linear complexity profile of nonlinear pseudorandom number generators. Appl. Algebra Eng. Commun. Comput. **13**, 499–508 (2003)

Drmota, M., Tichy, R.F.: Sequences, Descrepancies and Applications. Lecture Notes in Mathematics, vol. 1651. Springer, Berlin (1997)

Eichenauer, J., Lehn, J.: A non-linear congruential pseudorandom number generator. Stat. Hefte **27**, 315–326 (1986)

Eichenauer, J., Niederreiter, H.: On Marsaglia's lattice test for pseudorandom numbers. Manuscr. Math. **62**, 245–248 (1988)

Eichenauer, J., Grothe, H., Lehn, J., Topuzoğlu, A.: A multiple recursive nonlinear congruential pseudo random number generator. Manuscr. Math. **59**, 331–346 (1987)

Eichenauer, J., Grothe, H., Lehn, J.: Marsaglia's lattice test and non-linear congruential pseudo random number generators. Metrika **35**, 241–250 (1988a)

Eichenauer, J., Lehn, J., Topuzoğlu, A.: A nonlinear congruential pseudorandom number generator with power of two modulus. Math. Comput. **51**, 757–759 (1988b)

Eichenauer, J., Grothe, H., Lehn, J.: On the period length of pseudorandom vector sequences generated by matrix generators. Math. Comput. **52**, 145–148 (1989)

Eichenauer-Herrmann, J.: Inversive congruential pseudorandom numbers avoid the planes. Math. Comput. **56**, 297–301 (1991)

Eichenauer-Herrmann, J.: Inversive congruential pseudorandom numbers: a tutorial. Int. Stat. Rev. **60**, 167–176 (1992a)

Eichenauer-Herrmann, J.: On the autocorrelation structure of inversive congruential pseudorandom number sequences. Stat. Pap. **33**, 261–268 (1992b)

Eichenauer-Herrmann, J.: Construction of inversive congruential pseudorandom number generators with maximal period length. J. Comput. Appl. Math. **40**, 345–349 (1992c)

Eichenauer-Herrmann, J.: Statistical independence of a new class of inversive congruential pseudorandom numbers. Math. Comput. **60**, 375–384 (1993)

Eichenauer-Herrmann, J.: Pseudorandom number generation by nonlinear methods. Int. Stat. Rev. **63**, 245–255 (1995)

Eichenauer-Herrmann, J., Emmerich, F.: Compound inversive congruential pseudorandom numbers: an average-case analysis. Math. Comput. **65**, 215–225 (1996)

Eichenauer-Herrmann, J., Grothe, H.: A new inversive congruential pseudorandom number generator with power of two modulus. ACM Trans. Model. Comput. Simul. **2**, 1–11 (1992)

Eichenauer-Herrmann, J., Ickstadt, K.: Explicit inversive congruential pseudorandom numbers with power of two modulus. Math. Comput. **62**, 787–797 (1994)

Eichenauer-Herrmann, J., Niederreiter, H.: Digital inversive pseudorandom numbers. ACM Trans. Model. Comput. Simul. **4**, 339–349 (1994)

Eichenauer-Herrmann, J., Topuzoğlu, A.: On the period length of congruential pseudorandom number sequences generated by inversions. J. Comput. Appl. Math. **31**, 87–96 (1990)

Eichenauer-Herrmann, J., Herrmann, E., Wegenkittl, S.: A survey of quadratic and inversive congruential pseudorandom numbers. In: Niederreiter, H., et al. (eds.) Monte Carlo and Quasi-Monte Carlo Methods 1996. Lecture Notes in Statistics, vol. 127, pp. 66–97. Springer, New York (1998)

Emmerich, F.: Pseudorandom number and vector generation by compound inversive methods. PhD thesis. Technische Hochschule Darmstadt (1996)

Flahive, M., Niederreiter, H.: On inversive congruential generators for pseudorandom numbers. In: Mullen, G.L., Shiue, P.J.-S. (eds.) Finite Fields, Coding Theory, and Advances in Communications and Computing, Las Vegas, NV, 1991. Lecture Notes in Pure and Appl. Math., vol. 141, pp. 75–80. Marcel Dekker, New York (1993)

Golomb, S.W., Gong, G.: The status of Costas arrays. IEEE Trans. Inf. Theory **53**, 4260–4265 (2007)

Gutierrez, J., Shparlinski, I., Winterhof, A.: On the linear and nonlinear complexity profile of nonlinear pseudorandom number-generators. IEEE Trans. Inf. Theory **49**, 60–64 (2003)

Heegard, C., Wicker, S.B.: Turbo Coding. Kluwer Academic, Dordrecht (1999)

Hellekalek, P.: On the assessment of random and quasi-random point sets. In: Hellekalek, P., Larcher, G. (eds.) Random and Quasi-Random Point Sets. Lecture Notes in Statistics, vol. 138, pp. 49–108. Springer, Berlin (1998)

Larcher, G., Wolf, R., Eichenauer-Herrmann, J.: On the average discrepancy of successive tuples of pseudo-random numbers over parts of the period. Monatshefte Math. **127**, 141–154 (1999)

L'Ecuyer, P., Hellekalek, P.: Random number generators: selection criteria and testing. In: Hellekalek, P., Larcher, G. (eds.) Random and Quasi-Random Point Sets. Lecture Notes in Statistics, vol. 138, pp. 223–265. Springer, Berlin (1998)

Marsaglia, G.: Random numbers fall mainly in the planes. Proc. Natl. Acad. Sci. USA **61**, 25–28 (1968)

Massey, J.: Shift-register synthesis and BCH decoding. IEEE Trans. Inf. Theory **15**, 122–127 (1969)

Mauduit, C., Sarközi, A.: On finite pseudorandom binary sequences, I: measure of pseudorandomness, the Legendre symbol. Acta Arith. **82**, 365–377 (1997)

Meidl, W., Winterhof, A.: On the linear complexity profile of explicit nonlinear pseudorandom numbers. Inf. Process. Lett. **85**, 13–18 (2003)

Niederreiter, H.: Pseudo-random numbers and optimal coefficients. Adv. Math. **26**, 99–181 (1977)

Niederreiter, H.: Remarks on nonlinear congruential pseudorandom numbers. Metrika **35**, 321–328 (1988)

Niederreiter, H.: Random Number Generation and Quasi-Monte Carlo Methods. SIAM, Philadelphia (1992)

Niederreiter, H.: Pseudorandom vector generation by the inversive method. ACM Trans. Model. Comput. Simul. **4**, 191–212 (1994)

Niederreiter, H.: Linear complexity and related complexity measures for sequences. In: Johansson, T., Maitra, S. (eds.) Progress in Cryptology (INDOCRYPT 2003). Lecture Notes in Computer Science, vol. 2904, pp. 1–17. Springer, Berlin (2003)

Niederreiter, H., Rivat, J.: On the correlation of pseudorandom numbers generated by inversive methods. Monatshefte Math. **153**, 251–264 (2008)

Niederreiter, H., Shparlinski, I.: On the distribution and lattice structure of nonlinear congruential pseudorandom numbers. Finite Fields Appl. **5**, 246–253 (1999)

Niederreiter, H., Shparlinski, I.: On the distribution of inversive congruential pseudorandom numbers in parts of the period. Math. Comput. **70**, 1569–1574 (2001)

Niederreiter, H., Shparlinski, I.: Recent advances in the theory of nonlinear pseudorandom number generators. In: Fang, K.T., Hickernell, F.J., Niederreiter, H. (eds.) Monte Carlo and quasi-Monte Carlo methods, 2000, pp. 86–102. Springer, Berlin (2002a)

Niederreiter, H., Shparlinski, I.: On the average distribution of inversive pseudorandom numbers. Finite Fields Appl. **8**, 86–102 (2002b)

Niederreiter, H., Winterhof, A.: Incomplete exponential sums over finite fields and their applications to new inversive pseudorandom number generators. Acta Arith. **93**, 387–300 (2000)

Niederreiter, H., Winterhof, A.: On the distribution of some new explicit nonlinear congruential pseudorandom numbers. In: Helleseth, T., et al. (eds.) Proceedings of SETA 2004. Lecture Notes in Computer Science, vol. 3486, pp. 266–274. Springer, Berlin (2005)

Rubio, I.M., Corrada-Bravo, C.J.: Cyclic decomposition of permutations of finite fields obtained using monomials. In: Poli, A., Stichtenoth, H. (eds.) Proceedings of $\mathbb{F}_q 7$. Lecture Notes in Computer Science, vol. 2948, pp. 254–261. Springer, Berlin (2004)

Rubio, I.M., Mullen, G.L., Corrada, C.J., Castro, F.N.: Dickson permutation polynomials that decompose in cycles of the same length. In: Mullen, G.L., Panario, D., Shparlinski, I. (eds.) Proceedings of $\mathbb{F}_q 8$. Contemp. Math., vol. 461, pp. 229–239 (2008)

Shparlinski, I.: Cryptographic Applications of Analytic Number Theory. Progress in Computer Science and Applied Logic, vol. 22. Birkhäuser, Basel (2003)

Sloane, N.J.: On-line Encyclopedia of integer sequences. Published electronically at http://www.research.att.com/~njas/sequences

Sole, P., Zinoviev, D.: Inversive pseudorandom numbers over Galois rings. Eur. J. Comb. **30**, 458–467 (2009)

Topuzoğlu, A., Winterhof, A.: On the linear complexity profile of nonlinear congruential pseudorandom number generators of higher orders. Appl. Algebra Eng. Commun. Comput. **16**, 219–228 (2005)

Topuzoğlu, A., Winterhof, A.: Pseudorandom sequences. In: Garcia, A., Stichtenoth, H. (eds.) Topics in Geometry, Coding Theory and Cryptography. Algebra and Applications, vol. 6, pp. 135–166. Springer, Berlin (2007)

Winterhof, A.: On the distribution of some new explicit inversive pseudorandom numbers and vectors. In: Niederreiter, H., Talay, D. (eds.) Monte Carlo and Quasi-Monte Carlo Methods 2004, pp. 487–499. Springer, Berlin (2006)

Strong and Weak Approximation Methods for Stochastic Differential Equations—Some Recent Developments

Andreas Rößler

Abstract Some efficient stochastic Runge–Kutta (SRK) methods for the strong as well as for the weak approximation of solutions of stochastic differential equations (SDEs) with improved computational complexity are considered. Their convergence is analyzed by a concise colored rooted tree approach for both, Itô as well as Stratonovich SDEs. Further, order conditions for the coefficients of order 1.0 and 1.5 strong SRK methods as well as for order 2.0 weak SRK methods are given. As the main novelty, the computational complexity of the presented order 1.0 strong SRK method and the order 2.0 weak SRK method depends only linearly on the dimension of the driving Wiener process. This is a significant improvement compared to well known methods where the computational complexity depends quadratically on the dimension of the Wiener process.

1 Approximation of Solutions of Stochastic Differential Equations

Let $(\Omega, \mathcal{F}, \mathrm{P})$ be a complete probability space with a filtration $(\mathcal{F}_t)_{t\geq 0}$ fulfilling the usual conditions and let $\mathcal{I} = [t_0, T]$ for some $0 \leq t_0 < T < \infty$. We denote by $X = (X_t)_{t\in\mathcal{I}}$ the solution of the d-dimensional SDE system

$$X_t = X_{t_0} + \int_{t_0}^{t} a(s, X_s)\,\mathrm{d}s + \sum_{j=1}^{m} \int_{t_0}^{t} b^j(s, X_s) * \mathrm{d}W_s^j \tag{1}$$

with an m-dimensional driving Wiener process $(W_t)_{t\geq 0} = ((W_t^1, \dots, W_t^m)^T)_{t\geq 0}$ w.r.t. $(\mathcal{F}_t)_{t\geq 0}$ for $d, m \geq 1$ and $t \in \mathcal{I}$. We write $*\mathrm{d}W_s^j = \mathrm{d}W_s^j$ in the case of an

Andreas Rößler, Department Mathematik, Universität Hamburg, Bundesstraße 55, 20146 Hamburg, Germany
e-mail: andreas.roessler@math.uni-hamburg.de

L. Devroye et al. (eds.), *Recent Developments in Applied Probability and Statistics*,
DOI 10.1007/978-3-7908-2598-5_6,

Itô stochastic integral and $*dW_s^j = \circ dW_s^j$ for a Stratonovich stochastic integral. Suppose that $a : \mathcal{I} \times \mathbb{R}^d \to \mathbb{R}^d$ and $b : \mathcal{I} \times \mathbb{R}^d \to \mathbb{R}^{d \times m}$ are continuous functions which fulfill a global Lipschitz condition and denote by b^j the jth column of the $d \times m$-matrix function $b = (b^{i,j})$ for $j = 1, \ldots, m$. Let $X_{t_0} \in L^2(\Omega)$ be the $\mathcal{F}_{t_0}$-measurable initial value. In the following, we suppose that the conditions of the Existence and Uniqueness Theorem (cf., e.g., Kloeden and Platen 1999) are fulfilled for SDE (1) and we denote by $\| \cdot \|$ the Euclidean norm. Let $C_P^l(\mathbb{R}^d, \mathbb{R})$ denote the space of all $g \in C^l(\mathbb{R}^d, \mathbb{R})$ with polynomial growth, see e.g. Kloeden and Platen (1999) or Rößler (2006a, 2009) for details. Then g belongs to $C_P^{k,l}(\mathcal{I} \times \mathbb{R}^d, \mathbb{R})$ if $g \in C^{k,l}(\mathcal{I} \times \mathbb{R}^d, \mathbb{R})$ and $g(t, \cdot) \in C_P^l(\mathbb{R}^d, \mathbb{R})$ is fulfilled uniformly in $t \in \mathcal{I}$.

For the numerical approximation let a discretization $\mathcal{I}_h = \{t_0, t_1, \ldots, t_N\}$ with $t_0 < t_1 < \cdots < t_N = T$ of the time interval $\mathcal{I}$ with step sizes $h_n = t_{n+1} - t_n$ for $n = 0, 1, \ldots, N-1$ be given. Further, let $h = \max_{0 \le n < N} h_n$ denote the maximum step size. If one is interested in a good pathwise approximation of the solution of SDE (1), then strong approximation methods converging in the mean-square sense are applied. Note that mean-square convergence implies strong convergence.

Definition 1. A sequence of approximation processes $Y^h = (Y(t))_{t \in \mathcal{I}_h}$ converges in the mean-square sense with order p to the solution X of SDE (1) at time T if there exists a constant $C > 0$ and some $\delta_0 > 0$ such that for each $h \in]0, \delta_0]$

$$(\mathrm{E}(\|X_T - Y^h(T)\|^2))^{1/2} \le C\, h^p \,. \tag{2}$$

However, if one is interested in the approximation of some distributional characteristics of the solution of SDE (1), then weak approximation methods are applied.

Definition 2. A sequence of approximation processes $Y^h = (Y(t))_{t \in \mathcal{I}_h}$ converges in the weak sense with order p to the solution X of SDE (1) at time T if for each $f \in C_P^{2(p+1)}(\mathbb{R}^d, \mathbb{R})$ exists a constant C_f and some $\delta_0 > 0$ such that for each $h \in]0, \delta_0]$

$$|\,\mathrm{E}(f(X_T)) - \mathrm{E}(f(Y^h(T)))| \le C_f\, h^p \,. \tag{3}$$

2 A General Class of Stochastic Runge–Kutta Methods

For the approximation of the solution X of SDE (1) we consider the universal class of stochastic Runge–Kutta (SRK) methods introduced in Rößler (2006b): Let $\mathcal{M}$ be an arbitrary finite set of multi-indices with $\kappa = |\mathcal{M}|$ elements, let $\theta_\iota^{(k)}(h) \in L^2(\Omega)$ for $\iota \in \mathcal{M}$ and $0 \le k \le m$ be some suitable random variables. Further, define $b^0(t, x) := a(t, x)$. Then, an s-stages SRK method is given by $Y_0 = X_{t_0}$ and

$$Y_{n+1} = Y_n + \sum_{i=1}^{s} \sum_{k=0}^{m} \sum_{\nu \in \mathcal{M}} z_i^{(k),(\nu)}\; b^k\left(t_n + c_i^{(\nu)} h_n,\, H_i^{(\nu)}\right) \tag{4}$$

for $n = 0, 1, \ldots, N-1$ with $Y_n = Y(t_n)$, $t_n \in \mathcal{I}_h$, and with stages

$$H_i^{(\nu)} = Y_n + \sum_{j=1}^{s} \sum_{l=0}^{m} \sum_{\mu \in \mathcal{M}} Z_{ij}^{(\nu),(l),(\mu)} \, b^l \left(t_n + c_j^{(\mu)} h_n, H_j^{(\mu)} \right)$$

for $i = 1, \ldots, s$ and $\nu \in \mathcal{M}$. Here, let $0 \in \mathcal{M}$ and let for $i, j = 1, \ldots, s$

$$z_i^{(k),(\nu)} = \sum_{\iota \in \mathcal{M}} \gamma_i^{(\iota)(k),(\nu)} \, \theta_\iota^{(k)}(h_n), \qquad Z_{ij}^{(\nu),(l),(\mu)} = \sum_{\iota \in \mathcal{M}} C_{ij}^{(\iota)(\nu),(l),(\mu)} \, \theta_\iota^{(l)}(h_n)$$

with $\theta_0^{(0)}(h_n) = h_n$ and the coefficients $\gamma_i^{(\iota)(k),(\nu)}, C_{ij}^{(\iota)(\nu),(l),(\mu)} \in \mathbb{R}$ of the SRK method. In the following, we use the notation $z^{(k),(\nu)} = (z_i^{(k),(\nu)})_{1 \le i \le s} \in \mathbb{R}^s$ and $Z^{(\nu),(l),(\mu)} = (Z_{ij}^{(\nu),(l),(\mu)})_{1 \le i,j \le s} \in \mathbb{R}^{s \times s}$. The vector of weights can be defined by

$$c^{(\nu)} = \sum_{\mu \in \mathcal{M}} C_{ij}^{(0)(\nu),(0),(\mu)} e \tag{5}$$

with $e = (1, \ldots, 1)^T \in \mathbb{R}^s$. If $C_{ij}^{(\iota)(\nu),(l),(\mu)} = 0$ for $j \ge i$ then (4) is called an explicit SRK method, otherwise it is called implicit. We assume that the random variables $\theta_\iota^{(k)}(h)$ satisfy the moment condition

$$\mathrm{E}\left(\prod_{k=0}^{m} ((\theta_{\iota_1}^{(k)}(h))^{p_1^k} \cdot \ldots \cdot (\theta_{\iota_\kappa}^{(k)}(h))^{p_\kappa^k}) \right) = O\left(h^{p_1^0 + \ldots + p_\kappa^0 + \sum_{k=1}^m (p_1^k + \ldots + p_\kappa^k)/2} \right) \tag{6}$$

for all $p_i^k \in \mathbb{N}_0$, $k = 0, 1, \ldots, m$, and $\iota_i \in \mathcal{M}$, $1 \le i \le \kappa$. Further, we assume that in the case of an implicit method each random variable can be expressed as $\theta_\iota^{(0)}(h) = h \cdot \vartheta_\iota^{(0)}$ and $\theta_\iota^{(k)}(h) = \sqrt{h} \cdot \vartheta_\iota^{(k)}$, $1 \le k \le m$, for $\iota \in \mathcal{M}$ with suitable bounded random variables $\vartheta_\iota^{(0)}, \vartheta_\iota^{(k)} \in L^2(\Omega)$ such that each stage can be solved w.r.t. $H_i^{(\nu)}$ for sufficiently small h. These conditions are not necessary in the case of explicit SRK methods (see also Rößler 2006b or Milstein and Tretyakov 2004).

3 Colored Rooted Tree Analysis

In the following, we present a concise rooted tree analysis for the convergence of the general class of SRK methods (4). For simplicity, we restrict our investigations without loss of generality to the autonomous SDE (1) in this section. We denote by TS the set of all stochastic trees, see also Rößler (2004b, 2010a), which have a root $\tau_\gamma = \otimes$ and which can furthermore be composed of deterministic nodes $\tau_0 = \bullet$ and stochastic nodes $\tau_j = \bigcirc_j$ with some $j \in \{1, \ldots, m\}$. The index j is associated with the jth component of the m-dimensional driving Wiener process of the considered SDE. Some examples of trees in TS are presented in Fig. 1. Let $d(\mathbf{t})$ denote the number of deterministic nodes τ_0 and let $s(\mathbf{t})$ denote the number of stochastic nodes τ_j with $j \in \{1, \ldots, m\}$ of the tree $\mathbf{t} \in TS$. The order $\rho(\mathbf{t})$ of the tree $\mathbf{t} \in TS$ is

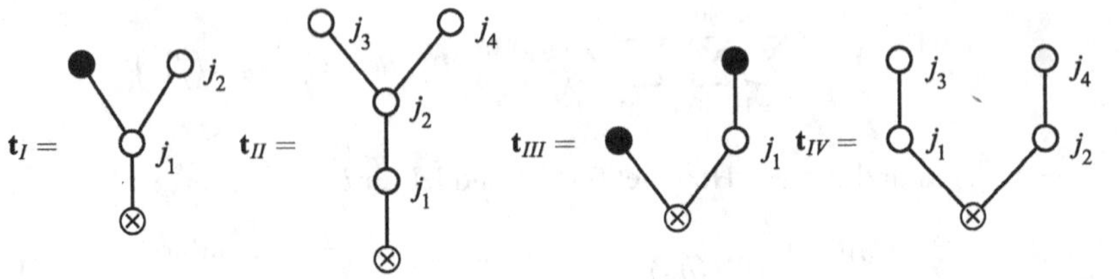

Fig. 1 Four elements of TS with some $j_1, j_2, j_3, j_4 \in \{1, \dots, m\}$

defined as $\rho(\tau_\gamma) = 0$ and $\rho(\mathbf{t}) = d(\mathbf{t}) + \frac{1}{2}s(\mathbf{t})$. As an example, for the trees in Fig. 1 we have $\rho(\mathbf{t}_I) = \rho(\mathbf{t}_{II}) = \rho(\mathbf{t}_{IV}) = 2$ and $\rho(\mathbf{t}_{III}) = 2.5$.

Every tree can be written by a combination of brackets: If $\mathbf{t}_1, \dots, \mathbf{t}_k$ are colored subtrees then we denote by $[\mathbf{t}_1, \dots, \mathbf{t}_k]_j$ the tree in which $\mathbf{t}_1, \dots, \mathbf{t}_k$ are each joined by a single branch to the node τ_j for some $j \in \{\gamma, 0, 1, \dots, m\}$. Therefore proceeding recursively, for the trees in Fig. 1 we obtain $\mathbf{t}_I = [[\tau_0, \tau_{j_2}]_{j_1}]_\gamma$, $\mathbf{t}_{II} = [[[\tau_{j_3}, \tau_{j_4}]_{j_2}]_{j_1}]_\gamma$, $\mathbf{t}_{III} = [\tau_0, [\tau_0]_{j_1}]_\gamma$ and $\mathbf{t}_{IV} = [[\tau_{j_3}]_{j_1}, [\tau_{j_4}]_{j_2}]_\gamma$.

Next, we assign to each tree $\mathbf{t} \in TS$ an elementary differential which is defined recursively by $F(\tau_\gamma)(x) = f(x)$, $F(\tau_j)(x) = b^j(x)$ and

$$F(\mathbf{t})(x) = \begin{cases} f^{(k)}(x) \cdot (F(\mathbf{t}_1)(x), \dots, F(\mathbf{t}_k)(x)) & \text{for } \mathbf{t} = [\mathbf{t}_1, \dots, \mathbf{t}_k]_\gamma, \\ b^{j\,(k)}(x) \cdot (F(\mathbf{t}_1)(x), \dots, F(\mathbf{t}_k)(x)) & \text{for } \mathbf{t} = [\mathbf{t}_1, \dots, \mathbf{t}_k]_j \end{cases} \tag{7}$$

for $j \in \{0, 1, \dots, m\}$. Here $f^{(k)}$ and $b^{j\,(k)}$ define a symmetric k-linear differential operator, and one can choose the sequence of subtrees $\mathbf{t}_1, \dots, \mathbf{t}_k$ in an arbitrary order.

Finally, we assign to every tree a multiple stochastic integral. Let $(Z_t)_{t \geq t_0}$ be a progressively measurable stochastic process. Then, we define for $\mathbf{t} \in TS$ the corresponding multiple stochastic integral recursively by

$$I_{\mathbf{t};t_0,t}[Z.] = \begin{cases} (\prod_{i=1}^k I_{\mathbf{t}_i;t_0,t})[Z.] & \text{if } \mathbf{t} = [\mathbf{t}_1, \dots, \mathbf{t}_k]_\gamma, \\ (\int_{t_0}^t \prod_{i=1}^k I_{\mathbf{t}_i;t_0,s} * \mathrm{d}W_s^j)[Z.] & \text{if } \mathbf{t} = [\mathbf{t}_1, \dots, \mathbf{t}_k]_j,\ j \in \{0, 1, \dots, m\} \end{cases} \tag{8}$$

with $*\mathrm{d}W_s^0 = \mathrm{d}s$, $I_{\tau_j;t_0,t}[Z.] = \int_{t_0}^t Z_s * \mathrm{d}W_s^j$, $I_{\tau_\gamma;t_0,t}[Z.] = Z_t$, $I_{\mathbf{t};t_0,t} = I_{\mathbf{t};t_0,t}[1]$ provided that the stochastic integral exists and by using the notation

$$\begin{aligned} \left(\int_{t_0}^t \int_{t_0}^{s_n} \cdots \int_{t_0}^{s_2} *\mathrm{d}W_{s_1}^{j_1} * \mathrm{d}W_{s_2}^{j_2} \cdots * \mathrm{d}W_{s_n}^{j_n} \right)[Z.] &= I_{(j_1, j_2, \dots, j_n)}[Z.]_{t_0,t} \\ = \int_{t_0}^t \int_{t_0}^{s_n} \cdots \int_{t_0}^{s_2} Z_{s_1} * \mathrm{d}W_{s_1}^{j_1} * \mathrm{d}W_{s_2}^{j_2} \cdots * \mathrm{d}W_{s_n}^{j_n} & \end{aligned} \tag{9}$$

in (8). The product of two stochastic integrals can be written as a sum (cf., e.g., Kloeden and Platen 1999)

- $$\int_{t_0}^{t} X_s * \mathrm{d}W_s^i \int_{t_0}^{t} Y_s * \mathrm{d}W_s^j$$
$$= \int_{t_0}^{t} X_s\, Y_s\, \mathbf{1}_{\{i=j\neq 0 \wedge *\neq\circ\}}\, \mathrm{d}s + \int_{t_0}^{t} X_s \left(\int_{t_0}^{s} Y_u * \mathrm{d}W_u^j \right) * \mathrm{d}W_s^i$$
$$+ \int_{t_0}^{t} \left(\int_{t_0}^{s} X_u * \mathrm{d}W_u^i \right) Y_s * \mathrm{d}W_s^j \tag{10}$$

for $0 \leq i, j \leq m$, where the first summand on the right hand side appears only in the case of Itô calculus. E.g., we calculate for $\mathbf{t}_I$ and $\mathbf{t}_{II}$

$$I_{\mathbf{t}_I;t_0,t}[1] = \int_{t_0}^{t} I_{\tau_0;t_0,s}\, I_{\tau_{j_2};t_0,s} * \mathrm{d}W_s^{j_1}[1] = I_{(0,j_2,j_1)}[1]_{t_0,t} + I_{(j_2,0,j_1)}[1]_{t_0,t}\,,$$

$$I_{\mathbf{t}_{II};t_0,t}[1] = \int_{t_0}^{t} \int_{t_0}^{s} I_{\tau_{j_3};t_0,u}\, I_{\tau_{j_4};t_0,u} * \mathrm{d}W_u^{j_2} * \mathrm{d}W_s^{j_1}$$
$$= I_{(j_3,j_4,j_2,j_1)}[1]_{t_0,t} + I_{(j_4,j_3,j_2,j_1)}[1]_{t_0,t} + I_{(0,j_2,j_1)}[\mathbf{1}_{\{j_3=j_4\neq 0\wedge *\neq\circ\}}]_{t_0,t}$$

where the last summand for $I_{\mathbf{t}_{II};t_0,t}[1]$ only appears in the case of Itô calculus.

Let $\mathbf{t} \in TS$ with $\mathbf{t} = [\mathbf{t}_1, \ldots, \mathbf{t}_1, \mathbf{t}_2, \ldots, \mathbf{t}_2, \ldots, \mathbf{t}_k, \ldots, \mathbf{t}_k]_j = [\mathbf{t}_1^{n_1}, \mathbf{t}_2^{n_2}, \ldots, \mathbf{t}_k^{n_k}]_j$, $j \in \{\gamma, 0, 1, \ldots, m\}$, where $\mathbf{t}_1, \ldots, \mathbf{t}_k$ are distinct subtrees with multiplicities $n_1, \ldots, n_k$, respectively. Then the symmetry factor σ is recursively defined by $\sigma(\tau_j) = 1$ and

$$\sigma(\mathbf{t}) = \prod_{i=1}^{k} n_i!\, \sigma(\mathbf{t}_i)^{n_i}\,. \tag{11}$$

For the trees in Fig. 1, we obtain $\sigma(\mathbf{t}_I) = \sigma(\mathbf{t}_{III}) = 1$. For the tree $\mathbf{t}_{II}$ we have to consider two cases: If $j_3 \neq j_4$ we have $\sigma(\mathbf{t}_{II}) = 1$. However, in the case of $j_3 = j_4$ we have some symmetry and thus we calculate $\sigma(\mathbf{t}_{II}) = 2$. Further, for tree $\mathbf{t}_{IV}$ we get $\sigma(\mathbf{t}_{IV}) = 2$ if $j_1 = j_2$ and $j_3 = j_4$ and $\sigma(\mathbf{t}_{IV}) = 1$ otherwise. E.g., all trees up to order 1.5 and the corresponding multiple integrals are presented in Table 1.

Next, we define the coefficient function Φ_S which assigns to every tree an elementary weight. For every $\mathbf{t} \in TS$ the function Φ_S is defined by $\Phi_S(\tau_\gamma) = 1$ and

$$\Phi_S(\mathbf{t}) = \begin{cases} \prod_{i=1}^{k} \Phi_S(\mathbf{t}_i) & \text{if } \mathbf{t} = [\mathbf{t}_1, \ldots, \mathbf{t}_k]_\gamma, \\ \sum_{\nu\in\mathcal{M}} z^{(j),(\nu)^T} \prod_{i=1}^{k} \Psi^{(\nu)}(\mathbf{t}_i) & \text{if } \mathbf{t} = [\mathbf{t}_1, \ldots, \mathbf{t}_k]_j,\ j \in \{0, 1, \ldots, m\} \end{cases} \tag{12}$$

where $\Psi^{(\nu)}(\emptyset) = e$ with the representation $\tau_j = [\emptyset]_j$ and for each subtree $\mathbf{t} = [\mathbf{t}_1, \ldots, \mathbf{t}_q]_l$ with some $l \in \{0, 1, \ldots, m\}$ we recursively define

$$\Psi^{(\nu)}(\mathbf{t}) = \sum_{\mu\in\mathcal{M}} Z^{(\nu),(l),(\mu)} \prod_{i=1}^{q} \Psi^{(\mu)}(\mathbf{t}_i). \tag{13}$$

Table 1 All trees $\mathbf{t} \in TS$ of order $\rho(\mathbf{t}) \leq 1.5$ with $j_1, j_2, j_3 \in \{1, \dots, m\}$ arbitrarily eligible

$\mathbf{t}$	tree	$I_{\mathbf{t};t_0,t}$	$\sigma(\mathbf{t})$	$\rho(\mathbf{t})$
$\mathbf{t}_{0,1}$	τ_γ	1	1	0
$\mathbf{t}_{0.5,1}$	$[\tau_{j_1}]_\gamma$	$I_{(j_1)}[1]_{t_0,t}$	1	0.5
$\mathbf{t}_{1,1}$	$[\tau_0]_\gamma$	$I_{(0)}[1]_{t_0,t}$	1	1
$\mathbf{t}_{1,2}$	$[\tau_{j_1}, \tau_{j_2}]_\gamma$	$I_{(j_1,j_2)}[1]_{t_0,t} + I_{(j_2,j_1)}[1]_{t_0,t}$ $+ I_{(0)}[\mathbf{1}_{\{j_1=j_2 \wedge * \neq \circ\}}]_{t_0,t}$	$1 + \mathbf{1}_{\{j_1=j_2\}}$	1
$\mathbf{t}_{1,3}$	$[[\tau_{j_2}]_{j_1}]_\gamma$	$I_{(j_2,j_1)}[1]_{t_0,t}$	1	1
$\mathbf{t}_{1.5,1}$	$[[\tau_{j_1}]_0]_\gamma$	$I_{(j_1,0)}[1]_{t_0,t}$	1	1.5
$\mathbf{t}_{1.5,2}$	$[[\tau_0]_{j_1}]_\gamma$	$I_{(0,j_1)}[1]_{t_0,t}$	1	1.5
$\mathbf{t}_{1.5,3}$	$[\tau_0, \tau_{j_1}]_\gamma$	$I_{(0,j_1)}[1]_{t_0,t} + I_{(j_1,0)}[1]_{t_0,t}$	1	1.5
$\mathbf{t}_{1.5,4}$	$[\tau_{j_1}, \tau_{j_2}, \tau_{j_3}]_\gamma$	$I_{(j_1,j_2,j_3)}[1]_{t_0,t} + I_{(j_1,j_3,j_2)}[1]_{t_0,t}$ $+ I_{(j_2,j_1,j_3)}[1]_{t_0,t} + I_{(j_2,j_3,j_1)}[1]_{t_0,t}$ $+ I_{(j_3,j_1,j_2)}[1]_{t_0,t} + I_{(j_3,j_2,j_1)}[1]_{t_0,t}$ $+ (I_{(j_1,0)}[1]_{t_0,t} + I_{(0,j_1)}[1]_{t_0,t})\mathbf{1}_{\{j_2=j_3 \wedge * \neq \circ\}}$ $+ (I_{(j_2,0)}[1]_{t_0,t} + I_{(0,j_2)}[1]_{t_0,t})\mathbf{1}_{\{j_1=j_3 \wedge * \neq \circ\}}$ $+ (I_{(j_3,0)}[1]_{t_0,t} + I_{(0,j_3)}[1]_{t_0,t})\mathbf{1}_{\{j_1=j_2 \wedge * \neq \circ\}}$	$1 + \mathbf{1}_{\{j_1=j_2 \neq j_3\}}$ $+ \mathbf{1}_{\{j_1=j_3 \neq j_2\}}$ $+ \mathbf{1}_{\{j_2=j_3 \neq j_1\}}$ $+ 5 \cdot \mathbf{1}_{\{j_1=j_2=j_3\}}$	1.5
$\mathbf{t}_{1.5,5}$	$[[\tau_{j_2}]_{j_1}, \tau_{j_3}]_\gamma$	$I_{(j_2,j_3,j_1)}[1]_{t_0,t} + I_{(j_3,j_2,j_1)}[1]_{t_0,t}$ $+ I_{(j_2,j_1,j_3)}[1]_{t_0,t} + I_{(0,j_1)}[\mathbf{1}_{\{j_2=j_3 \wedge * \neq \circ\}}]_{t_0,t}$ $+ I_{(j_2,0)}[\mathbf{1}_{\{j_1=j_3 \wedge * \neq \circ\}}]_{t_0,t}$	1	1.5
$\mathbf{t}_{1.5,6}$	$[[\tau_{j_2}, \tau_{j_3}]_{j_1}]_\gamma$	$I_{(j_3,j_2,j_1)}[1]_{t_0,t} + I_{(j_2,j_3,j_1)}[1]_{t_0,t}$ $+ I_{(0,j_1)}[\mathbf{1}_{\{j_2=j_3 \wedge * \neq \circ\}}]_{t_0,t}$	$1 + \mathbf{1}_{\{j_2=j_3\}}$	1.5
$\mathbf{t}_{1.5,7}$	$[[[\tau_{j_3}]_{j_2}]_{j_1}]_\gamma$	$I_{(j_3,j_2,j_1)}[1]_{t_0,t}$	1	1.5

Here $e = (1, \dots, 1)^T$ and the product of vectors in the definition of $\Psi^{(\nu)}$ is defined by component-wise multiplication, i.e. with $(a_1, \dots, a_n)(b_1, \dots, b_n) = (a_1 b_1, \dots, a_n b_n)$. In the following, we also write $\Phi_S(\mathbf{t}; t, t+h) = \Phi_S(\mathbf{t})$ in order to emphasize the dependency on the current time step with step size h.

Now, the following local Taylor expansions can be proved: For the solution X of SDE (1) and for $p \in \frac{1}{2}\mathbb{N}_0$ with $f \in C^{2p+2}(\mathbb{R}^d, \mathbb{R})$ and $a, b^j \in C^{2p+1}(\mathbb{R}^d, \mathbb{R}^d)$ for $j = 1, \dots, m$, we obtain the expansion (see Rößler 2004b, 2010a)

$$f(X_t) = \sum_{\substack{\mathbf{t} \in TS \\ \rho(\mathbf{t}) \leq p}} F(\mathbf{t})(X_{t_0}) \frac{I_{\mathbf{t};t_0,t}}{\sigma(\mathbf{t})} + \mathcal{R}_p^*(t, t_0) \tag{14}$$

P-a.s. with remainder term $\mathcal{R}_p^*(t, t_0)$ provided all multiple Itô integrals exist. For the approximation Y by the SRK method (4) and for $p \in \frac{1}{2}\mathbb{N}_0$ with $f \in C^{2p+1}(\mathbb{R}^d, \mathbb{R})$ and $a, b^j \in C^{2p}(\mathbb{R}^d, \mathbb{R}^d)$, $j = 1, \dots, m$, we get the expansion (see Rößler 2006b, 2009)

$$f(Y(t)) = \sum_{\substack{\mathbf{t} \in TS \\ \rho(\mathbf{t}) \leq p}} F(\mathbf{t})(Y(t_0)) \frac{\Phi_S(\mathbf{t}; t_0, t)}{\sigma(\mathbf{t})} + \mathcal{R}_p^{\Delta}(t, t_0) \tag{15}$$

P-a.s. with remainder term $\mathcal{R}_p^{\Delta}(t, t_0)$.

4 Order Conditions for Stochastic Runge–Kutta Methods

Using the colored rooted tree analysis, we obtain order conditions for the random variables and the coefficients of the SRK method (4) if it is applied to SDE (1). The following results can be applied for the development of SRK methods for the Itô as well as the Stratonovich version of SDE (1). First, we consider conditions for strong convergence with some order $p \in \frac{1}{2}\mathbb{N}$ due to Rößler (2010b). Therefore, let TS^* denote the set of trees $\mathbf{t} \in TS$ which have only one ramification at the root node τ_γ, i.e. which are of type $[[\ldots]_j]_\gamma$ for some $j \in \{0, 1, \ldots, m\}$. The reason is, that we are interested in the approximation of X, thus we have to choose $f(x) = x$. However, in this case all elementary differentials vanish except for the trees in TS^*. For example, the trees $\mathbf{t}_{1,2}$, $\mathbf{t}_{1.5,3}$, $\mathbf{t}_{1.5,4}$ and $\mathbf{t}_{1.5,5}$ in Table 1 as well as the trees $\mathbf{t}_{III}$ and $\mathbf{t}_{IV}$ in Fig. 1 do not belong to TS^*. A comparison of the Taylor expansions (14) and (15) results in the following two theorems.

Theorem 1. *Let $p \in \frac{1}{2}\mathbb{N}_0$ and $a, b^j \in C^{\lceil p \rceil, 2p+1}(\mathcal{I} \times \mathbb{R}^d, \mathbb{R}^d)$ for $j = 1, \ldots, m$. Then, the SRK method* (4) *has mean–square order of accuracy p if the conditions*

(a) *for all $\mathbf{t} \in TS^*$ with $\rho(\mathbf{t}) \leq p$*

$$I_{\mathbf{t};t,t+h} = \Phi_S(\mathbf{t}; t, t+h) \quad \text{P-a.s.,} \tag{16}$$

(b) *for all $\mathbf{t} \in TS^*$ with $\rho(\mathbf{t}) = p + \frac{1}{2}$*

$$\mathrm{E}(I_{\mathbf{t};t,t+h}) = \mathrm{E}(\Phi_S(\mathbf{t}; t, t+h)), \tag{17}$$

are fulfilled for arbitrary $t, t+h \in \mathcal{I}$ and if (5) *and* (6) *hold.*

For the proof of Theorem 1 we refer to Rößler (2009). Next, we give conditions for the weak convergence of the SRK method (4) based on trees in TS having also multiple ramifications at the root node (see Theorem 6.4 in Rößler 2006b).

Theorem 2. *Let $p \in \mathbb{N}$ and $a, b^j \in C_P^{p+1,2p+2}(\mathcal{I} \times \mathbb{R}^d, \mathbb{R}^d)$ for $j = 1, \ldots, m$. Then the SRK method* (4) *is of weak order p if for all $\mathbf{t} \in TS$ with $\rho(\mathbf{t}) \leq p + \frac{1}{2}$ the order conditions*

$$\mathrm{E}(I_{\mathbf{t};t,t+h}) = \mathrm{E}(\Phi_S(\mathbf{t}; t, t+h)) \tag{18}$$

are fulfilled for arbitrary $t, t+h \in \mathcal{I}$, provided that (5) *and* (6) *apply and that the approximation Y has uniformly bounded moments w.r.t. the number N of steps.*

For the proof of Theorem 2 we refer to Rößler (2006b).

Remark 1. The approximation Y by the SRK method (4) has uniformly bounded moments if bounded random variables are used by the method, if (6) is fulfilled and if $\mathrm{E}(z^{(k,\nu)^T} e) = 0$ for $1 \leq k \leq m$ and $\nu \in \mathcal{M}$ (see Rößler 2006b for details). Further, Theorem 2 provides uniform weak convergence with order p in the case of a non-random time discretization $\mathcal{I}_h$.

5 Strong Approximation of SDEs

For higher order strong numerical approximation methods for SDEs, the simulation of multiple stochastic integrals is necessary in general. Therefore, for $t_n, t_{n+1} \in \mathcal{I}_h$ and $1 \leq i, j \leq m$ let

$$I_{(i),n} = \int_{t_n}^{t_{n+1}} \mathrm{d}W_s^i, \qquad I_{(i,j),n} = \int_{t_n}^{t_{n+1}} \int_{t_n}^{s} \mathrm{d}W_u^i \, \mathrm{d}W_s^j,$$

denote the multiple Itô stochastic integrals. For convenience we write e.g. $I_{(i)} = I_{(i),n}$ if n is obvious from the context. The increments of the Wiener process $I_{(i),n}$ are independent $N(0, h_n)$ distributed with $h_n = t_{n+1} - t_n$. From (10) follows that $I_{(0,i)} = h_n I_{(i)} - I_{(i,0)}$. In the case of $i = j$, formula (10) results in $I_{(i,i)} = \frac{1}{2}(I_{(i)}^2 - h_n)$. Further, let $I_{(i,i,i)} = \frac{1}{6}(I_{(i)}^3 - 3 I_{(0)} I_{(i)})$. In the following, the multiple integrals $I_{(i,0)}$ can be simulated by $I_{(i,0)} = \frac{1}{2} h_n (I_{(i)} + \frac{1}{\sqrt{3}} \zeta_i)$ with some independent $N(0, h_n)$ distributed random variables ζ_i which are independent from $I_{(j)}$ for all $1 \leq j \leq m$ (cf., e.g., Kloeden and Platen 1999 or Milstein 1995). However, since the exact distribution and thus the exact simulation of the multiple stochastic integrals $I_{(i,j)}$ for $1 \leq i, j \leq m$ with $i \neq j$ is not known, we substitute them in our numerical experiments by sufficiently exact and efficient approximations as recently proposed by Wiktorsson (2001). Further, let (p_D, p_S) with $p_D \geq p_S$ denote the order of convergence of the considered SRK scheme if it is applied to a deterministic or stochastic differential equation, respectively.

5.1 Order 1.0 Strong SRK Methods

Firstly, we consider an efficient order 1.0 strong SRK method for Itô SDEs (1). Yet, known derivative free order 1.0 strong approximation methods suffer from an inefficiency in the case of an m-dimensional driving Wiener process. For example, the derivative free scheme (11.1.7) in Kloeden and Platen (1999) needs one evaluation of the drift coefficient a, however $m + 1$ evaluations of each diffusion coefficient b^j, $j = 1, \ldots, m$, each step. Thus, the computational complexity grows quadratically in m which is a significant drawback especially for high dimensional problems. Therefore, efficient SRK methods were firstly proposed in Rößler (2009) where the number of necessary evaluations of each drift and each diffusion coefficient is independent of the dimension m of the driving Wiener process.

For the multi-dimensional Itô SDE (1) with $d, m \geq 1$, the efficient s-stages order 1.0 strong SRK method due to Rößler (2009) is given by $Y_0 = X_{t_0}$ and

$$Y_{n+1} = Y_n + \sum_{i=1}^{s} \alpha_i \, a(t_n + c_i^{(0)} h_n, H_i^{(0)}) \, h_n + \sum_{k=1}^{m} \sum_{i=1}^{s} \left(\beta_i^{(1)} I_{(k)} + \beta_i^{(2)} \sqrt{h_n}\right) b^k(t_n + c_i^{(1)} h_n, H_i^{(k)}) \tag{19}$$

for $n = 0, 1, \ldots, N-1$ with stages

$$\begin{aligned} H_i^{(0)} =& Y_n + \sum_{j=1}^{s} A_{ij}^{(0)} \, a(t_n + c_j^{(0)} h_n, H_j^{(0)}) \, h_n \\ &+ \sum_{l=1}^{m} \sum_{j=1}^{s} B_{ij}^{(0)} \, b^l(t_n + c_j^{(1)} h_n, H_j^{(l)}) \, I_{(l)}, \\ H_i^{(k)} =& Y_n + \sum_{j=1}^{s} A_{ij}^{(1)} \, a(t_n + c_j^{(0)} h_n, H_j^{(0)}) \, h_n \\ &+ \sum_{l=1}^{m} \sum_{j=1}^{s} B_{ij}^{(1)} \, b^l(t_n + c_j^{(1)} h_n, H_j^{(l)}) \, \frac{I_{(l,k)}}{\sqrt{h_n}} \end{aligned} \tag{20}$$

for $i = 1, \ldots, s$ and $k = 1, \ldots, m$. A modified version of the efficient SRK method (19) suitable for Stratonovich SDEs can be found in Rößler (2009). The SRK method (19) can be characterized by its coefficients given by an extended Butcher tableau:

$$\begin{array}{c|c|c|c} c^{(0)} & A^{(0)} & B^{(0)} & \\ \hline c^{(1)} & A^{(1)} & B^{(1)} & \\ \hline & \alpha^T & \beta^{(1)^T} & \beta^{(2)^T} \end{array} \tag{21}$$

Here, the class of SRK methods (4) is applied with $\mathcal{M} = \{\nu : 0 \leq \nu \leq m\}$ and

$$z_i^{(0),(0)} = \alpha_i \, h_n \,, \qquad Z_{ij}^{(0),(0),(0)} = A_{ij}^{(0)} \, h_n \,, \qquad Z_{ij}^{(0),(k),(k)} = B_{ij}^{(0)} \, I_{(k)} \,,$$
$$z_i^{(k),(k)} = \beta_i^{(1)} I_{(k)} + \beta_i^{(2)} \sqrt{h_n} \,, \qquad Z_{ij}^{(k),(0),(0)} = A_{ij}^{(1)} h_n \,,$$
$$Z_{ij}^{(k),(l),(l)} = B_{ij}^{(1)} \frac{I_{(l,k)}}{\sqrt{h_n}} \,,$$

for $1 \leq k, l \leq m$ and all other coefficients in (4) are set equal to zero. Thus, the presented SRK method (19) belongs to the general class (4). The application of the rooted tree analysis and Theorem 1 gives order conditions up to strong order 1.0 for the coefficients of the SRK method (19), see also Rößler (2009).

Table 2 Coefficients for the strong SRK schemes SRI1 of order (1.0, 1.0) on the left hand side and SRI2 of order (2.0, 1.0) on the right hand side

SRI1:

0			
0	0	0	
0	0 0	0 0	
0			
0	0	1	
0	0 0	−1 0	
	1 0 0	1 0 0	0 $\frac{1}{2}$ $-\frac{1}{2}$

SRI2:

0			
1	1	0	
0	0 0	0 0	
0			
1	1	1	
1	1 0	−1 0	
	$\frac{1}{2}$ $\frac{1}{2}$ 0	1 0 0	0 $\frac{1}{2}$ $-\frac{1}{2}$

Theorem 3. *Let $a, b^j \in C^{1,2}(\mathcal{I} \times \mathbb{R}^d, \mathbb{R}^d)$ for $j = 1, \ldots, m$. If the coefficients of the SRK method* (19) *fulfill the equations*

$$1.\quad \alpha^T e = 1 \qquad 2.\quad \beta^{(1)^T} e = 1 \qquad 3.\quad \beta^{(2)^T} e = 0$$

then the method attains order 0.5 for the strong approximation of the solution of the Itô SDE (1). If $a, b^j \in C^{1,3}(\mathcal{I} \times \mathbb{R}^d, \mathbb{R}^d)$ for $j = 1, \ldots, m$ and if in addition the equations

$$4.\quad \beta^{(1)^T} B^{(1)} e = 0 \qquad 5.\quad \beta^{(2)^T} B^{(1)} e = 1 \qquad 6.\quad \beta^{(2)^T} A^{(1)} e = 0$$
$$7.\quad \beta^{(2)^T} (B^{(1)} e)^2 = 0 \qquad 8.\quad \beta^{(2)^T} (B^{(1)} (B^{(1)} e)) = 0$$

are fulfilled and if $c^{(i)} = A^{(i)} e$ for $i = 0, 1$, then the SRK method (19) *attains order* 1.0 *for the strong approximation of the solution of the Itô SDE* (1).

For the detailed proof of Theorem 3 we refer to Rößler (2009). The Euler–Maruyama scheme EM is the basic explicit order 0.5 strong SRK scheme with $s = 1$ stage, $\alpha_1 = \beta_1^{(1)} = 1$ and $\beta_1^{(2)} = A_{1,1}^{(0)} = A_{1,1}^{(1)} = B_{1,1}^{(0)} = B_{1,1}^{(1)} = 0$. As an example for some explicit order 1.0 strong SRK schemes, the coefficients presented in Table 2 define the order (1.0, 1.0) strong SRK scheme SRI1 and the order (2.0, 1.0) strong SRK scheme SRI2. As the main advantage, the scheme SRI1 needs one evaluation of the drift coefficient a and only 3 evaluations of each diffusion coefficient b^j, $j = 1, \ldots, m$, each step. Thus, the number of evaluations of the drift and diffusion coefficients is independent of the dimension m of the Wiener process.

5.2 Order 1.5 Strong SRK Methods for SDEs with Scalar Noise

In contrast to the multi-dimensional Wiener process case, higher order 1.5 strong approximation methods can be applied if the driving Wiener process is scalar. E.g., order 1.5 strong SRK methods for Stratonovich SDEs with a scalar Wiener process have been proposed by Burrage and Burrage (1996, 2000). On the other hand, for Itô

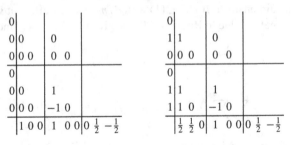

SDEs with a scalar Wiener process order 1.5 strong SRK methods have been proposed by Kaneko (1995) and by Kloeden and Platen (1999). However, the scheme due to Kaneko (1995) is not efficient because it needs 4 evaluations of the drift coefficient a, 12 evaluations of the diffusion coefficient b and the simulation of two independent normally distributed random variables for each step. On the other hand, the scheme (11.2.1) in Kloeden and Platen (1999) due to Platen needs 3 evaluations of the drift coefficient a, 5 evaluations of the diffusion b and also the simulation of two independent normally distributed random variables each step. In contrast to this, we consider the order 1.5 strong SRK method for Itô SDEs with less computational complexity proposed in Rößler (2009).

For the Itô SDE (1) with $d \geq 1$ and $m = 1$ the efficient order 1.5 strong SRK method due to Rößler (2009) is defined by $Y_0 = X_{t_0}$ and

$$\begin{aligned} Y_{n+1} = Y_n &+ \sum_{i=1}^{s} \alpha_i \, a(t_n + c_i^{(0)} h_n, H_i^{(0)}) \, h_n \\ &+ \sum_{i=1}^{s} \left(\beta_i^{(1)} I_{(1)} + \beta_i^{(2)} \frac{I_{(1,1)}}{\sqrt{h_n}} + \beta_i^{(3)} \frac{I_{(1,0)}}{h_n} + \beta_i^{(4)} \frac{I_{(1,1,1)}}{h_n} \right) \\ &\times b(t_n + c_i^{(1)} h_n, H_i^{(1)}) \end{aligned} \tag{22}$$

for $n = 0, 1, \ldots, N-1$ with stages

$$\begin{aligned} H_i^{(0)} &= Y_n + \sum_{j=1}^{s} A_{ij}^{(0)} a(t_n + c_j^{(0)} h_n, H_j^{(0)}) \, h_n + \sum_{j=1}^{s} B_{ij}^{(0)} \, b(t_n + c_j^{(1)} h_n, H_j^{(1)}) \\ &\quad \times \frac{I_{(1,0)}}{h_n}, \\ H_i^{(1)} &= Y_n + \sum_{j=1}^{s} A_{ij}^{(1)} \, a(t_n + c_j^{(0)} h_n, H_j^{(0)}) \, h_n + \sum_{j=1}^{s} B_{ij}^{(1)} \, b(t_n + c_j^{(1)} h_n, H_j^{(1)}) \sqrt{h_n} \end{aligned} \tag{23}$$

for $i = 1, \ldots, s$. A more general version of the order 1.5 strong SRK method (22) for SDEs with diagonal noise and a simplified version for additive noise can be found in Rößler (2009). The SRK method (22) is characterized by the Butcher tableau:

$$\begin{array}{c|c|c|c} c^{(0)} & A^{(0)} & B^{(0)} & \\ \hline c^{(1)} & A^{(1)} & B^{(1)} & \\ \hline & \alpha^T & \beta^{(1)^T} & \beta^{(2)^T} \\ \hline & & \beta^{(3)^T} & \beta^{(4)^T} \end{array} \tag{24}$$

For the SRK method (22) we choose $\mathcal{M} = \{0, 1\}$ and we then define

$$z_i^{(0),(0)} = \alpha_i \, h_n \, , \qquad z_i^{(1),(1)} = \beta_i^{(1)} \, I_{(1)} + \beta_i^{(2)} \, \frac{I_{(1,1)}}{\sqrt{h_n}} + \beta_i^{(3)} \, \frac{I_{(1,0)}}{h_n} + \beta_i^{(4)} \, \frac{I_{(1,1,1)}}{h_n} \, ,$$

$$Z_{ij}^{(0),(0),(0)} = A_{ij}^{(0)} \, h_n \, , \qquad Z_{ij}^{(0),(1),(1)} = B_{ij}^{(0)} \, \frac{I_{(1,0)}}{h_n} \, ,$$

$$Z_{ij}^{(1),(0),(0)} = A_{ij}^{(1)} \, h_n \, , \qquad Z_{ij}^{(1),(1),(1)} = B_{ij}^{(1)} \, \sqrt{h_n} \, ,$$

with all remaining coefficients in (4) defined equal to zero. Then, the SRK method (22) is also covered by the class (4) of SRK methods. Thus, we can apply Theorem 1 with $p = 1.5$ to obtain strong order 1.5 conditions, see Rößler (2009) for details.

Theorem 4. *Let $a, b \in C^{1,2}(\mathcal{J} \times \mathbb{R}^d, \mathbb{R}^d)$. If the coefficients of the SRK method* (22) *fulfill the equations*

$$1. \quad \alpha^T e = 1 \qquad 2. \quad {\beta^{(1)}}^T e = 1 \qquad 3. \quad {\beta^{(2)}}^T e = 0$$

$$4. \quad {\beta^{(3)}}^T e = 0 \qquad 5. \quad {\beta^{(4)}}^T e = 0$$

then the method attains order 0.5 *for the strong approximation of the solution of the Itô SDE* (1). *If $a, b \in C^{1,3}(\mathcal{J} \times \mathbb{R}^d, \mathbb{R}^d)$ and if in addition the equations*

$$6. \quad {\beta^{(1)}}^T B^{(1)} e = 0 \qquad 7. \quad {\beta^{(2)}}^T B^{(1)} e = 1$$

$$8. \quad {\beta^{(3)}}^T B^{(1)} e = 0 \qquad 9. \quad {\beta^{(4)}}^T B^{(1)} e = 0$$

are fulfilled and if $c^{(i)} = A^{(i)} e$ for $i = 0, 1$, then the SRK method (22) *attains order* 1.0 *for the strong approximation of the solution of the Itô SDE* (1) *with scalar noise. If $a, b \in C^{2,4}(\mathcal{J} \times \mathbb{R}^d, \mathbb{R}^d)$ and if in addition the equations*

$$10. \quad \alpha^T A^{(0)} e = \frac{1}{2} \qquad 11. \quad \alpha^T B^{(0)} e = 1$$

$$12. \quad \alpha^T (B^{(0)} e)^2 = \frac{3}{2} \qquad 13. \quad {\beta^{(1)}}^T A^{(1)} e = 1$$

$$14. \quad {\beta^{(2)}}^T A^{(1)} e = 0 \qquad 15. \quad {\beta^{(3)}}^T A^{(1)} e = -1$$

$$16. \quad {\beta^{(4)}}^T A^{(1)} e = 0 \qquad 17. \quad {\beta^{(1)}}^T (B^{(1)} e)^2 = 1$$

$$18. \quad {\beta^{(2)}}^T (B^{(1)} e)^2 = 0 \qquad 19. \quad {\beta^{(3)}}^T (B^{(1)} e)^2 = -1$$

$$20. \quad {\beta^{(4)}}^T (B^{(1)} e)^2 = 2 \qquad 21. \quad {\beta^{(1)}}^T (B^{(1)} (B^{(1)} e)) = 0$$

$$22. \quad {\beta^{(2)}}^T (B^{(1)} (B^{(1)} e)) = 0 \qquad 23. \quad {\beta^{(3)}}^T (B^{(1)} (B^{(1)} e)) = 0$$

$$24. \quad {\beta^{(4)}}^T (B^{(1)} (B^{(1)} e)) = 1$$

$$25. \quad \frac{1}{2} {\beta^{(1)}}^T (A^{(1)} (B^{(0)} e)) + \frac{1}{3} {\beta^{(3)}}^T (A^{(1)} (B^{(0)} e)) = 0$$

Table 3 Strong SRK scheme SRI1W1 of order (2.0, 1.5) on the left hand side and SRI2W1 of order (3.0, 1.5) on the right hand side

$$
\begin{array}{c|cccc|cccc|cccc}
0 &&&&&&&&&&&& \\
\frac{3}{4} & \frac{3}{4} &&&& \frac{3}{2} &&&&&&& \\
0 & 0 & 0 &&& 0 & 0 &&&&&& \\
0 & 0 & 0 & 0 && 0 & 0 & 0 &&&&& \\
\hline
0 &&&&&&&&&&&& \\
\frac{1}{4} & \frac{1}{4} &&&& \frac{1}{2} &&&&&&& \\
1 & 1 & 0 &&& -1 & 0 &&&&&& \\
\frac{1}{4} & 0 & 0 & \frac{1}{4} && -5 & 3 & \frac{1}{2} &&&&& \\
\hline
& \frac{1}{3} & \frac{2}{3} & 0 & 0 & -1 & \frac{4}{3} & \frac{2}{3} & 0 & -1 & \frac{4}{3} & -\frac{1}{3} & 0 \\
\hline
&&&&& 2 & -\frac{4}{3} & -\frac{2}{3} & 0 & -2 & \frac{5}{3} & -\frac{2}{3} & 1
\end{array}
\qquad
\begin{array}{c|cccc|cccc|cccc}
0 &&&&&&&&&&&& \\
1 & 1 &&&& 0 &&&&&&& \\
\frac{1}{2} & \frac{1}{4} & \frac{1}{4} &&& 1 & \frac{1}{2} &&&&&& \\
0 & 0 & 0 & 0 && 0 & 0 & 0 &&&&& \\
\hline
0 &&&&&&&&&&&& \\
\frac{1}{4} & \frac{1}{4} &&&& -\frac{1}{2} &&&&&&& \\
1 & 1 & 0 &&& 1 & 0 &&&&&& \\
\frac{1}{4} & 0 & 0 & \frac{1}{4} && 2 & -1 & \frac{1}{2} &&&&& \\
\hline
& \frac{1}{6} & \frac{1}{6} & \frac{2}{3} & 0 & -1 & \frac{4}{3} & \frac{2}{3} & 0 & 1 & -\frac{4}{3} & \frac{1}{3} & 0 \\
\hline
&&&&& 2 & -\frac{4}{3} & -\frac{2}{3} & 0 & -2 & \frac{5}{3} & -\frac{2}{3} & 1
\end{array}
$$

are fulfilled and if $c^{(i)} = A^{(i)}e$ *for* $i = 0, 1$, *then the SRK method* (22) *attains order* 1.5 *for the strong approximation of the solution of the Itô SDE* (1) *in the case of scalar noise.*

For a proof of Theorem 4 we refer to Rößler (2009). Coefficients for the order 1.5 strong SRK schemes SRI1W1 of order (2.0, 1.5) and SRI2W1 of order (3.0, 1.5) are given in Table 3. The SRK scheme SRI1W1 needs only 2 evaluations of the drift coefficient, 4 evaluations of the diffusion coefficient b and the simulation of two independent normally distributed random variables for each step. Note that the explicit 2-stages SRK method (22) with coefficients $\alpha_1 = \beta_1^{(1)} = \beta_2^{(2)} = A_{2,1}^{(1)} = B_{2,1}^{(1)} = 1$, $\beta_1^{(2)} = -1$ and $\alpha_2 = \beta_2^{(1)} = A_{2,1}^{(0)} = B_{2,1}^{(0)} = \beta_1^{(3)} = \beta_2^{(3)} = \beta_1^{(4)} = \beta_2^{(4)} = 0$ coincides with the order 1.0 strong scheme (11.1.3) in Kloeden and Platen (1999).

5.3 Numerical Results

The presented efficient SRK methods are applied to some test SDEs in order to analyze their performance. Let EM denote the order 0.5 strong Euler–Maruyama scheme and let MIL denote the order 1.0 strong Milstein scheme in Milstein (1995). Further, the order 1.0 strong scheme (11.1.7) denoted as SPLI and the order 1.5 strong scheme (11.2.1) called SPLIW1 for Itô SDEs with scalar noise in Kloeden and Platen (1999) are applied. As a measure for the computational effort, we take the number of evaluations of the drift and diffusion coefficients as well as the number of realizations of (normally distributed) random variables needed each step. If the approximation method needs the random variables $I_{(i,j)}$ for $1 \leq i, j \leq m$ with $i \neq j$, then $I_{(i,j)}$ is simulated by the method due to Wiktorsson (2001) and we need to simulate $\frac{1}{2}m(m-1) + 2m\,q$ independent normally distributed random variables

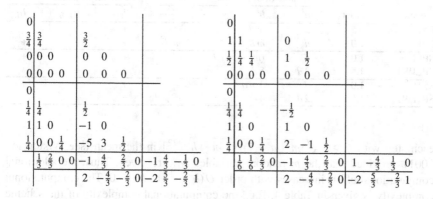

Table 4 Computational complexity of some schemes per step for a d-dimensional SDE system with a m-dimensional Wiener process ($m = 1$ for SPLIW1 and SRI1W1)

		number of evaluations			random variables		
scheme	order	a^k	$b^{k,j}$	$\frac{\partial b^{k,j}}{\partial x^l}$	$I_{(j)}$	$I_{(j,0)}$	$I_{(i,j)}$
EM	0.5	d	$m\,d$	–	+	–	–
MIL	1.0	d	$m\,d$	$m\,d^2$	+	–	+
SPLI	1.0	d	$(m^2+m)\,d$	–	+	–	+
SPLIW1	1.5	$3\,d$	$5\,d$	–	+	+	–
SRI1	1.0	d	$3\,m\,d$	–	+	–	+
SRI1W1	1.5	$2\,d$	$4\,d$	–	+	+	–

each step with $q \geq \lceil \sqrt{5m^2(m-1)/(24\pi^2)}\, h^{-1/2} \rceil$ in the mean (see Wiktorsson 2001), provided that the m random variables $I_{(i)}$ are given. Thus, the additional computational effort increases with order $O(h^{-1/2})$ as $h \to 0$. The computational complexity is given in Table 4. E.g., the computational complexity of the scheme MIL is $d+md+md^2+m+\frac{1}{2}m(m-1)+2m\,q$ whereas scheme SRI1 has only complexity $d+3md+m+\frac{1}{2}m(m-1)+2m\,q$ each step. Thus, the scheme SRI1 has lower computational complexity than the Milstein scheme MIL in the case of $d > 2$ and $m \geq 1$ even if we neglect the effort for the calculation of the derivatives of b^j needed by the Milstein scheme. Further, the scheme SRI1 has also lower computational complexity than the scheme SPLI1 due to Platen in the case of $d \geq 1$ and $m > 2$.

We simulate 2000 trajectories and take the mean of the attained errors at $T = 1$ as an estimator for the expectation in (2). Then, we analyze the mean–square errors versus the computational effort as well as versus the step size in log–log-diagrams with base two. We denote by p_{eff} the effective order of convergence which is the slope of the resulting line in the mean–square errors versus effort diagrams. Considering the effective order may cause an order reduction such that an strong order 1.0 scheme attains the effective order $p_{\text{eff}} = 2/3$ as $h \to 0$. This is due to the effort for the simulation of the multiple integrals $I_{(i,j)}$ which depends on h. Dotted order lines with slope 0.5, 1.0, 2/3 and 1.5 are plotted as a reference. Clearly, a more efficient method to simulate the multiple integrals $I_{(i,j)}$ would result in a higher effective order. However, compared to the Euler–Maruyama scheme EM with $p_{\text{eff}} = 0.5$, there is still a significantly improved convergence for the order 1.0 methods. As a result of this, the order 1.0 strong approximation methods are superior to the order 0.5 strong Euler–Maruyama scheme, which is also confirmed by the simulation results.

As the first example, consider for $d = m = 1$ the nonlinear Itô SDE

$$dX_t = -\left(\frac{1}{10}\right)^2 \sin(X_t)\cos^3(X_t)\,dt + \frac{1}{10}\cos^2(X_t)\,dW_t\,, \qquad X_0 = 1\,, \tag{25}$$

with solution $X_t = \arctan(\frac{1}{10}W_t + \tan(X_0))$ in Kloeden and Platen (1999). The results for $h = 2^0, \ldots, 2^{-16}$ are plotted on the left of Fig. 2. Scheme SRI1W1 has

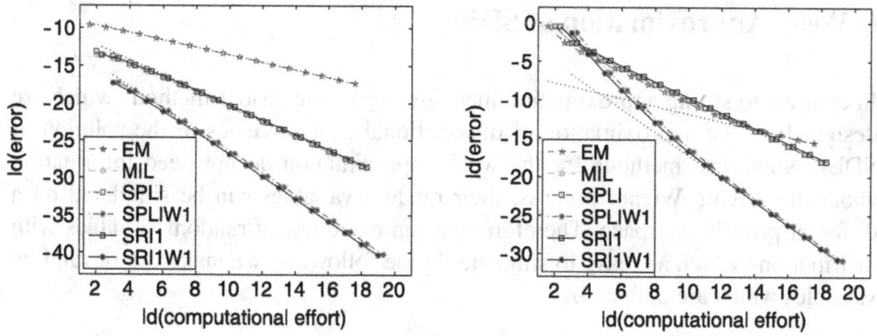

Fig. 2 Errors vs. effort for SDE (25) and SDE (26) with $d = m = 1$

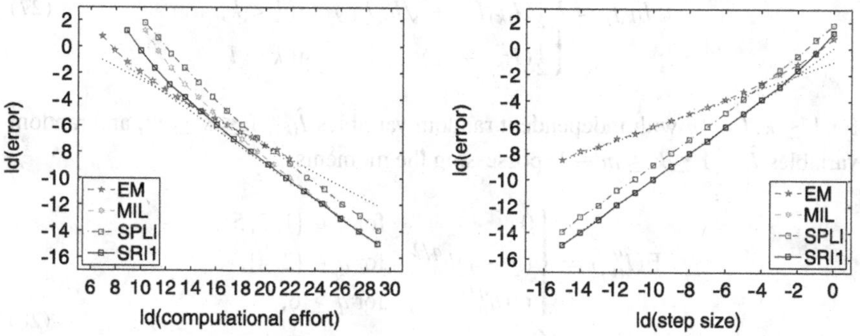

Fig. 3 Errors vs. effort for SDE (26) and errors vs. step sizes for SDE (26) with $d = m = 10$

effective order 1.5 and performs better than the other schemes due to its reduced complexity.

In order to consider also a multi-dimensional Itô SDE with $d, m \geq 1$, we define $A \in \mathbb{R}^{d \times d}$ as a matrix with entries $A_{ij} = \frac{1}{20}$ if $i \neq j$ and $A_{ii} = -\frac{3}{2}$ for $1 \leq i, j \leq d$. Further, define $B^k \in \mathbb{R}^{d \times d}$ by $B^k_{ij} = \frac{1}{100}$ for $i \neq j$ and $B^k_{ii} = \frac{1}{5}$ for $1 \leq i, j \leq d$ and $k = 1, \ldots, m$. Then, we consider the Itô SDE

$$\mathrm{d}X_t = AX_t \,\mathrm{d}t + \sum_{k=1}^{m} B^k X_t \,\mathrm{d}W_t^k, \qquad X_0 = (1, \ldots, 1)^T \in \mathbb{R}^d, \tag{26}$$

with solution $X_t = X_0 \exp((A - \frac{1}{2} \sum_{k=1}^{m} (B^k)^2)\, t + \sum_{k=1}^{m} B^k W_t^k)$. For the case of $d = m = 1$ the numerical results for $h = 2^0, \ldots, 2^{-16}$ are presented on the right of Fig. 2 where the scheme SRI1W1 has the best performance. On the other hand, for the case of $d = m = 10$ the effective and the strong orders are analyzed for $h = 2^0, \ldots, 2^{-15}$ in Fig. 3. Here, the schemes MIL, SPLI, and SRI1 have strong order 1.0 while the Euler–Maruyama scheme EM has order $1/2$. Further, due to the effort for the simulation of the multiple integrals, all order 1.0 strong schemes attain the effective order $2/3$ and thus perform significantly better than the Euler–Maruyama scheme EM with effective order $1/2$. The scheme SRI1 shows the best performance, especially compared to the Milstein scheme MIL and the scheme SPLI.

6 Weak Approximation of SDEs

In contrast to strong approximation methods, we now consider methods which are designed for the approximation of distributional characteristics of the solution of SDEs. Numerical methods for the weak approximation do not need information about the driving Wiener process, their random variables can be simulated on a different probability space. Therefore, we can make use of random variables with distributions which are easy to simulate. In the following, we make use of random variables which are defined by

$$\hat{I}_{(k,l)} = \begin{cases} \frac{1}{2}(\hat{I}_{(k)}\hat{I}_{(l)} - \sqrt{h_n}\tilde{I}_{(k)}) & \text{if } k < l, \\ \frac{1}{2}(\hat{I}_{(k)}\hat{I}_{(l)} + \sqrt{h_n}\tilde{I}_{(l)}) & \text{if } l < k, \\ \frac{1}{2}(\hat{I}_{(k)}^2 - h_n) & \text{if } k = l \end{cases} \tag{27}$$

for $1 \leq k, l \leq m$ with independent random variables $\hat{I}_{(k)}$, $1 \leq k \leq m$, and random variables $\tilde{I}_{(k)}$, $1 \leq k \leq m-1$, possessing the moments

$$\begin{aligned} \mathrm{E}(\hat{I}_{(k)}^q) &= \begin{cases} 0 & \text{for } q \in \{1,3,5\}, \\ (q-1)h_n^{q/2} & \text{for } q \in \{2,4\}, \\ O(h_n^{q/2}) & \text{for } q \geq 6, \end{cases} \\ \mathrm{E}(\tilde{I}_{(k)}^q) &= \begin{cases} 0 & \text{for } q \in \{1,3\}, \\ h_n & \text{for } q = 2, \\ O(h_n^{q/2}) & \text{for } q \geq 4. \end{cases} \end{aligned} \tag{28}$$

Thus, only $2m-1$ independent random variables are needed for each step $n = 0, 1, \ldots, N-1$. For example, we can choose $\hat{I}_{(k)}$ as three point distributed random variables with $\mathrm{P}(\hat{I}_{(k)} = \pm\sqrt{3h_n}) = \frac{1}{6}$ and $\mathrm{P}(\hat{I}_{(k)} = 0) = \frac{2}{3}$. The random variables $\tilde{I}_{(k)}$ can be defined by a two point distribution with $\mathrm{P}(\tilde{I}_{(k)} = \pm\sqrt{h_n}) = \frac{1}{2}$.

6.1 Order 2.0 Weak SRK Methods

We consider the class of efficient SRK methods introduced in Rößler (2009) for the weak approximation of the solution of the Itô SDE (1) where the number of stages s is independent of the dimension m of the driving Wiener process. A similar class of second order SRK methods for the Stratonovich version of SDE (1) can be found in Rößler (2007). For the Itô SDE (1) the d-dimensional SRK approximation Y with $Y_n = Y(t_n)$ for $t_n \in \mathcal{I}_h$ due to Rößler (2009) is defined by $Y_0 = x_0$ and

$$Y_{n+1} = Y_n + \sum_{i=1}^{s} \alpha_i \, a(t_n + c_i^{(0)} h_n, H_i^{(0)}) \, h_n$$

$$
\begin{aligned}
&+\sum_{i=1}^{s}\sum_{k=1}^{m}\beta_i^{(1)}\,b^k(t_n+c_i^{(1)}h_n,H_i^{(k)})\hat{I}_{(k)}\\
&+\sum_{i=1}^{s}\sum_{k=1}^{m}\beta_i^{(2)}\,b^k(t_n+c_i^{(1)}h_n,H_i^{(k)})\frac{\hat{I}_{(k,k)}}{\sqrt{h_n}}\\
&+\sum_{i=1}^{s}\sum_{k=1}^{m}\beta_i^{(3)}\,b^k(t_n+c_i^{(2)}h_n,\hat{H}_i^{(k)})\hat{I}_{(k)}\\
&+\sum_{i=1}^{s}\sum_{k=1}^{m}\beta_i^{(4)}\,b^k(t_n+c_i^{(2)}h_n,\hat{H}_i^{(k)})\sqrt{h_n}
\end{aligned}
\tag{29}
$$

for $n = 0, 1, \ldots, N-1$ with stage values

$$
\begin{aligned}
H_i^{(0)} &= Y_n+\sum_{j=1}^{s}A_{ij}^{(0)}\,a(t_n+c_j^{(0)}h_n,H_j^{(0)})\,h_n\\
&\quad+\sum_{j=1}^{s}\sum_{l=1}^{m}B_{ij}^{(0)}\,b^l(t_n+c_j^{(1)}h_n,H_j^{(l)})\,\hat{I}_{(l)},\\
H_i^{(k)} &= Y_n+\sum_{j=1}^{s}A_{ij}^{(1)}\,a(t_n+c_j^{(0)}h_n,H_j^{(0)})\,h_n\\
&\quad+\sum_{j=1}^{s}B_{ij}^{(1)}\,b^k(t_n+c_j^{(1)}h_n,H_j^{(k)})\,\sqrt{h_n},\\
\hat{H}_i^{(k)} &= Y_n+\sum_{j=1}^{s}A_{ij}^{(2)}\,a(t_n+c_j^{(0)}h_n,H_j^{(0)})\,h_n\\
&\quad+\sum_{j=1}^{s}\sum_{\substack{l=1\\ l\neq k}}^{m}B_{ij}^{(2)}\,b^l(t_n+c_j^{(1)}h_n,H_j^{(l)})\,\frac{\hat{I}_{(k,l)}}{\sqrt{h_n}}
\end{aligned}
$$

for $i = 1, \ldots, s$ and $k = 1, \ldots, m$. In the case of a scalar driving Wiener process, i.e. for $m = 1$, the SRK method (29) reduces to the SRK method proposed in Rößler (2006b). The coefficients of the SRK method (29) can be represented by an extended Butcher array:

$$
\begin{array}{c|c|c|c}
c^{(0)} & A^{(0)} & B^{(0)} & \\
\hline
c^{(1)} & A^{(1)} & B^{(1)} & \\
\hline
c^{(2)} & A^{(2)} & B^{(2)} & \\
\hline
 & \alpha^T & \beta^{(1)^T} & \beta^{(2)^T} \\
\hline
 & & \beta^{(3)^T} & \beta^{(4)^T}
\end{array}
$$

Applying the rooted tree analysis and Theorem 2 with $p = 2$, we obtain order two conditions for the SRK method (29) which were calculated in Rößler (2009).

Theorem 5. *Let $a^i, b^{i,j} \in C_P^{2,4}(\mathcal{J} \times \mathbb{R}^d, \mathbb{R})$ for $1 \leq i \leq d$, $1 \leq j \leq m$. If the coefficients of the stochastic Runge–Kutta method (29) fulfill the equations*

1. $\alpha^T e = 1$
2. ${\beta^{(4)}}^T e = 0$
3. ${\beta^{(3)}}^T e = 0$
4. $({\beta^{(1)}}^T e)^2 = 1$
5. ${\beta^{(2)}}^T e = 0$
6. ${\beta^{(1)}}^T B^{(1)} e = 0$
7. ${\beta^{(4)}}^T A^{(2)} e = 0$
8. ${\beta^{(3)}}^T B^{(2)} e = 0$
9. ${\beta^{(4)}}^T (B^{(2)} e)^2 = 0$

then the method attains order 1 for the weak approximation of the solution of the Itô SDE (1). Further, if $a^i, b^{i,j} \in C_P^{3,6}(\mathcal{J} \times \mathbb{R}^d, \mathbb{R})$ for $1 \leq i \leq d$, $1 \leq j \leq m$ and if in addition the equations

10. $\alpha^T A^{(0)} e = \frac{1}{2}$
11. $\alpha^T (B^{(0)} e)^2 = \frac{1}{2}$
12. $({\beta^{(1)}}^T e)(\alpha^T B^{(0)} e) = \frac{1}{2}$
13. $({\beta^{(1)}}^T e)({\beta^{(1)}}^T A^{(1)} e) = \frac{1}{2}$
14. ${\beta^{(3)}}^T A^{(2)} e = 0$
15. ${\beta^{(2)}}^T B^{(1)} e = 1$
16. ${\beta^{(4)}}^T B^{(2)} e = 1$
17. $({\beta^{(1)}}^T e)({\beta^{(1)}}^T (B^{(1)} e)^2) = \frac{1}{2}$
18. $({\beta^{(1)}}^T e)({\beta^{(3)}}^T (B^{(2)} e)^2) = \frac{1}{2}$
19. ${\beta^{(1)}}^T (B^{(1)} (B^{(1)} e)) = 0$
20. ${\beta^{(3)}}^T (B^{(2)} (B^{(1)} e)) = 0$
21. ${\beta^{(3)}}^T (B^{(2)} (B^{(1)} (B^{(1)} e))) = 0$
22. ${\beta^{(1)}}^T (A^{(1)} (B^{(0)} e)) = 0$
23. ${\beta^{(3)}}^T (A^{(2)} (B^{(0)} e)) = 0$
24. ${\beta^{(4)}}^T (A^{(2)} e)^2 = 0$
25. ${\beta^{(4)}}^T (A^{(2)} (A^{(0)} e)) = 0$
26. $\alpha^T (B^{(0)} (B^{(1)} e)) = 0$
27. ${\beta^{(2)}}^T A^{(1)} e = 0$
28. ${\beta^{(1)}}^T ((A^{(1)} e)(B^{(1)} e)) = 0$
29. ${\beta^{(3)}}^T ((A^{(2)} e)(B^{(2)} e)) = 0$
30. ${\beta^{(4)}}^T (A^{(2)} (B^{(0)} e)) = 0$
31. ${\beta^{(2)}}^T (A^{(1)} (B^{(0)} e)) = 0$
32. ${\beta^{(4)}}^T ((B^{(2)} e)^2 (A^{(2)} e)) = 0$
33. ${\beta^{(4)}}^T (A^{(2)} (B^{(0)} e)^2) = 0$
34. ${\beta^{(2)}}^T (A^{(1)} (B^{(0)} e)^2) = 0$
35. ${\beta^{(1)}}^T (B^{(1)} (A^{(1)} e)) = 0$
36. ${\beta^{(3)}}^T (B^{(2)} (A^{(1)} e)) = 0$
37. ${\beta^{(2)}}^T (B^{(1)} e)^2 = 0$
38. ${\beta^{(4)}}^T (B^{(2)} (B^{(1)} e)) = 0$
39. ${\beta^{(2)}}^T (B^{(1)} (B^{(1)} e)) = 0$
40. ${\beta^{(1)}}^T (B^{(1)} e)^3 = 0$
41. ${\beta^{(3)}}^T (B^{(2)} e)^3 = 0$
42. ${\beta^{(1)}}^T (B^{(1)} (B^{(1)} e)^2) = 0$
43. ${\beta^{(3)}}^T (B^{(2)} (B^{(1)} e)^2) = 0$
44. ${\beta^{(4)}}^T (B^{(2)} e)^4 = 0$
45. ${\beta^{(4)}}^T (B^{(2)} (B^{(1)} e))^2 = 0$
46. ${\beta^{(4)}}^T ((B^{(2)} e)(B^{(2)} (B^{(1)} e))) = 0$
47. $\alpha^T ((B^{(0)} e)(B^{(0)} (B^{(1)} e))) = 0$
48. ${\beta^{(1)}}^T ((A^{(1)} (B^{(0)} e))(B^{(1)} e)) = 0$
49. ${\beta^{(3)}}^T ((A^{(2)} (B^{(0)} e))(B^{(2)} e)) = 0$
50. ${\beta^{(1)}}^T (A^{(1)} (B^{(0)} (B^{(1)} e))) = 0$
51. ${\beta^{(3)}}^T (A^{(2)} (B^{(0)} (B^{(1)} e))) = 0$
52. ${\beta^{(4)}}^T ((B^{(2)} (A^{(1)} e))(B^{(2)} e)) = 0$
53. ${\beta^{(1)}}^T (B^{(1)} (A^{(1)} (B^{(0)} e))) = 0$
54. ${\beta^{(3)}}^T (B^{(2)} (A^{(1)} (B^{(0)} e))) = 0$
55. ${\beta^{(1)}}^T ((B^{(1)} e)(B^{(1)} (B^{(1)} e))) = 0$
56. ${\beta^{(3)}}^T ((B^{(2)} e)(B^{(2)} (B^{(1)} e))) = 0$
57. ${\beta^{(1)}}^T (B^{(1)} (B^{(1)} (B^{(1)} e))) = 0$

58. $\beta^{(4)^T}((B^{(2)}e)(B^{(2)}(B^{(1)}(B^{(1)}e)))) = 0$

59. $\beta^{(4)^T}((B^{(2)}e)(B^{(2)}(B^{(1)}e)^2)) = 0$

are fulfilled and if $c^{(i)} = A^{(i)}e$ *for* $i = 0, 1, 2$, *then the stochastic Runge–Kutta method* (29) *attains order* 2 *for the weak approximation of the solution of the Itô SDE* (1).

Proof. We only give a sketch of the proof and refer to Rößler (2009) for the detailed proof. Calculating the order conditions by Theorem 2, it turns out that there are some trees which restrict the class of efficient SRK methods significantly and which give a deep insight to the necessary structure of such methods. Therefore, we concentrate our investigation to the trees

$$\mathbf{t}_{2,12} = [\tau_{j_1}, \tau_{j_2}, [\tau_{j_4}]_{j_3}]_\gamma, \qquad \mathbf{t}_{2,15} = [[\tau_{j_2}]_{j_1}, [\tau_{j_4}]_{j_3}]_\gamma, \tag{30}$$

with some $j_1, j_2, j_3, j_4 \in \{1, \ldots, m\}$. Then, we have $l(\mathbf{t}_{2,12}) = l(\mathbf{t}_{2,15}) = 5$, $\rho(\mathbf{t}_{2,12}) = \rho(\mathbf{t}_{2,15}) = 2$ and $s(\mathbf{t}_{2,12}) = s(\mathbf{t}_{2,15}) = 4$. Now, for the SRK method (29) we choose $\mathcal{M} = \{(0), (\nu), (\nu, 0), (\nu, 1) : 1 \le \nu \le m\}$ and

$$z_i^{(0),(0)} = \alpha_i \, h_n \,, \qquad z_i^{(k),(k,0)} = \beta_i^{(1)} \, \hat{I}_{(k)} + \beta_i^{(2)} \, \frac{\hat{I}_{(k,k)}}{\sqrt{h_n}} \,,$$

$$z_i^{(k),(k,1)} = \beta_i^{(3)} \, \hat{I}_{(k)} + \beta_i^{(4)} \, \sqrt{h_n} \,,$$

$$Z_{ij}^{(0),(0),(0)} = A_{ij}^{(0)} \, h_n \,, \qquad Z_{ij}^{(k,0),(0),(0)} = A_{ij}^{(1)} \, h_n \,, \qquad Z_{ij}^{(k,1),(0),(0)} = A_{ij}^{(2)} \, h_n \,,$$

$$Z_{ij}^{(0),(k),(k,0)} = B_{ij}^{(0)} \, \hat{I}_{(k)} \,, \qquad Z_{ij}^{(k,0),(k),(k,0)} = B_{ij}^{(1)} \sqrt{h_n},$$

$$Z_{ij}^{(k,1),(l),(l,0)} = B_{ij}^{(2)} \, \frac{\hat{I}_{(k,l)}}{\sqrt{h_n}} \,,$$

for $1 \le k, l \le m$ with $k \ne l$ and with $H_i^{(k,0)} = H_i^{(k)}$ and $H_i^{(k,1)} = \hat{H}_i^{(k)}$ for $1 \le i, j \le s$. Thus, the class of SRK methods is covered by the general class (4). Then, the coefficient function (12) yields

$$\begin{aligned}
\Phi_S(\mathbf{t}_{2,12}) = {} & (z^{(j_1),(j_1,0)^T} e + z^{(j_1),(j_1,1)^T} e)(z^{(j_2),(j_2,0)^T} e + z^{(j_2),(j_2,1)^T} e) \\
& \times (z^{(j_3),(j_3,0)^T} Z^{(j_3,0),(j_4),(j_4,0)} e + z^{(j_3),(j_3,1)^T} Z^{(j_3,1),(j_4),(j_4,0)} e) \,, \\
\Phi_S(\mathbf{t}_{2,15}) = {} & (z^{(j_1),(j_1,0)^T} Z^{(j_1,0),(j_2),(j_2,0)} e + z^{(j_1),(j_1,1)^T} Z^{(j_1,1),(j_2),(j_2,0)} e) \\
& \times (z^{(j_3),(j_3,0)^T} Z^{(j_3,0),(j_4),(j_4,0)} e + z^{(j_3),(j_3,1)^T} Z^{(j_3,1),(j_4),(j_4,0)} e) \,,
\end{aligned} \tag{31}$$

for $j_1, j_2, j_3, j_4 \in \{1, \ldots, m\}$. Further, the multiple stochastic integrals are

$$\begin{aligned}
I_{\mathbf{t}_{2,12};t,t+h} = {} & I_{(j_4,j_3,j_2,j_1);t,t+h} + I_{(j_4,j_3,j_1,j_2);t,t+h} \\
& + I_{(j_1,j_4,j_3,j_2);t,t+h} + I_{(j_4,j_1,j_3,j_2);t,t+h}
\end{aligned}$$

$$
\begin{aligned}
&+ I_{(0,j_3,j_2);t,t+h}[\mathbf{1}_{\{j_1=j_4\}}] + I_{(j_4,0,j_2);t,t+h}[\mathbf{1}_{\{j_1=j_3\}}] \\
&+ I_{(j_4,j_3,0);t,t+h}[\mathbf{1}_{\{j_1=j_2\}}] \\
&+ I_{(j_4,j_2,j_3,j_1);t,t+h} + I_{(j_4,j_2,j_1,j_3);t,t+h} + I_{(j_1,j_4,j_2,j_3);t,t+h} \\
&+ I_{(j_4,j_1,j_2,j_3);t,t+h} \\
&+ I_{(0,j_2,j_3);t,t+h}[\mathbf{1}_{\{j_1=j_4\}}] + I_{(j_4,0,j_3);t,t+h}[\mathbf{1}_{\{j_1=j_2\}}] \\
&+ I_{(j_4,j_2,0);t,t+h}[\mathbf{1}_{\{j_1=j_3\}}] \\
&+ I_{(j_2,j_4,j_3,j_1);t,t+h} + I_{(j_2,j_4,j_1,j_3);t,t+h} + I_{(j_1,j_2,j_4,j_3);t,t+h} \\
&+ I_{(j_2,j_1,j_4,j_3);t,t+h} \\
&+ I_{(0,j_4,j_3);t,t+h}[\mathbf{1}_{\{j_1=j_2\}}] + I_{(j_2,0,j_3);t,t+h}[\mathbf{1}_{\{j_1=j_4\}}] \\
&+ I_{(j_2,j_4,0);t,t+h}[\mathbf{1}_{\{j_1=j_3\}}] \\
&+ I_{(0,j_3,j_1);t,t+h}[\mathbf{1}_{\{j_2=j_4\}}] + I_{(j_1,0,j_3);t,t+h}[\mathbf{1}_{\{j_2=j_4\}}] \\
&+ I_{(0,j_1,j_3);t,t+h}[\mathbf{1}_{\{j_2=j_4\}}] \\
&+ I_{(j_4,0,j_1);t,t+h}[\mathbf{1}_{\{j_2=j_3\}}] + I_{(j_1,j_4,0);t,t+h}[\mathbf{1}_{\{j_2=j_3\}}] \\
&+ I_{(j_4,j_1,0);t,t+h}[\mathbf{1}_{\{j_2=j_3\}}] \\
&+ I_{(0,0);t,t+h}[\mathbf{1}_{\{j_2=j_4\}}\mathbf{1}_{\{j_1=j_3\}}] + I_{(0,0);t,t+h}[\mathbf{1}_{\{j_2=j_3\}}\mathbf{1}_{\{j_1=j_4\}}]
\end{aligned}
$$

and

$$
\begin{aligned}
I_{\mathbf{t}_{2,15};t,t+h} = {} & I_{(j_4,j_3,j_2,j_1);t,t+h} + I_{(j_4,j_2,j_3,j_1);t,t+h} + I_{(j_2,j_4,j_3,j_1);t,t+h} \\
&+ I_{(j_2,j_1,j_4,j_3);t,t+h} \\
&+ I_{(0,j_3,j_1);t,t+h}[\mathbf{1}_{\{j_2=j_4\}}] + I_{(j_4,0,j_1);t,t+h}[\mathbf{1}_{\{j_2=j_3\}}] \\
&+ I_{(j_2,0,j_3);t,t+h}[\mathbf{1}_{\{j_1=j_4\}}] \\
&+ I_{(j_2,j_4,j_1,j_3);t,t+h} + I_{(j_4,j_2,j_1,j_3);t,t+h} + I_{(0,0);t,t+h}[\mathbf{1}_{\{j_1=j_3\}}\mathbf{1}_{\{j_2=j_4\}}] \\
&+ I_{(j_4,j_2,0);t,t+h}[\mathbf{1}_{\{j_1=j_3\}}] + I_{(j_2,j_4,0);t,t+h}[\mathbf{1}_{\{j_1=j_3\}}] \\
&+ I_{(0,j_1,j_3);t,t+h}[\mathbf{1}_{\{j_2=j_4\}}].
\end{aligned}
$$

If we apply Theorem 2 to $\mathbf{t}_{2,12}$ and $\mathbf{t}_{2,15}$, then we have to consider the cases $j_k = j_l$ and $j_k \neq j_l$ for $1 \leq k < l \leq 4$. In the case of $j_1 = j_2 = j_3 = j_4$ we obtain $\sigma(\mathbf{t}_{2,12}) = 2$ and $\mathrm{E}(I_{\mathbf{t}_{2,12};t,t+h}) = h^2$. The order condition (18) yields that $\mathrm{E}(\Phi_S(\mathbf{t}_{2,12}; t, t+h)) = h^2$ has to be fulfilled. Applying (31) and taking into account the order conditions ${\beta^{(4)}}^T e = 0$ and ${\beta^{(2)}}^T e = 0$ due to the trees $\mathbf{t}_{0.5,1} = [\tau_{j_1}]_\gamma$ and $\mathbf{t}_{1.5,4} = [\tau_{j_1}, \tau_{j_2}, \tau_{j_3}]_\gamma$ (see Rößler 2009 for details) yields

$$
\begin{aligned}
\mathrm{E}(\Phi_S(\mathbf{t}_{2,12})) = \mathrm{E}\Bigg(& \Bigg(\Bigg({\beta^{(1)}}^T e\, \hat{I}_{(j_1)} + {\beta^{(2)}}^T e\, \frac{\hat{I}_{(j_1,j_1)}}{\sqrt{h}} \Bigg) \\
& + ({\beta^{(3)}}^T e\, \hat{I}_{(j_1)} + {\beta^{(4)}}^T e \sqrt{h}) \Bigg)^2 \\
& \times \Bigg({\beta^{(1)}}^T B^{(1)} e\, \hat{I}_{(j_1)} \sqrt{h} + {\beta^{(2)}}^T B^{(1)} e\, \frac{\hat{I}_{(j_1,j_1)}}{\sqrt{h}} \sqrt{h} \Bigg) \Bigg)
\end{aligned}
$$

$$= ({\beta^{(1)}}^T e + {\beta^{(3)}}^T e)^2 ({\beta^{(2)}}^T B^{(1)} e) \, \mathrm{E}(\hat{I}_{(j_1)}^2 \hat{I}_{(j_1,j_1)}) \, .$$

Due to $\mathrm{E}(\hat{I}_{(j_1)}^2 \hat{I}_{(j_1,j_1)}) = h^2$, the order condition is fulfilled if for the coefficients holds $({\beta^{(1)}}^T e + {\beta^{(3)}}^T e)^2 ({\beta^{(2)}}^T B^{(1)} e) = 1$. In the case of $j_1 = j_3 \neq j_2 = j_4$ we calculate with $\sigma(\mathbf{t}_{2,12}) = 2$ and $\mathrm{E}(I_{\mathbf{t}_{2,12};t,t+h}) = \frac{1}{2}h^2$ from (18) the order condition $\mathrm{E}(\Phi_S(\mathbf{t}_{2,12}; t, t+h)) = \frac{1}{2}h^2$. Then, we obtain for the SRK method (29)

$$\begin{aligned}
\mathrm{E}(\Phi_S(\mathbf{t}_{2,12})) = \mathrm{E}\Bigg(& \left(\left({\beta^{(1)}}^T e \, \hat{I}_{(j_1)} + {\beta^{(2)}}^T e \, \frac{\hat{I}_{(j_1,j_1)}}{\sqrt{h}} \right) + ({\beta^{(3)}}^T e \, \hat{I}_{(j_1)} + {\beta^{(4)}}^T e \sqrt{h}) \right) \\
& \times \left(\left({\beta^{(1)}}^T e \, \hat{I}_{(j_2)} + {\beta^{(2)}}^T e \, \frac{\hat{I}_{(j_2,j_2)}}{\sqrt{h}} \right) + ({\beta^{(3)}}^T e \, \hat{I}_{(j_2)} + {\beta^{(4)}}^T e \sqrt{h}) \right) \\
& \times \left({\beta^{(3)}}^T B^{(2)} e \, \hat{I}_{(j_1)} \frac{\hat{I}_{(j_1,j_2)}}{\sqrt{h}} + {\beta^{(4)}}^T B^{(2)} e \sqrt{h} \frac{\hat{I}_{(j_1,j_2)}}{\sqrt{h}} \right) \Bigg) \\
= & ({\beta^{(1)}}^T e + {\beta^{(3)}}^T e)^2 ({\beta^{(4)}}^T B^{(2)} e) \, \mathrm{E}(\hat{I}_{(j_1)} \hat{I}_{(j_2)} \hat{I}_{(j_1,j_2)}) \, .
\end{aligned}$$

Now, we can calculate that $\mathrm{E}(\hat{I}_{(j_1)} \hat{I}_{(j_2)} \hat{I}_{(j_1,j_2)}) = \frac{1}{2}h^2$. Thus, the order condition is fulfilled if $({\beta^{(1)}}^T e + {\beta^{(3)}}^T e)^2 ({\beta^{(4)}}^T B^{(2)} e) = 1$.

For $\mathbf{t}_{2,15}$, we calculate in the case of $j_1 = j_2 = j_3 = j_4$ with $\sigma(\mathbf{t}_{2,15}) = 2$ and $\mathrm{E}(I_{\mathbf{t}_{2,15};t,t+h}) = \frac{1}{2}h^2$ from (18) the order condition $\mathrm{E}(\Phi_S(\mathbf{t}_{2,15}; t, t+h)) = \frac{1}{2}h^2$. Again, applying (31) results in

$$\begin{aligned}
\mathrm{E}(\Phi_S(\mathbf{t}_{2,15})) &= \mathrm{E}\left(\left({\beta^{(1)}}^T B^{(1)} e \, \hat{I}_{(j_1)} \sqrt{h} + {\beta^{(2)}}^T B^{(1)} e \, \frac{\hat{I}_{(j_1,j_1)}}{\sqrt{h}} \sqrt{h} \right)^2 \right) \\
&= ({\beta^{(1)}}^T B^{(1)} e)^2 \, \mathrm{E}(\hat{I}_{(j_1)}^2) h + ({\beta^{(2)}}^T B^{(1)} e)^2 \, \mathrm{E}(\hat{I}_{(j_1,j_1)}^2) \, .
\end{aligned}$$

Now, due to $\mathrm{E}(\hat{I}_{(j_1)}^2) = h$ and $\mathrm{E}(\hat{I}_{(j_1,j_1)}^2) = \frac{1}{2}h^2$ the order condition is $({\beta^{(1)}}^T B^{(1)} e)^2 + \frac{1}{2}({\beta^{(2)}}^T B^{(1)} e)^2 = \frac{1}{2}$. On the other hand, in the case of $j_1 = j_3 \neq j_2 = j_4$ with $\sigma(\mathbf{t}_{2,15}) = 2$ and $\mathrm{E}(I_{\mathbf{t}_{2,15};t,t+h}) = \frac{1}{2}h^2$, we get from (18) that $\mathrm{E}(\Phi_S(\mathbf{t}_{2,15}; t, t+h)) = \frac{1}{2}h^2$ has to be fulfilled. Now, we obtain with (31) that

$$\begin{aligned}
\mathrm{E}(\Phi_S(\mathbf{t}_{2,15})) &= \mathrm{E}\left(\left({\beta^{(3)}}^T B^{(2)} e \, \hat{I}_{(j_1)} \frac{\hat{I}_{(j_1,j_2)}}{\sqrt{h}} + {\beta^{(4)}}^T B^{(2)} e \sqrt{h} \frac{\hat{I}_{(j_1,j_2)}}{\sqrt{h}} \right)^2 \right) \\
&= ({\beta^{(3)}}^T B^{(2)} e)^2 \, \mathrm{E}(\hat{I}_{(j_1)}^2 \hat{I}_{(j_1,j_2)}^2) h^{-1} + ({\beta^{(4)}}^T B^{(2)} e)^2 \, \mathrm{E}(\hat{I}_{(j_1,j_2)}^2) \, .
\end{aligned}$$

Table 5 Weak SRK scheme RI5 of order $p_D = 3$ and $p_S = 2$ and RI6 of order $p_D = p_S = 2$

0									
1	1			$\frac{1}{3}$					
$\frac{5}{12}$	$\frac{25}{144}$	$\frac{35}{144}$		$-\frac{5}{6}$	0				
0									
$\frac{1}{4}$	$\frac{1}{4}$			$\frac{1}{2}$					
$\frac{1}{4}$	$\frac{1}{4}$	0		$-\frac{1}{2}$	0				
0									
0	0			1					
0	0	0		-1	0				
	$\frac{1}{10}$	$\frac{3}{14}$	$\frac{24}{35}$	1	-1	-1	0	1	-1
				$\frac{1}{2}$	$-\frac{1}{4}$	$-\frac{1}{4}$	0	$\frac{1}{2}$	$-\frac{1}{2}$

0									
1	1			1					
0	0	0		0	0				
0									
1	1			1					
1	1	0		-1	0				
0									
0	0			1					
0	0	0		-1	0				
	$\frac{1}{2}$	$\frac{1}{2}$	0	$\frac{1}{2}$	$\frac{1}{4}$	$\frac{1}{4}$	0	$\frac{1}{2}$	$-\frac{1}{2}$
				$-\frac{1}{2}$	$\frac{1}{4}$	$\frac{1}{4}$	0	$\frac{1}{2}$	$-\frac{1}{2}$

Due to $\mathrm{E}(\hat{I}^2_{(j_1)} \hat{I}^2_{(j_1,j_2)}) = h^3$ and $\mathrm{E}(\hat{I}^2_{(j_1,j_2)}) = \frac{1}{2}h^2$, we finally get the order condition $(\beta^{(3)^T} B^{(2)} e)^2 + \frac{1}{2}(\beta^{(4)^T} B^{(2)} e)^2 = \frac{1}{2}$.

For all remaining cases of type $j_k = j_l$ or $j_k \neq j_l$ for $1 \leq k < l \leq 4$, we have $\mathrm{E}(I_{\mathbf{t}_{2,12};t,t+h}) = \mathrm{E}(I_{\mathbf{t}_{2,15};t,t+h}) = 0$ and we also calculate that $\mathrm{E}(\Phi_S(\mathbf{t}_{2,12}; t, t+h)) = \mathrm{E}(\Phi_S(\mathbf{t}_{2,15}; t, t+h)) = 0$. Therefore, (18) is fulfilled in these cases without any additional restrictions for the coefficients. Applying the rooted tree analysis and Theorem 2 to all remaining rooted trees up to order 2.5, we can calculate the complete order two conditions for the SRK method (29), see Rößler (2009). □

Remark 2. In the case of $m = 1$ and if we choose $A^{(2)}_{ij} = 0$ for $1 \leq i, j \leq s$ then the 59 conditions of Theorem 5 reduce to 28 conditions (see also Rößler 2006b). For an explicit SRK method of type (29) $s \geq 3$ is needed due to conditions 4., 6. and 17. Further, in the case of commutative noise significantly simplified SRK methods have been developed in Rößler (2004b).

For example, the well known Euler-Maruyama scheme EM belongs to the introduced class of SRK methods having weak order 1 with $s = 1$ stage and with coefficients $\alpha_1 = \beta^{(1)}_1 = 1$, $\beta^{(2)}_1 = \beta^{(3)}_1 = \beta^{(4)}_1 = 0$, $A^{(0)}_{11} = A^{(1)}_{11} = 0$ and $B^{(0)}_{11} = B^{(1)}_{11} = 0$. We refer to Debrabant and Rößler (2009b) for a detailed analysis of the solution space of the order conditions in Theorem 5 and for some coefficients which minimize the error constants of the SRK method (29). The SRK scheme RI5 presented on the left hand side of Table 5, is of order $p_S = 2$ and $p_D = 3$, while the SRK scheme RI6 on the right hand side of Table 5 is of order $p_D = p_S = 2$. Considering the computational complexity of the efficient SRK schemes RI5 and RI6, we take again the number of evaluations of the drift and diffusion functions and the number of random numbers needed in each step as a measure for the complexity of the schemes. Then, the SRK scheme RI5 needs 3 evaluations of the drift a while the SRK scheme RI6 needs 2 evaluations of a. Furthermore, we have to point out that only 5 evaluations of each diffusion function b^k for $k = 1, \ldots, m$ are needed by both SRK schemes RI5 and RI6. This is due to the fact that the number of stages $s = 3$ does not depend on the dimension m of the driving Wiener process and because of $H_1^{(k)} = \hat{H}_1^{(k)}$, which saves one evaluation of each b^k in the case of explicit SRK schemes. As a further feature, only $2m-1$ independent random numbers have to be generated for the new SRK schemes in each step. Thus, the scheme RI6 has computational complexity $2d + 5md + 2m - 1$ while e.g. the order 2.0 weak SRK method PL1WM due to Platen (see Kloeden and Platen 1999 or Tocino and Vigo-Aguiar 2002) has computational complexity $2d + (2m^2 + m)d + m + \frac{1}{2}m(m-1)$ which grows quadratically with the dimension m of the Wiener process. Thus, this is a significant reduction of complexity for the new SRK method (29) compared to well known SRK methods.

7 Numerical Results

We compare the schemes RI5 and RI6 with the order one Euler-Maruyama scheme EM, with the order 2.0 weak SRK scheme PL1WM due to Platen (see Kloeden and Platen 1999 or Tocino and Vigo-Aguiar 2002) and with the extrapolated Euler-Maruyama scheme ExEu due to Talay and Tubaro (1990) attaining order two. In the following, we approximate $\mathrm{E}(f(X_T))$ for $f(x^1, \ldots, x^d) = x^1$ by Monte Carlo simulation. Therefore, we estimate $\mathrm{E}(f(Y_T))$ by the sample average of M independently simulated realizations of the approximations $f(Y_{T,k})$, $k = 1, \ldots, M$, with $Y_{T,k}$ calculated by the scheme under consideration. The obtained errors at time $T = 1.0$ are plotted versus the corresponding step sizes or the corresponding computational effort with double logarithmic scale in order to analyze the empirical order of convergence and the performance of the schemes, respectively.

The first test equation is a non-linear SDE system for $d = m = 2$ with non-commutative noise given by

$$\mathrm{d}\begin{pmatrix} X_t^1 \\ X_t^2 \end{pmatrix} = \begin{pmatrix} -\frac{1}{2}X_t^1 + \frac{3}{2}X_t^2 \\ \frac{3}{2}X_t^1 - \frac{1}{2}X_t^2 \end{pmatrix} \mathrm{d}t + \begin{pmatrix} \sqrt{\frac{3}{4}(X_t^1)^2 - \frac{3}{2}X_t^1X_t^2 + \frac{3}{4}(X_t^2)^2 + \frac{3}{20}} \\ 0 \end{pmatrix} \mathrm{d}W_t^1 + \begin{pmatrix} -\sqrt{\frac{1}{4}(X_t^1)^2 - \frac{1}{2}X_t^1X_t^2 + \frac{1}{4}(X_t^2)^2 + \frac{1}{20}} \\ \sqrt{(X_t^1)^2 - 2X_t^1X_t^2 + X_t^2 + \frac{1}{5}} \end{pmatrix} \mathrm{d}W_t^2, \quad (32)$$

with initial value $X_0 = (\frac{1}{10}, \frac{1}{10})^T$. Then, we calculate the first moments as $\mathrm{E}(X_t^i) = \frac{1}{10}\exp(t)$ for $i = 1, 2$. Here, we choose $M = 10^9$ and the corresponding results are presented in Fig. 4.

Next, we consider a non-linear SDE with non-commutative noise and some higher dimension $d = 4$ which is given for $\lambda, \mu \in \{0, 1\}$ as

$$
\begin{aligned}
\mathrm{d}\begin{pmatrix} X_t^1 \\ X_t^2 \\ X_t^3 \\ X_t^4 \end{pmatrix} &= \begin{pmatrix} \frac{243}{154}X_t^1 - \frac{27}{77}X_t^2 + \frac{23}{154}X_t^3 - \frac{65}{154}X_t^4 \\ \frac{27}{77}X_t^1 - \frac{243}{154}X_t^2 + \frac{65}{154}X_t^3 - \frac{23}{154}X_t^4 \\ \frac{5}{154}X_t^1 - \frac{61}{154}X_t^2 + \frac{162}{77}X_t^3 - \frac{36}{77}X_t^4 \\ \frac{61}{154}X_t^1 - \frac{5}{154}X_t^2 + \frac{36}{77}X_t^3 - \frac{162}{77}X_t^4 \end{pmatrix} \mathrm{d}t \\
&\quad + \frac{1}{9}\sqrt{(X_t^2)^2 + (X_t^3)^2 + \frac{2}{23}} \begin{pmatrix} \frac{1}{13} \\ \frac{1}{14} \\ \frac{1}{13} \\ \frac{1}{15} \end{pmatrix} \mathrm{d}W_t^1 \\
&\quad + \frac{1}{8}\sqrt{(X_t^4)^2 + (X_t^1)^2 + \frac{1}{11}} \begin{pmatrix} \frac{1}{14} \\ \frac{1}{16} \\ \frac{1}{16} \\ \frac{1}{12} \end{pmatrix} \mathrm{d}W_t^2 \\
&\quad + \frac{\lambda}{12}\sqrt{(X_t^1)^2 + (X_t^2)^2 + \frac{1}{9}} \begin{pmatrix} \frac{1}{6} \\ \frac{1}{5} \\ \frac{1}{5} \\ \frac{1}{6} \end{pmatrix} \mathrm{d}W_t^3 \\
&\quad + \frac{\lambda}{14}\sqrt{(X_t^3)^2 + (X_t^4)^2 + \frac{3}{29}} \begin{pmatrix} \frac{1}{8} \\ \frac{1}{9} \\ \frac{1}{8} \\ \frac{1}{9} \end{pmatrix} \mathrm{d}W_t^4 \\
&\quad + \frac{\mu}{10}\sqrt{(X_t^1)^2 + (X_t^3)^2 + \frac{1}{13}} \begin{pmatrix} \frac{1}{11} \\ \frac{1}{15} \\ \frac{1}{13} \\ \frac{1}{11} \end{pmatrix} \mathrm{d}W_t^5 \\
&\quad + \frac{\mu}{11}\sqrt{(X_t^2)^2 + (X_t^4)^2 + \frac{2}{25}} \begin{pmatrix} \frac{1}{12} \\ \frac{1}{13} \\ \frac{1}{16} \\ \frac{1}{13} \end{pmatrix} \mathrm{d}W_t^6
\end{aligned} \tag{33}
$$

with initial value $X_0 = (\frac{1}{8}, \frac{1}{8}, 1, \frac{1}{8})^T$. Then, we have $m = 2 + 2\lambda + 2\mu$ as the dimension of the driving Wiener process. The moments of the solution can be calculated as $\mathrm{E}(X_T^i) = \frac{1}{8}\exp(2T)$ for $i = 1, 2, 4$ and $\mathrm{E}(X_T^3) = \exp(2T)$. We compare the performance of the considered schemes for the cases $m = 2$ with $\lambda = \mu = 0$, for $m = 4$ with $\lambda = 1$ and $\mu = 0$, and for $m = 6$ if $\lambda = \mu = 1$. Here, $M = 10^8$

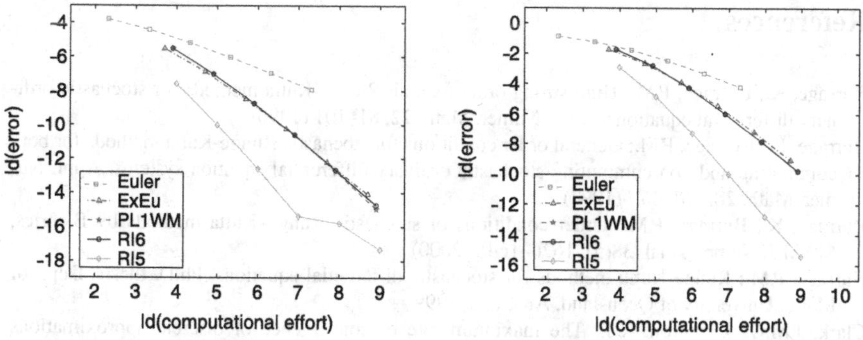

Fig. 4 Computational effort vs. error for the approximation of $\mathrm{E}(X_T^1)$ for SDE (32) in the left and for SDE (33) for $\lambda = \mu = 0$ with $m = 2$ in the right figure

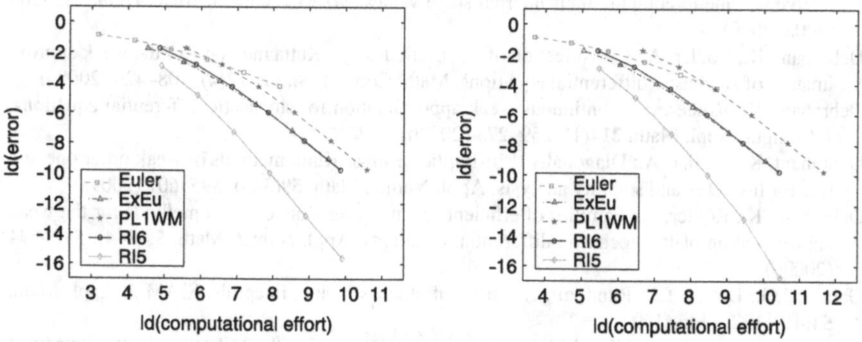

Fig. 5 Computational effort vs. error for the approximation of $\mathrm{E}(X_T^1)$ for SDE (33) for $\lambda = 1$, $\mu = 0$ with $m = 4$ in the left and for $\lambda = \mu = 1$ with $m = 6$ in the right figure

independent trajectories are simulated and the results are presented in Figs. 4–5. On the right hand side in Fig. 4 and in Fig. 5, we can see the performance of the considered schemes as the dimension m increases from 2 to 6. Comparing these results, we can see the significantly reduced complexity for the new SRK schemes RI5 and RI6 compared to the well known SRK scheme PL1WM in the case of $m > 2$. This benefit becomes more and more significant if we increase the dimension m of the driving Wiener process, which confirms our theoretical results. For the considered examples, we obtained very good results especially for the SRK scheme RI5 having order $p_D = 3$ and $p_S = 2$.

Further references relevant to the material in this paper include Burrage and Burrage (1998), Burrage (1999), Clark and Cameron (1980), Debrabant and Kværnø (2008–2009), Debrabant and Rößler (2008a, 2008b, 2009a, 2009b), Gaines and Lyons (1994), Giles (2008), Gilsing and Shardlow (2007), Hairer et al. (1993), Kloeden et al. (1995), Kloeden et al. (1992), Komori (2007), Komori et al. (1997), Küpper et al. (2007), Lehn et al. (2002), Moon et al. (2005), Newton (1991), Rößler (2004a), Rümelin (1982) and Talay (1990).

References

Burrage, K., Burrage, P.M.: High strong order explicit Runge–Kutta methods for stochastic ordinary differential equations. Appl. Numer. Math. **22**, 81–101 (1996)

Burrage, K., Burrage, P.M.: General order conditions for stochastic Runge-Kutta methods for both commuting and non-commuting stochastic ordinary differential equation systems. Appl. Numer. Math. **28**, 161–177 (1998)

Burrage, K., Burrage, P.M.: Order conditions of stochastic Runge–Kutta methods by B-series. SIAM J. Numer. Anal. **38**(5), 1626–1646 (2000)

Burrage, P.M.: Runge-Kutta methods for stochastic differential equations. Ph.D. thesis, Dept. of Math., University of Qeensland, Australia, 1999

Clark, J.M.C., Cameron, R.J.: The maximum rate of convergence of discrete approximations for stochastic differential equations. In: Lecture Notes in Control and Information Sciences, vol. 25, pp. 162–171. Springer, Berlin (1980)

Debrabant, K., Kværnø, A.: B-series analysis of stochastic Runge-Kutta methods that use an iterative scheme to compute their internal stage values. SIAM J. Numer. Anal. **47**(1), 181–203 (2008–2009)

Debrabant, K., Rößler, A.: Classification of stochastic Runge–Kutta methods for the weak approximation of stochastic differential equations. Math. Comput. Simul. **77**(4), 408–420 (2008a)

Debrabant, K., Rößler, A.: Continuous weak approximation for stochastic differential equations. J. Comput. Appl. Math. **214**(1), 259–273 (2008b)

Debrabant, K., Rößler, A.: Diagonally drift-implicit Runge-Kutta methods of weak order one and two for Itô SDEs and stability analysis. Appl. Numer. Math. **59**(3–4), 595–607 (2009a)

Debrabant, K., Rößler, A.: Families of efficient second order Runge–Kutta methods for the weak approximation of Itô stochastic differential equations. Appl. Numer. Math. **59**(3–4), 582–594 (2009b)

Gaines, J.G., Lyons, T.J.: Random generation of stochastic area integrals. SIAM J. Appl. Math. **54**(4), 1132–1146 (1994)

Giles, M.B.: Improved multilevel Monte Carlo convergence using the Milstein scheme. In: Monte Carlo and Quasi-Monte Carlo Methods 2006, pp. 343–358. Springer, Berlin (2008)

Gilsing, H., Shardlow, T.: SDELab: a package for solving stochastic differential equations in MATLAB. J. Comput. Appl. Math. **205**(2), 1002–1018 (2007)

Hairer, E., Nørsett, S.P., Wanner, G.: Solving Ordinary Differential Equations I. Springer, Berlin (1993)

Kaneko, J.: Explicit order 1.5 Runge-Kutta scheme for solutions of Itô stochastic differential equations. Various problems in stochastic numerical analysis II, Kyoto, Suri-Kaiseki-Kenkyusho-Kokyuroku, No. 932, 46-60 (1995)

Kloeden, P.E., Platen, E.: Numerical Solution of Stochastic Differential Equations. Springer, Berlin (1999)

Kloeden, P.E., Platen, E., Hofmann, N.: Extrapolation methods for the weak approximation of Ito diffusions. SIAM J. Numer. Anal. **32**(5), 1519–1534 (1995)

Kloeden, P.E., Platen, E., Wright, I.W.: The approximation of multiple stochastic integrals. Stoch. Anal. Appl. **10**(4), 431–441 (1992)

Komori, Y.: Weak second-order stochastic Runge-Kutta methods for non-commutative stochastic differential equations. J. Comput. Appl. Math. **206**(1), 158–173 (2007)

Komori, Y., Mitsui, T., Sugiura, H.: Rooted tree analysis of the order conditions of ROW-type scheme for stochastic differential equations. BIT **37**(1), 43–66 (1997)

Küpper, D., Lehn, J., Rößler, A.: A step size control algorithm for the weak approximation of stochastic differential equations. Numer. Algorithms **44**(4), 335–346 (2007)

Lehn, J., Rößler, A., Schein, O.: Adaptive schemes for the numerical solution of SDEs—a comparison. J. Comput. Appl. Math. **138**(2), 297–308 (2002)

Milstein, G.N.: Numerical Integration of Stochastic Differential Equations. Kluwer Academic, Dordrecht (1995)

Milstein, G.N., Tretyakov, M.V.: Stochastic Numerics for Mathematical Physics. Springer, Berlin (2004)

Moon, K.-S., Szepessy, A., Tempone, R., Zouraris, G.E.: Convergence rates for adaptive weak approximation of stochastic differential equations. Stoch. Anal. Appl. **23**, 511–558 (2005)

Newton, N.J.: Asymptotically efficient Runge–Kutta methods for a class of Itô and Stratonovich equations. SIAM J. Appl. Math. **51**(2), 542–567 (1991)

Rößler, A.: Runge–Kutta methods for Stratonovich stochastic differential equation systems with commutative noise. J. Comput. Appl. Math. **164–165**, 613–627 (2004a)

Rößler, A.: Stochastic Taylor expansions for the expectation of functionals of diffusion processes. Stoch. Anal. Appl. **22**(6), 1553–1576 (2004b)

Rößler, A.: Runge-Kutta methods for Itô stochastic differential equations with scalar noise. BIT **46**(1), 97–110 (2006a)

Rößler, A.: Rooted tree analysis for order conditions of stochastic Runge–Kutta methods for the weak approximation of stochastic differential equations. Stoch. Anal. Appl. **24**(1), 97–134 (2006b)

Rößler, A.: Second order Runge–Kutta methods for Stratonovich stochastic differential equations. BIT **47**(3), 657–680 (2007)

Rößler, A.: Stochastic Taylor expansions for functionals of diffusion processes. Stoch. Anal. Appl. **28**(3), 415–429 (2010a)

Rößler, A.: Runge–Kutta methods for the strong approximation of solutions of stochastic differential equations. SIAM J. Number. Anal. (2010b, to appear)

Rößler, A.: Second order Runge–Kutta methods for Itô stochastic differential equations. SIAM J. Numer. Anal. **47**(3), 1713–1738 (2009)

Rümelin, W.: Numerical treatment of stochastic differential equations. SIAM J. Numer. Anal. **19**(3), 604–613 (1982)

Talay, D.: Second-order discretization schemes of stochastic differential systems for the computation of the invariant law. Stoch. Stoch. Rep. **29**(1), 13–36 (1990)

Talay, D., Tubaro, L.: Expansion of the global error for numerical schemes solving stochastic differential equations. Stoch. Anal. Appl. **8**(4), 483–509 (1990)

Tocino, A., Vigo-Aguiar, J.: Weak second order conditions for stochastic Runge–Kutta methods. SIAM J. Sci. Comput. **24**(2), 507–523 (2002)

Wiktorsson, M.: Joint characteristic function and simultaneous simulation of iterated Itô integrals for multiple independent Brownian motions. Ann. Appl. Probab. **11**(2), 470–487 (2001)

On Robust Gaussian Graphical Modeling

Daniel Vogel and Roland Fried

Abstract The objective of this exposition is to give an overview of the existing approaches to robust Gaussian graphical modeling. We start by thoroughly introducing Gaussian graphical models (also known as *covariance selection models* or *concentration graph models*) and then review the established, likelihood-based statistical theory (estimation, testing and model selection). Afterwards we describe robust methods and compare them to the classical approaches.

1 Introduction

Graphical modeling is the analysis of conditional associations between random variables by means of graph theoretic methods. The graphical representation of the interrelation of several variables is an attractive data analytical tool. Besides allowing parsimonious modeling of the data it facilitates the understanding and the interpretation of the data generating process. The importance of considering *conditional* rather than marginal associations for assessing the dependence structure of several variables is vividly exemplified by Simpson's paradox, see e.g. Edwards (2000), Chap. 1.4. The statistical literature knows several different types of graphical models, differing in the type of relation coded by an edge, in the type of data and hence in the statistical methodology. In this chapter we deal with undirected graphs only, that is, the type of association we consider is mutual. Precisely, we are going to define partial correlation graphs in Sect. 2.2.

Undirected models are in a sense closer to the data. A directed association suggests a causal relationship. Even though it can often be justified, e.g. by chronol-

Daniel Vogel, Fakultät Statistik, Technische Universität Dortmund, Dortmund, Germany
e-mail: daniel.vogel@tu-dortmund.de

Roland Fried, Fakultät Statistik, Technische Universität Dortmund, Dortmund, Germany
e-mail: fried@statistik.tu-dortmund.de

L. Devroye et al. (eds.), *Recent Developments in Applied Probability and Statistics*,
DOI 10.1007/978-3-7908-2598-5_7, © Springer-Verlag Berlin Heidelberg 2010

ogy or knowledge about the physiological process, the direction of the effect is an additional assumption. Undirected models constitute the simplest case, the understanding of which is crucial for the study of directed models and models with both, directed and undirected edges.

Furthermore we restrict our attention to continuous data, which are assumed to stem from a multivariate Gaussian distribution. Conditional independence in the normal model is nicely expressed through its second order characteristics, cf. Sect. 2.3. This fact, along with its general predominant role in multivariate statistics (largely due to the Central limit theorem justification), is the reason for the almost exclusive use of the multivariate normal distribution in graphical models for continuous data.

With rapidly increasing data sizes, and on the other hand computer hardware available to process them, the need for robust methods becomes more and more important. The sample covariance matrix possesses good statistical properties in the normal model and is very fast to compute, but highly non-robust, cf. Sect. 4.1. We are going to survey robust alternatives to the classical Gaussian graphical modeling, which is based on the sample covariance matrix.

The paper is organized as follows. Section 2 introduces Gaussian graphical models (GGMs). We start by studying partial correlations, a purely moment based relation, without any distributional assumption and then examine the special case of the normal distribution where partial uncorrelatedness coincides with conditional independence. The better transferability of the former concept to more general data situations is the reason for taking this route. Section 3 reviews the classical, non-robust, likelihood-based statistical theory for Gaussian graphical models. Each step is motivated, and important points are emphasized. Sections 2 and 3 thus serve as a self-contained introduction to GGMs. The basis for this first part are the books Whittaker (1990) and Lauritzen (1996). Other standard volumes on graphical models in statistics are Cox and Wermuth (1996) and Edwards (2000), both with a stronger emphasis on applications. Section 4 deals with robust Gaussian graphical modeling. We focus on the use of robust affine equivariant scatter estimators, since the robust estimators proposed for GGMs in the past belong to this class. As an important robustness measure we consider the influence function and give the general form of the influence functions of affine equivariant scatter estimators and derived partial correlation estimators.

We close this section by introducing some of the mathematical notation we are going to use. Bold letters $\mathbf{b}$, $\boldsymbol{\mu}$, etc., denote vectors, capital letters X, Y, etc., indicate (univariate) random variables and bold capital letters $\mathbf{X}$, $\mathbf{Y}$, etc., random vectors. We view vectors, by default, neither as a column nor as a row, but just as an ordered collection of elements of the same type. This makes $(\mathbf{X}, \mathbf{Y})$ again a vector and not a two-column matrix. However, if matrix notation, such as $(\cdot)^T$, is applied to vectors, they are always interpreted as $n \times 1$ matrices.

Matrices are also denoted by non-bold capital letters, and the corresponding small letter is used for an element of the matrix, e.g., the $p \times p$ matrix Σ is the collection of all $\sigma_{i,j}$, $i, j = 1, \ldots, p$. Alternatively, if matrices are denoted by more complicated compound symbols (e.g. if they carry subscripts already) square brackets will be used to refer to individual elements, e.g. $[\hat{\Sigma}_G^{-1}]_{i,j}$. Throughout the paper upright small Greek letters will denote index sets. Subvectors and submatrices are referenced by subscripts, e.g. for $\alpha, \beta \subseteq \{1, \ldots, p\}$ the $|\alpha| \times |\beta|$ matrix $\Sigma_{\alpha,\beta}$ is obtained from Σ by deleting all rows that are not in α and all columns that are not in β. Similarly, the $p \times p$ matrix $[\Sigma_{\alpha,\beta}]^p$ is obtained from Σ by putting all rows not in α and all columns not in β to zero. We want to view this matrix operation as two operations performed sequentially: first $(\cdot)_{\alpha,\beta}$ extracting the submatrix and then $[\cdot]^p$ writing it back on a "blank" matrix at the coordinates specified by α and β. Of course, the latter is not well defined without the former, but this allows us e.g. to write $[(\Sigma_{\alpha,\beta})^{-1}]^p$.

We adopt the general convention that subscripts have stronger ties than superscripts, for instance, we write $\Sigma_{\alpha,\beta}^{-1}$ for $(\Sigma_{\alpha,\beta})^{-1}$. Let $\mathcal{S}_p$ and $\mathcal{S}_p^+$ be the sets of all symmetric, respectively positive definite $p \times p$ matrices, and define for any $A \in \mathcal{S}_p^+$

$$\operatorname{Corr}(A) = A_D^{-\frac{1}{2}} A A_D^{-\frac{1}{2}}, \tag{1}$$

where A_D denotes the diagonal matrix having the same diagonal as A. Recall the important inversion formula for partitioned matrices. Let $r \in \{1, \ldots, p-1\}$, $\alpha = \{1, \ldots, r\}$ and $\beta = \{r+1, \ldots, p\}$. Then

$$\begin{pmatrix} \Sigma_{\alpha,\alpha} & \Sigma_{\alpha,\beta} \\ \Sigma_{\beta,\alpha} & \Sigma_{\beta,\beta} \end{pmatrix}^{-1} = \begin{pmatrix} \Omega^{-1} & -\Omega^{-1}\Sigma_{\alpha,\beta}\Sigma_{\beta,\beta}^{-1} \\ -\Sigma_{\beta,\beta}^{-1}\Sigma_{\beta,\alpha}\Omega^{-1} & \Sigma_{\beta,\beta}^{-1} + \Sigma_{\beta,\beta}^{-1}\Sigma_{\beta,\alpha}\Omega^{-1}\Sigma_{\alpha,\beta}\Sigma_{\beta,\beta}^{-1} \end{pmatrix}, \tag{2}$$

where the $r \times r$ matrix $\Omega = \Sigma_{\alpha,\alpha} - \Sigma_{\alpha,\beta}\Sigma_{\beta,\beta}^{-1}\Sigma_{\beta,\alpha}$ is called the *Schur complement* of $\Sigma_{\beta,\beta}$. The inverse exists if and only if Ω and $\Sigma_{\beta,\beta}$ are both invertible. Note that, by simultaneously re-ordering rows and columns, the formula is valid for any partition $\{\alpha, \beta\}$ of $\{1, \ldots, p\}$.

Finally, the Kronecker product $A \otimes B$ of two matrices $A, B \in \mathbb{R}^{p\times p}$ is defined as the $p^2 \times p^2$ matrix with entry $a_{i,j}b_{k,l}$ at position $(i(p-1)+k, j(p-1)+l)$. Let $\mathbf{e}_1, \ldots, \mathbf{e}_p$ be the unit vectors in $\mathbb{R}^p$ and $\mathbf{1}_p$ the p vector consisting only of ones. Define further the following matrices:

$$J_p = \sum_{i=1}^{p} \mathbf{e}_i\mathbf{e}_i^T \otimes \mathbf{e}_i\mathbf{e}_i^T, \qquad K_p = \sum_{i=1}^{p}\sum_{j=1}^{p} \mathbf{e}_i\mathbf{e}_j^T \otimes \mathbf{e}_j\mathbf{e}_i^T, \qquad M_p = \frac{1}{2}\left(I_{p^2} + K_p\right),$$

where I_{p^2} denotes the $p^2 \times p^2$ identity matrix. K_p is also called the *commutation matrix*. Let vec(A) be the p^2 vector obtained by stacking the columns of $A \in \mathbb{R}^{p\times p}$ from left to right underneath each other. More on these concepts and their properties can be found in Magnus and Neudecker (1999).

2 Partial Correlation Graphs and Properties of the Gaussian Distribution

This section explains the basic concepts of Gaussian graphical models: We define the terms *partial variance* and *partial correlation* (Sect. 2.1), review basic graph theory terms and explain the merit of a *partial correlation graph* (Sect. 2.2). Gaussianity enters in Sect. 2.3, where we deduce the conditional independence interpretation of a partial correlation graph which is valid under normality. Statistics is deferred to Sect. 3.

2.1 Partial Variance

Let $\mathbf{X} = (X_1, \ldots, X_p)$ be a random vector in $\mathbb{R}^p$ with distribution F and positive definite variance matrix $\Sigma = \Sigma_{\mathbf{X}} \in \mathbb{R}^{p\times p}$. The inverse of Σ is called *concentration matrix* (or *precision matrix*) of $\mathbf{X}$ and shall be denoted by K or $K_{\mathbf{X}}$.

Now let $\mathbf{X}$ be partitioned into $\mathbf{X} = (\mathbf{Y}, \mathbf{Z})$, where $\mathbf{Y}$ and $\mathbf{Z}$ are subvectors of lengths q and r, respectively. The corresponding index sets shall be called α and β, i.e. $\alpha = \{1, \ldots, q\}$ and $\beta = \{q+1, \ldots, q+r\}$.

The variance matrix of $\mathbf{Y}$ is $\Sigma_{\mathbf{Y}} = \Sigma_{\alpha,\alpha} \in \mathbb{R}^{q\times q}$ and its concentration matrix $K_{\mathbf{Y}} = \Sigma_{\alpha,\alpha}^{-1} = (K_{\mathbf{X}}^{-1})_{\alpha,\alpha}^{-1}$. The covariance matrix of $\mathbf{Y}$ and $\mathbf{Z}$ is $\Sigma_{\alpha,\beta} \in \mathbb{R}^{q\times r}$. The orthogonal projection of $\mathbf{Y}$ onto the space of all affine linear functions of $\mathbf{Z}$ shall be denoted by $\hat{\mathbf{Y}}(\mathbf{Z})$ and is given by

$$\hat{\mathbf{Y}}(\mathbf{Z}) = \mathbb{E}\mathbf{Y} + \Sigma_{\alpha,\beta}\Sigma_{\beta,\beta}^{-1}(\mathbf{Z} - \mathbb{E}\mathbf{Z}). \tag{3}$$

This is the best linear prediction of $\mathbf{Y}$ from $\mathbf{Z}$, in the sense that the squared prediction error $\mathbb{E}\|\mathbf{Y} - h(\mathbf{Z})\|^2$ is uniquely minimized by $h = \hat{\mathbf{Y}}(\cdot)$ among all (affine) linear functions h. The *partial variance of* $\mathbf{Y}$ *given* $\mathbf{Z}$ is the variance of the residual $\mathbf{Y} - \hat{\mathbf{Y}}(\mathbf{Z})$. It shall be denoted by $\Sigma_{\mathbf{Y}\bullet\mathbf{Z}}$, i.e.

$$\Sigma_{\mathbf{Y}\bullet\mathbf{Z}} = \mathrm{Var}\big(\mathbf{Y} - \hat{\mathbf{Y}}(\mathbf{Z})\big) = \Sigma_{\alpha,\alpha} - \Sigma_{\alpha,\beta}\Sigma_{\beta,\beta}^{-1}\Sigma_{\beta,\alpha}. \tag{4}$$

The notation $\mathbf{Y}\bullet\mathbf{Z}$ is intended to resemble $\mathbf{Y}\,|\,\mathbf{Z}$, that is, we look at $\mathbf{Y}$ in dependence on $\mathbf{Z}$, but instead of conditioning $\mathbf{Y}$ on $\mathbf{Z}$ the type of connection we consider here is a linear regression. In particular, $\Sigma_{\mathbf{Y}\bullet\mathbf{Z}}$ is—contrary to a conditional variance—a fixed parameter and not random.

If $\mathbf{Y}$ is at least two-dimensional, we partition it further into $\mathbf{Y} = (\mathbf{Y}_1, \mathbf{Y}_2)$ with corresponding index sets $\alpha_1 \cup \alpha_2 = \alpha$ and lengths $q_1 + q_2 = q$, and define

$$\Sigma_{\mathbf{Y}_1,\mathbf{Y}_2\bullet\mathbf{Z}} = (\Sigma_{\mathbf{Y}\bullet\mathbf{Z}})_{\alpha_1,\alpha_2} = \Sigma_{\alpha_1,\alpha_2} - \Sigma_{\alpha_1,\beta}\Sigma_{\beta,\beta}^{-1}\Sigma_{\beta,\alpha_2}$$

as the *partial covariance between* $\mathbf{Y}_1$ *and* $\mathbf{Y}_2$ *given* $\mathbf{Z}$. If $\Sigma_{\mathbf{Y}_1,\mathbf{Y}_2\bullet\mathbf{Z}} = \mathbf{0}$, we say $\mathbf{Y}_1$ *and* $\mathbf{Y}_2$ *are partially uncorrelated given* $\mathbf{Z}$ and write

$$\mathbf{Y}_1 \perp \mathbf{Y}_2 \bullet \mathbf{Z}.$$

Furthermore, if $\mathbf{Y}_1 = Y_1$ and $\mathbf{Y}_2 = Y_2$ are both one-dimensional, $\Sigma_{\mathbf{Y} \bullet \mathbf{Z}}$ is a positive definite 2×2 matrix. The correlation coefficient computed from this matrix, i.e. the $(1, 2)$ element of $\mathrm{Corr}(\Sigma_{\mathbf{Y} \bullet \mathbf{Z}})$, cf. (1), is called the *partial correlation (coefficient) of* Y_1 *and* Y_2 *given* $\mathbf{Z}$ and denoted by $\varrho_{Y_1, Y_2 \bullet \mathbf{Z}}$. This is nothing but the correlation between the residuals $Y_1 - \hat{Y}_1(\mathbf{Z})$ and $Y_2 - \hat{Y}_2(\mathbf{Z})$ and may be interpreted as a measure of the linear association between Y_1 and Y_2 after the linear effects of $\mathbf{Z}$ have been removed. For $\alpha_1 = \{i\}$ and $\alpha_2 = \{j\}$, $i \neq j$, we use the simplified notation $\varrho_{i,j\bullet}$ for $\varrho_{X_i, X_j \bullet \mathbf{X}_{\setminus \{i,j\}}}$.

The simple identity (4) is fundamental and the actual starting point for all following considerations. We recognize $\Sigma_{\mathbf{Y} \bullet \mathbf{Z}}$ as the Schur complement of $\Sigma_{\mathbf{Z}}$ in $\Sigma_{\mathbf{X}}$, cf. (2), implying that

$$\Sigma_{\mathbf{Y} \bullet \mathbf{Z}}^{-1} = K_{\alpha, \alpha}. \qquad (5)$$

In words: the concentration matrix of $\mathbf{Y} - \hat{\mathbf{Y}}(\mathbf{Z})$ is the submatrix of $K_{\mathbf{X}}$ corresponding to $\mathbf{Y}$, or—very roughly put—while marginalizing means partitioning the covariance matrix, partializing means partitioning its inverse. This has some immediate implications about the interpretation of K, which explain why K, rather than Σ, is of interest in graphical modeling.

Proposition 1. *The partial correlation* $\varrho_{i,j\bullet}$ *between* X_i *and* X_j, $1 \leq i < j \leq p$, *given all remaining variables* $\mathbf{X}_{\setminus \{i,j\}}$ *is*

$$\varrho_{i,j\bullet} = -\frac{k_{i,j}}{\sqrt{k_{i,i} k_{j,j}}}.$$

Another way of phrasing this assertion is to say, the matrix $P = -\mathrm{Corr}(K)$ contains the partial correlations (of each pair of variables given the respective remainder) as its off-diagonal elements. We call P the *partial correlation matrix of* $\mathbf{X}$. Proposition 1 is a direct consequence of (5) involving the inversion of a 2×2 matrix. For a detailed derivation see Whittaker (1990), Chap. 5.

2.2 Partial Correlation Graph

The partial correlation structure of the random variable $\mathbf{X}$ can be coded in a graph, which originates the term *graphical model*. An undirected graph $G = (V, E)$, where V is the vertex set and E the edge set, is constructed the following way: the variables $X_1, \ldots, X_p$ are the vertices, and an (undirected) edge is drawn between X_i and X_j, $i \neq j$, if and only if $\varrho_{i,j\bullet} \neq 0$. The thus obtained graph G is called the *partial correlation graph (PCG)* of $\mathbf{X}$. Formally we set $V = \{1, \ldots, p\}$ and write the elements of E as unordered pairs $\{i, j\}$, $1 \leq i < j \leq p$. Before we dwell on the benefits of this graphical representation, let us briefly recall some terms from graph theory. We only consider undirected graphs with a single type of nodes.

If $\{a, b\} \in E$, the vertices a and b are called *adjacent* or *neighbors*. The set of neighbors of the vertex $a \in V$ is denoted by ne(a). An alternative notation is bd(a), which stands for *boundary*, but keep in mind that in graphs containing directed edges the set of neighbors and the boundary of a node are generally different.

A *path of length k*, $k \geq 1$, is a sequence $(a_1, \ldots, a_{k+1})$ of distinct vertices such that $\{a_i, a_{i+1}\} \in E$, $i = 1, \ldots, k$. If $k \geq 2$ and additionally $\{a_1, a_{k+1}\} \in E$, then the sequence $(a_1, \ldots, a_{k+1}, a_1)$ is called a *cycle of length* $k + 1$ or a $(k + 1)$*-cycle*. Note that the length, in both cases, refers to the number of edges.

The n-cycle $(a_1, \ldots, a_n, a_1)$ is *chordless*, if no other than successive vertices in the cycle are adjacent, i.e. $\{a_i, a_j\} \in E \Rightarrow |i - j| \in \{1, n - 1\}$. Otherwise the cycle possesses a *chord*. All cycles of length 3 are chordless.

The graph is called *complete*, if it contains all possible edges. Every subset $\alpha \subset V$ induces a *subgraph* $G_\alpha = (\alpha, E_\alpha)$, where E_α contains those edges in E with both endpoints in α, i.e. $E_\alpha = E \cap (\alpha \times \alpha)$. A subset $\alpha \subset V$, for which G_α is complete, but adding another vertex would render it incomplete, is called a *clique*. Thus the cliques identify the maximal complete subgraphs.

The set $\gamma \subset V$ is said to *separate* the sets $\alpha, \beta \subset V$ in G, if α, β, γ are mutually disjoint and every path from a vertex in α to a vertex in β contains a node from γ. The set γ may be empty.

Definition 1. A partition (α, β, γ) of V is a *decomposition* of the graph G, if

(1) α, β are both non-empty,
(2) γ separates α and β,
(3) G_γ is complete.

If such a decomposition exists, G is called *reducible* (otherwise *irreducible*). It can then be *decomposed into* or *reduced to* the components $G_{\alpha\cup\gamma}$ and $G_{\beta\cup\gamma}$.

Our terminology is in concordance with Whittaker (1990), Chap. 12, however, there are different definitions around. For instance, Lauritzen (1996) calls a decomposition in the above sense a "proper weak decomposition". Also be aware that the expression "G is decomposable", which is defined below, denotes something different than "there exists a decomposition of G", for which the term "reducible" is used.

Definition 1 suggests a recursive application of decompositions until ultimately the graph is fully decomposed into irreducible components, which then are viewed as atomic building blocks of the graph. It is not at all obvious, if such atomic components exist or are well defined, since, at least in principle, any sequence of decompositions may lead to different irreducible components, cf. Example 12.3.1 in Whittaker (1990). With an additional constraint, the irreducible components of a given graph are indeed well defined.

Definition 2. The system of subsets $\{\alpha_1, \ldots, \alpha_k\} \subset 2^{|V|}$ is called the (set of) *maximal irreducible components of* G, if

(1) G_{α_i} is irreducible, $i = 1, \ldots, k$,

(2) α_i and α_j are mutually incomparable, i.e. α_i is not a proper subset of α_j and vice versa, $1 \le i < j \le k$, and
(3) $\bigcup_{i=1}^{k} \alpha_i = V$.

The maximal irreducible components of any graph G are unique and can be obtained by first fully decomposing the graph into irreducible components (by any sequence of decompositions) and then deleting those that are a proper subset of another one—the *maximal* irreducible components remain.

Definition 3. The graph G is *decomposable*, if all of its maximal irreducible components are complete.

Decomposability also admits the following recursive definition: G is decomposable, if it is complete or there exists a decomposition (α, β, γ) into decomposable subgraphs $G_{\alpha\cup\gamma}$ and $G_{\beta\cup\gamma}$. Another characterization is to say, a decomposable graph can be decomposed into its cliques. Figure 1 shows two reducible graphs and their respective maximal irreducible components. The decomposability of a graph is a

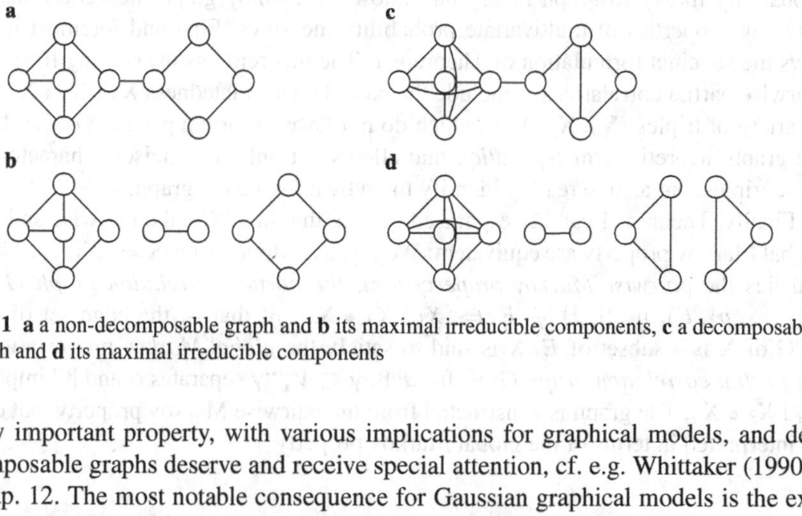

Fig. 1 **a** a non-decomposable graph and **b** its maximal irreducible components, **c** a decomposable graph and **d** its maximal irreducible components

very important property, with various implications for graphical models, and decomposable graphs deserve and receive special attention, cf. e.g. Whittaker (1990), Chap. 12. The most notable consequence for Gaussian graphical models is the existence of closed form maximum likelihood estimates, cf. Sect. 3.1. The recursive nature of Definition 3 makes it hard to determine whether a given graph is decomposable or not. Several equivalent characterizations of decomposability are given e.g. in Lauritzen (1996). We want to name one, which is helpful for spotting decomposable graphs.

Definition 4. The graph G is *triangulated*, if every cycle of length greater than 3 has a chord.

Proposition 2. *A graph G is decomposable if and only if it is triangulated.*

For a proof see Lauritzen (1996), p. 9, or Whittaker (1990), p. 390.

We close this subsection by giving a motivation for partial correlation graphs. Clearly, the information in the graph is fully contained in Σ and can directly be read off its inverse K: a zero off-diagonal element at position (i, j) signifies the absence of an edge between the corresponding nodes. Of course, graphs in general are helpful visual tools. This argument is valid for representing any type of association between variables by means of a graph and is not the sole justification for partial correlation graphs. The purpose of a PCG is explained by the following theorem, which lies at the core of graphical models.

Theorem 1 (Separation Theorem for PCGs). *For a random vector* $\mathbf{X}$ *with positive definite covariance matrix* Σ *and partial correlation graph* G *the following is true:* γ *separates* α *and* β *in* G *if and only if* $\mathbf{X}_\alpha \perp \mathbf{X}_\beta \bullet \mathbf{X}_\gamma$.

This result is not trivial, but its proof can be accomplished by matrix manipulation. It is also a corollary of Theorem 3.7 in Lauritzen (1996) by exploiting the equivalence of partial uncorrelatedness and conditional independence in the normal model, cf. Sect. 2.3. The theorem roughly tells that the association "partial uncorrelatedness" (of two random vectors given a third one) exhibits the same properties as the association "separation" (of two sets of vertices by a third one). Thus it links probability theory to graph theory and allows to employ graph theoretic tools in studying properties of multivariate probability measures. First and foremost it allows the succinct formulation of Theorem 1. The theorem lets us, starting from the pairwise partial correlations, conclude the partial uncorrelatedness $\mathbf{X}_\alpha \perp \mathbf{X}_\beta \bullet \mathbf{X}_\gamma$ for a variety of triples $(\mathbf{X}_\alpha, \mathbf{X}_\beta, \mathbf{X}_\gamma)$ (which do not have to form a partition of $\mathbf{X}$). It is the graph theoretic term *separation* that allows not only to concisely characterize these triples, but also to readily identify them by drawing the graph.

Finally, Theorem 1 can be re-phrased, saying that in a PCG the pairwise and the global Markov property are equivalent: We say, a random vector $\mathbf{X} = (X_1, \ldots, X_p)$ satisfies the *pairwise Markov property w.r.t. the partial correlation graph* $G = (\{1, \ldots, p\}, E)$, if $\{i, j\} \notin E \Rightarrow X_i \perp X_j \bullet \mathbf{X}_{\setminus\{i,j\}}$, that is, the edge set of the PCG of $\mathbf{X}$ is a subset of E. $\mathbf{X}$ is said to satisfy the *global Markov property w.r.t. the partial correlation graph* G, if, for $\alpha, \beta, \gamma \subset V$, "$\gamma$ separates α and β" implies $\mathbf{X}_\alpha \perp \mathbf{X}_\beta \bullet \mathbf{X}_\gamma$. The graph is constructed from the pairwise Markov property, but can be interpreted in terms of the global Markov property.

2.3 The Multivariate Normal Distribution and Conditional Independence

We want to make further assumptions on the distribution F of $\mathbf{X}$. A random vector $\mathbf{X} = (X_1, \ldots, X_p)$ is said to have a *regular* p*-variate normal* (or *Gaussian*) distribution, denoted by $\mathbf{X} \sim N_p(\boldsymbol{\mu}, \Sigma)$, if it possesses a Lebesgue density of the form

$$f_{\mathbf{X}}(\mathbf{x}) = (2\pi)^{-\frac{p}{2}} (\det \Sigma)^{-\frac{1}{2}} \exp\left\{-\frac{1}{2}(\mathbf{x}-\boldsymbol{\mu})\Sigma^{-1}(\mathbf{x}-\boldsymbol{\mu})\right\}, \quad \mathbf{x} \in \mathbb{R}^p, \tag{6}$$

for some $\mu \in \mathbb{R}^p$ and $\Sigma \in \mathcal{S}_p^+$. Then $\mathbb{E}\mathbf{X} = \mu$ and $\mathrm{Var}(\mathbf{X}) = \Sigma$. The term *regular* refers to the positive definiteness of the variance matrix. We will only deal with regular normal distributions—which allow the density characterization given above—without necessarily stressing the regularity.

The multivariate normal (*MVN*) distribution is a well studied object, it is treated e.g. in Bilodeau and Brenner (1999) or any other book on multivariate statistics. Of the properties of the MVN distribution the following three are of particular interest to us. Let, as before, $\mathbf{X}$ be partitioned into $\mathbf{X} = (\mathbf{Y}, \mathbf{Z})$. Then we have:

(I) The (marginal) distribution of $\mathbf{Y}$ is $N_q(\mu_\alpha, \Sigma_{\alpha,\alpha})$.
(II) $\mathbf{Y}$ and $\mathbf{Z}$ are independent (in notation $\mathbf{Y} \perp\!\!\!\perp \mathbf{Z}$) if and only if $\Sigma_{\alpha,\beta} = \mathbf{0}$ (which is equivalent to $K_{\alpha,\beta} = \mathbf{0}$).
(III) The conditional distribution of $\mathbf{Y}$ given $\mathbf{Z} = \mathbf{z}$ is

$$N_q\left(\mathbb{E}\mathbf{Y} + \Sigma_{\alpha,\beta}\Sigma_{\beta,\beta}^{-1}(\mathbf{z} - \mathbb{E}\mathbf{Z}),\ \Sigma_{\mathbf{Y}\bullet\mathbf{Z}}\right).$$

These fundamental properties of the MVN distribution can be proved by directly manipulating the density (6). We want to spare a few words about assertion (III). It can be phrased as to say, the multivariate normal model is closed under conditioning—just as (I) tells that it is closed under marginalizing. Moreover, (III) gives expressions for the conditional expectation and the conditional variance:

$$\mathbb{E}(\mathbf{Y}|\,\mathbf{Z}) = \hat{\mathbf{Y}}(\mathbf{Z}) \quad \text{and} \quad \mathrm{Var}(\mathbf{Y}|\,\mathbf{Z}) = \Sigma_{\mathbf{Y}\bullet\mathbf{Z}}.$$

In general, $\mathbb{E}(\mathbf{Y}|\,\mathbf{Z})$ and $\mathrm{Var}(\mathbf{Y}|\,\mathbf{Z})$ are random variables that can be expressed as functions of the conditioning variable $\mathbf{Z}$. Thus (III) tells us that in the MVN model $\mathbb{E}(\mathbf{Y}|\cdot)$ is a *linear* function, whereas $\mathrm{Var}(\mathbf{Y}|\cdot)$ is *constant*. Further, $\mathbb{E}(\mathbf{Y}|\,\mathbf{Z})$ is the best prediction of $\mathbf{Y}$ from $\mathbf{Z}$, in the sense that $\mathbb{E}\|\mathbf{Y} - h(\mathbf{Z})\|^2$ is uniquely minimized by $h = \hat{\mathbf{Y}}(\cdot)$ among *all* measurable functions h. Here this best prediction coincides with the best linear prediction $\hat{\mathbf{Y}}(\mathbf{Z})$ given in (3). Finally, $\mathrm{Var}(\mathbf{Y}|\,\mathbf{Z})$ being constant means that the accuracy gain for predicting $\mathbf{Y}$ that we get from knowing $\mathbf{Z}$ is the same no matter what value $\mathbf{Z}$ takes on. It is not least this linearity of the MVN distribution that makes it very appealing for statistical modeling.

The occupation with the conditional distribution is guided by our interest in conditional independence, which is—although it has not been mentioned yet—the actual primary object of study in graphical models. Let, as in Sect. 2.1, $\mathbf{Y} = (\mathbf{Y}_1, \mathbf{Y}_2)$ be further partitioned. $\mathbf{Y}_1$ and $\mathbf{Y}_2$ are *conditionally independent given* $\mathbf{Z}$—in writing: $\mathbf{Y}_1 \perp\!\!\!\perp \mathbf{Y}_2|\,\mathbf{Z}$—if the conditional distribution of $(\mathbf{Y}_1, \mathbf{Y}_2)$ given $\mathbf{Z} = \mathbf{z}$ is for (almost) all $\mathbf{z} \in \mathbb{R}^r$ a product measure with independent margins corresponding to $\mathbf{Y}_1$ and $\mathbf{Y}_2$. If $\mathbf{X}$ possesses a density $f_{\mathbf{X}} = f_{(\mathbf{Y}_1,\mathbf{Y}_2,\mathbf{Z})}$ w.r.t. some σ-finite measure, conditional independence admits the following characterization: $\mathbf{Y}_1 \perp\!\!\!\perp \mathbf{Y}_2|\,\mathbf{Z}$ if and only if there exist functions $g : \mathbb{R}^{q_1+r} \to \mathbb{R}$ and $h : \mathbb{R}^{q_2+r} \to \mathbb{R}$ such that

$$f_{(\mathbf{Y}_1,\mathbf{Y}_2,\mathbf{Z})}(\mathbf{y}_1, \mathbf{y}_2, \mathbf{z}) = g(\mathbf{y}_1, \mathbf{z})h(\mathbf{y}_2, \mathbf{z}) \quad \text{for almost all } (\mathbf{y}_1, \mathbf{y}_2, \mathbf{z}) \in \mathbb{R}^p.$$

This factorization criterion ought to be compared to its analogue for (marginal) independence. It shall serve as a definition here, saving us a proper introduction of the terms *conditional distribution* or *conditional density*.

We can construct for any random variable $\mathbf{X}$ in $\mathbb{R}^p$ a *conditional independence graph* (*CIG*) in an analogous way as before the partial correlation graph: We put an edge between nodes i and j unless $X_i \perp\!\!\!\perp X_j \,|\, \mathbf{X}_{\setminus\{i,j\}}$. Then, for "nice" distributions F—for instance, if F has a continuous, strictly positive density f (w.r.t. some σ-finite measure)—we have in analogy to Theorem 1 a separation property for CIGs: $\mathbf{X}_\alpha \perp\!\!\!\perp \mathbf{X}_\beta \,|\, \mathbf{X}_\gamma$ if and only if γ separates α and β in the CIG of $\mathbf{X}$.

Assertions (I) to (III) are the link from conditional independence to the analysis of the second moment characteristics in Sect. 2.1. A direct consequence is:

Proposition 3. *If* $\mathbf{X} = (\mathbf{Y}_1, \mathbf{Y}_2, \mathbf{Z}) \sim N_p(\mu, \Sigma)$, $\Sigma \in \mathcal{S}_p^+$, *then*

$$\mathbf{Y}_1 \perp \mathbf{Y}_2 \bullet \mathbf{Z} \quad \Longleftrightarrow \quad \mathbf{Y}_1 \perp\!\!\!\perp \mathbf{Y}_2 \,|\, \mathbf{Z}.$$

In other words, the PCG and the CIG of a regular normal vector coincide. It must be emphasized that this is a particular property of the Gaussian distribution. Conditional independence and partial uncorrelatedness are generally different, cf. Baba et al. (2004), and so are the respective association graphs.

3 Gaussian Graphical Models

We have defined the partial correlation graph of a random vector and have recalled some properties of the multivariate normal distribution. We have thus gathered the ingredients we need to deal with Gaussian graphical models.

We understand a *graphical model* as a family of probability distributions on $\mathbb{R}^p$ satisfying the pairwise zero partial correlations specified by a given (undirected) graph $G = (V, E)$, i.e. for all $i, j \in V$

$$\{i, j\} \notin E \Rightarrow \varrho_{i,j\bullet} = 0. \tag{7}$$

If the model consists of all (regular) p-variate normal distributions satisfying (7) we call it a *Gaussian graphical model* (*GGM*). Another equivalent term is *covariance selection model*, originated by Dempster (1972).

We write $\mathcal{M}(G)$ to denote the GGM induced by the graph G. The model $\mathcal{M}(G)$ is called *saturated* if G is complete. It is called *decomposable* if the graph is decomposable. A Gaussian graphical model is a parametric family, which may be succinctly described as follows. Let $\mathcal{S}_p^+(G)$ be the subset of $\mathcal{S}_p^+$ consisting of all positive definite matrices with zero entries at the positions specified by G, i.e.

$$K \in \mathcal{S}_p^+(G) \iff K \in \mathcal{S}_p^+ \text{ and } k_{i,j} = 0 \text{ for } i \neq j \text{ and } \{i, j\} \notin E.$$

Then

$$\mathcal{M}(G) = \left\{ N_p(\mu, \Sigma) \mid \mu \in \mathbb{R}^p,\ K = \Sigma^{-1} \in \mathcal{S}_p^+(G) \right\}. \tag{8}$$

In the context of GGMs it is more convenient to parametrize the normal model by (μ, K), which may be less common, but is quite intuitive considering that K directly appears in the density formula (6). The GGM $\mathcal{M}(G)$ is also specified by its parameter space $\mathbb{R}^p \times \mathcal{S}_p^+(G)$.

The term *graphical modeling* refers to the statistical task of deciding on a graphical model for given data and the collection of the statistical methods employed toward this end. Within the parametric family of Gaussian graphical models we have the powerful maximum likelihood theory available. We continue by stating the maximum likelihood estimates and some of their properties (Sect. 3.1), then review the properties of the likelihood ratio test for comparing two nested models (Sect. 3.2) and finally describe some model selection procedures (Sect. 3.3).

3.1 Estimation

Suppose we have i.i.d. observations $\mathbf{X}_1, \ldots, \mathbf{X}_n$ sampled from the normal distribution $N_p(\mu, \Sigma)$ with $\Sigma \in \mathcal{S}_p^+$. Let furthermore $\mathbb{X}_n = (\mathbf{X}_1^T, \ldots, \mathbf{X}_n^T)^T$ be the $n \times p$ data matrix containing the data points as rows. We will make use of the following matrix notation. For an undirected graph $G = (V, E)$ and an arbitrary square matrix A define the matrix $A(G)$ by

$$[A(G)]_{i,j} = \begin{cases} a_{i,j} & \text{if } i = j \text{ or } \{i, j\} \in E, \\ 0 & \text{if } i \neq j \text{ and } \{i, j\} \notin E. \end{cases}$$

The Saturated Model

We start with the saturated model, i.e. there is no further restriction on K. The main quantities of interest in Gaussian graphical models are the concentration matrix K and the partial correlation matrix P. Their computation ought to be part of any initial explorative data analysis. Both are functions of the covariance matrix Σ, thus we start with the latter.

Proposition 4. *If $n > p$, the maximum likelihood estimator (MLE) of Σ in the multivariate normal model (with unknown location μ) is*

$$\hat{\Sigma} = \frac{1}{n}\sum_{i=1}^{n}(\mathbf{X}_i - \bar{\mathbf{X}})(\mathbf{X}_i - \bar{\mathbf{X}})^T = \frac{1}{n}\mathbb{X}_n^T H_n \mathbb{X}_n,$$

where $H_n = I_n - \frac{1}{n}\mathbf{1}_n\mathbf{1}_n^T$ is an idempotent matrix of rank $n-1$. The MLEs of K and P are $\hat{K} = \hat{\Sigma}^{-1}$ and $\hat{P} = -\mathrm{Corr}(\hat{K})$, respectively.

Apparently $\mathbb{X}_n^T H_n \mathbb{X}_n$ has to be non-singular in order to be able to compute $\hat{K}$ and $\hat{P}$. It should be noted that this is also necessary for the MLE to exist in the sense that the ML equations have a unique solution. If n is strictly larger than p, this is almost surely true, but never if $n \leq p$.

We want to review some properties of these estimators. The strong law of large numbers, the continuous mapping theorem, the central limit theorem and the delta method yield the following asymptotic results, cf. Vogel (2009).

Proposition 5. *In the MVN model $\hat{\Sigma}$, $\hat{K}$ and $\hat{P}$ are strongly consistent estimators of Σ, K and P, respectively. Furthermore,*

(1) $\sqrt{n}\mathrm{vec}(\hat{\Sigma} - \Sigma) \xrightarrow{\mathscr{L}} N_{p^2}(\mathbf{0}, 2M_p(\Sigma \otimes \Sigma))$,

(2) $\sqrt{n}\mathrm{vec}(\hat{K} - K) \xrightarrow{\mathscr{L}} N_{p^2}(\mathbf{0}, 2M_p(K \otimes K))$,

(3) $\sqrt{n}\mathrm{vec}(\hat{P} - P) \xrightarrow{\mathscr{L}} N_{p^2}(\mathbf{0}, 2\Gamma M_p(K \otimes K)\Gamma^T)$,

where $\Gamma = (K_D^{-\frac{1}{2}} \otimes K_D^{-\frac{1}{2}}) + M_p(P \otimes K_D^{-1})J_p$.

Since the normal distribution and the empirical covariance matrix are of such utter importance, the exact distribution of the MLEs has also been the subject of study.

Proposition 6. *In the MVN model, if $n > p$, $\hat{\Sigma}$ has a Wishart distribution with parameter $\frac{1}{n}\Sigma$ and $n - 1$ degrees of freedom, for which we use the notation $\hat{\Sigma} \sim W_p(n-1, \frac{1}{n}\Sigma)$.*

For a definition and properties of the Wishart distribution see e.g. Bilodeau and Brenner (1999), Chap. 7, or Srivastava and Khatri (1979), Chap. 3. It is also treated in most textbook on multivariate statistics. The distribution of $\hat{K}$ is then called an *inverse Wishart distribution*. Of the various results on Wishart and related distributions we want to name the following three, but remark that more general results are available.

Proposition 7. *In the MVN model with $n > p$ we have*

(1) $\mathbb{E}\hat{\Sigma} = \frac{n-1}{n}\Sigma$ *and*

(2) $\mathrm{Var}(\mathrm{vec}\,\hat{\Sigma}) = \frac{2}{n}M_p(\Sigma \otimes \Sigma)$.

(3) *If furthermore* $\varrho_{i,j\bullet} = 0$, *then*

$$\sqrt{n-p}\frac{\hat{\varrho}_{i,j\bullet}}{\sqrt{1-\hat{\varrho}_{i,j\bullet}^2}} \sim t_{n-p}, \quad \text{which implies} \quad \hat{\varrho}_{i,j\bullet}^2 \sim \mathrm{Beta}\left(\frac{1}{2}, \frac{n-p}{2}\right),$$

where t_{n-p} denotes Student's t-distribution with $n - p$ degrees of freedom and Beta(c, d) the beta distribution with parameters $c, d > 0$ and density

$$b(x) = \frac{\Gamma(c+d)}{\Gamma(c)\Gamma(d)} x^{c-1}(1-x)^{d-1}\mathbb{1}_{[0,1]}(x).$$

The last assertion (3) ought to be compared to the analogous results for the empirical correlation coefficient $\hat{\varrho}_{i,j} = \hat{\sigma}_{i,j}/\sqrt{\hat{\sigma}_{i,j}\hat{\sigma}_{j,j}}$: if the true correlation is zero, then

$$\sqrt{n-2}\frac{\hat{\varrho}_{i,j}}{\sqrt{1-\hat{\varrho}_{i,j}^2}} \sim t_{n-2} \quad \text{and} \quad \hat{\varrho}_{i,j}^2 \sim \text{Beta}\left(\frac{1}{2}, \frac{n-2}{2}\right).$$

Estimation under a Given Graphical Model

We have dealt so far with unrestricted estimators of Σ, K and the partial correlation matrix P. Small absolute values of the estimated partial correlations suggest that the corresponding true partial correlations may be zero. However assuming a non-saturated model, using unrestricted estimates for the remaining parameters is no longer optimal. The estimation efficiency generally decreases with the number of parameters to estimate. Also, for stepwise model selection procedures, as described in Sect. 3.3, which successively compare the appropriateness of different GGMs, estimates under model constraints are necessary.

Consider the graph $G = (V, E)$ with $|V| = p$ and $|E| = m$, and let $\mathbf{X}_1, \ldots, \mathbf{X}_n$ be an i.i.d. sample from the model $\mathscr{M}(G)$ given in (8). Keep in mind that K is then an element of the $(m + p)$-dimensional vector space $\mathscr{S}_p(G)$, where m may range from 0 to $p(p-1)/2$. The matrix Σ is fully determined by the $m + p$ values $k_{1,1}, \ldots, k_{p,p}$ and $k_{i,j}$, $\{i, j\} \in E$ (which have to meet the further restriction that K is positive definite) and in this sense has to be regarded as an $(m + p)$-dimensional object.

Theorem 2.

(1) *The ML estimate $\hat{\Sigma}_G$ of Σ in the model $\mathscr{M}(G)$ exists if $\hat{\Sigma} = \frac{1}{n}\mathbb{X}_n^T H_n \mathbb{X}_n$ is positive definite, which happens with probability one if $n > p$.*

(2) *If the ML estimate $\hat{\Sigma}_G$ exists, it is the unique solution of the following system of equations*

$$[\hat{\Sigma}_G]_{i,j} = \hat{\sigma}_{i,j}, \quad \{i, j\} \in E \text{ or } i = j,$$

$$[\hat{\Sigma}_G^{-1}]_{i,j} = 0, \quad \{i, j\} \notin E \text{ and } i \neq j,$$

which may be succinctly formulated as

$$\hat{\Sigma}_G(G) = \hat{\Sigma}(G) \quad \text{and} \quad \hat{K}_G = \hat{K}_G(G), \tag{9}$$

where $\hat{K}_G = \hat{\Sigma}_G^{-1}$.

This result follows from general maximum likelihood theory for exponential models. The key is to observe that a GGM is a regular exponential model, cf. Lauritzen (1996), p. 133. It is important to note that, contrary to the saturated case, the positive definiteness of $\mathbb{X}_n^T H_n \mathbb{X}_n$ is sufficient but not necessary. In a decomposable model, for instance, it suffices that n is larger than the number of nodes of the largest clique, cf. Proposition 8. Generally this condition is necessary but not sufficient. Details on stricter conditions on the existence of the ML estimate in the general case can be found in Buhl (1993) or Lauritzen (1996), p. 148.

Theorem 2 gives instructive information about the structure of $\hat{\Sigma}_G$, in particular, that it is a function of the sample covariance matrix $\hat{\Sigma}$. The relation between $\hat{\Sigma}_G$

and $\hat{\Sigma}_G$ is specified by (9), and Theorem 2 states furthermore that these equations always have a unique solution $\hat{\Sigma}_G$, if $\hat{\Sigma}$ is positive definite. What remains unclear is how to compute $\hat{\Sigma}_G$ from $\hat{\Sigma}$. This is accomplished by the *iterative proportional scaling* (*IPS*) algorithm, sometimes also referred to as *iterative proportional fitting*, which is explained in the following.

Iterative Proportional Scaling

The IPS algorithm generally solves the problem of fitting a multivariate density that obeys a given interaction structure to specified marginal densities. Another application is the computation of the ML estimate in log-linear models, i.e. graphical models for discrete data. In the statistical literature the IPS algorithm can be traced back to at least Deming and Stephan (1940). In the case of multivariate normal densities the IPS procedure comes down to an iterative matrix manipulation. The IPS algorithm for GGMs, as it is described in the following, is mainly due to Speed and Kiiveri (1986).

Suppose we are given a graph G with cliques $\gamma_1, \ldots, \gamma_c$ and an unrestricted ML estimate $\hat{\Sigma} \in \mathcal{S}_p$. Then define for every clique γ the following matrix operator $T_\gamma : \mathcal{S}_p \to \mathcal{S}_p$:

$$T_\gamma(K) = K + \left[(\hat{\Sigma}_{\gamma,\gamma})^{-1}\right]^p - \left[(K^{-1})^{-1}_{\gamma,\gamma}\right]^p .$$

The operator T_γ has the following properties:

(I) If $K \in \mathcal{S}_p^+(G)$, then so is $T_\gamma K$.
(II) $(T_\gamma K)^{-1}_{\gamma,\gamma} = \hat{\Sigma}_{\gamma,\gamma}$, i.e. if the updated matrix $T_\gamma K$ is the concentration matrix of a random vector, $\mathbf{X}$ say, then $\mathbf{X}_\gamma$ has covariance matrix $\hat{\Sigma}_{\gamma,\gamma}$.

Apparently T_γ preserves the zero pattern of G. That it also preserves the positive definiteness and assertion (II) is not as straightforward, but both can be deduced by applying (2) to K^{-1}, cf. Lauritzen (1996), p. 135. The IPS algorithm then goes as follows: choose any $K_0 \in \mathcal{S}_p^+$, for instance $K_0 = I_p$, and repeat

$$K_{n+1} = T_{\gamma_1} T_{\gamma_2} \cdots T_{\gamma_c} K_n$$

until convergence is reached. If the ML estimate $\hat{\Sigma}_G$ exists (for which $\hat{\Sigma} \in \mathcal{S}_p^+$ is sufficient but not necessary), then (K_n) converges to $\hat{K}_G = \hat{\Sigma}_G^{-1}$, where $\hat{\Sigma}_G$ is the solution of (9), see again Lauritzen (1996), p. 135. Thus the IPS algorithm cycles through the cliques of G, in each step updating the concentration matrix K such that the clique has marginal covariance $\hat{\Sigma}_{\gamma,\gamma}$ while preserving the zero pattern specified by G.

Decomposable models

As mentioned before, in the case of decomposable models the ML estimate can be given in explicit form, and we do not have to resort to iterative approximations. As a decomposable graph can be *decomposed* into its cliques, the ML estimate of a decomposable model can be *composed* from the (unconstrained) MLEs of the cliques. Let $G = (V, E)$ be a decomposable graph with cliques $\gamma_1, \ldots, \gamma_c$ and $c > 1$. Define the sequence $(\delta_1, \ldots, \delta_{c-1})$ of subsets of V by

$$\delta_k = (\gamma_1 \cup \cdots \cup \gamma_k) \cap \gamma_{k+1}, \quad k = 1, \ldots, c-1.$$

We assume that the numbering $\gamma_1, \ldots, \gamma_c$ is such that for every $k \in \{1, \ldots, c-1\}$ there is a $j \le k$ with $\delta_k \subseteq \gamma_j$. It is always possible to order the cliques of a decomposable graph in such a way, cf. (Lauritzen 1996), p. 18. The sequence $(\gamma_1, \ldots, \gamma_c)$ is then said to be *perfect*, and it corresponds to a reversed sequence of successive decompositions. The δ_k do not have to be distinct. For instance, the graph in Fig. 2 has four cliques and, for any numbering of the cliques, $\delta_i = \{3\}, i = 1, 2, 3$.

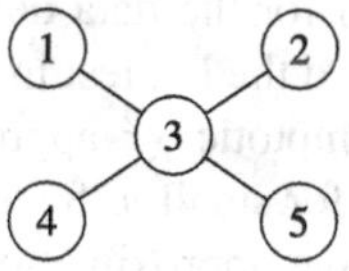

Fig. 2 Example graph

Proposition 8.

(1) *The ML estimate $\hat{\Sigma}_G$ of Σ in the decomposable model $\mathcal{M}(G)$ exists with probability one if and only if $n > \max_{k=1,\ldots,c} |\gamma_k|$.*

(2) *If the ML estimate $\hat{\Sigma}_G = \hat{K}_G$ exists, then it is given by*

$$\hat{K}_G = \sum_{k=1}^{c} \left[(\hat{\Sigma}_{\gamma_k,\gamma_k})^{-1} \right]^p - \sum_{k=1}^{c-1} \left[(\hat{\Sigma}_{\delta_k,\delta_k})^{-1} \right]^p .$$

See Lauritzen (1996), p. 146, for a proof. Results on the asymptotic distribution of the restrained ML-estimator $\hat{\Sigma}_G$ in the decomposable as well as the general case can be found in Lauritzen (1996), Chap. 5. The exact, non-asymptotic distribution of $\hat{\Sigma}_G$ has also been studied. For decomposable G, it is known as the *hyper Wishart distribution* (Dawid and Lauritzen 1993), and the distribution of $\hat{K}_G$ as *inverse hyper Wishart distribution* (Roverato 2000).

3.2 Testing

We want to test a graphical model against a larger one, possibly but not necessarily the saturated model. Consider two graphs $G = (V, E)$ and $G_0 = (V, E_0)$ with $E_0 \subset E$, or equivalently $\mathcal{M}(G_0) \subset \mathcal{M}(G)$. Then the likelihood ratio for testing $\mathcal{M}(G_0)$ against the larger model $\mathcal{M}(G)$ based on the observation $\mathbb{X}_n$ reduces to

$$\mathrm{LR}(G_0, G) = \left(\frac{\det \hat{\Sigma}_G}{\det \hat{\Sigma}_{G_0}} \right)^{\frac{n}{2}},$$

small values of which suggest to dismiss $\mathcal{M}(G_0)$ in favor of $\mathcal{M}(G)$. It follows by the general theory for LR tests that the test statistic

$$D_n(G_0, G) = -2 \ln \mathrm{LR}(G_0, G) = n \left(\ln \det \hat{\Sigma}_{G_0} - \ln \det \hat{\Sigma}_G \right) \tag{10}$$

is asymptotically χ^2 distributed with $|E| - |E_0|$ degrees of freedom under the model $\mathcal{M}(G_0)$. The test statistic D_n may be interpreted as a measure of how much the appropriateness of model $\mathcal{M}(G_0)$ for the data deviates from that of $\mathcal{M}(G)$. It is thus also referred to as *deviance* and the LR test in GGMs is called *deviance test.*

It has been noted that the asymptotic χ^2 approximation of the distribution of D_n is generally not very accurate for small n. Several suggestions have been made on how to improve the finite sample approximation. One approach is to apply the Bartlett correction to the LR test statistic (Porteous 1989). Another approximation, which is considerably better than the asymptotic distribution, is given by the exact distribution for decomposable models in Proposition 9 (Eriksen 1996).

Decomposable Models

Again decomposable models play a special role. We are able to give the exact distribution of the deviance if both models compared are decomposable. Thus assume in the following that G and G_0 are decomposable. Then one can find a sequence of decomposable models $G_0 \subset G_1 \subset \cdots \subset G_k = G$ such that each successive pair (G_{i-1}, G_i) differs by exactly one edge e_i, $i = 1, \ldots, k$, cf. Lauritzen (1996), p. 20. Let a_i denote the number of common neighbors of both endpoints of e_i in the graph G_i.

Proposition 9. *If G_0 and G are decomposable and $G_0 \subset G$, then*

$$\frac{\det \hat{\Sigma}_G}{\det \hat{\Sigma}_{G_0}} = \exp\left(-\frac{D_n}{n} \right) \sim B_1 B_2 \cdots B_k,$$

where the B_i are independent random variables with $B_i \sim$ $\mathrm{Beta}(\frac{n-a_i-2}{2}, \frac{1}{2})$.

Since a complete graph and a graph with exactly one missing edge are both decomposable, the test of conditional independence of two components of a random vector is a special case of Proposition 9. If we let G_0 be the graph with all edges but $\{i, j\}$, some matrix calculus yields (cf. Lauritzen 1996, p. 150)

$$\frac{\det \hat{\Sigma}}{\det \hat{\Sigma}_{G_0}} = 1 - \hat{\varrho}^2_{i,j\bullet}.$$

By Proposition 9 this has a Beta$(\frac{n-p}{2}, \frac{1}{2})$ distribution, which is in concordance with Proposition 7 (3).

3.3 Model Selection

Contrary to estimation and statistical testing in GGMs there is no generally agreed-upon, optimal way to select a model. Statistical theory gives a relatively precise answer to the question if a certain model fits the data or not, but not which model to choose among those that fit. There are many model selection procedures (MSPs), and comparing them is rather difficult, since many aspects play a role—computing time being just one of them. Furthermore, theoretic results are usually hard to derive. For most MSPs, consistency can be shown, but distributional results are seldom available. Selecting a graphical model means to decide, based on the data, which partial correlations should be set to zero and which should be estimated freely. This decision, of course, heavily depends on the nature of the problem at hand, for example, if too few or too many edges are judged more severe. Ultimately, the choice of the MSP is a matter of personal taste, and the model selection has to be tailored to the specific situation. Expert knowledge should be incorporated to obtain sensible and interpretable models, especially when it comes to choosing from several equally adequate models.

The total number of p-dimensional GGMs is $2^{\binom{p}{2}}$, and only for very small p an evaluation of all possible models, based on some model selection criterion like AIC or BIC, is feasible. With respect to interpretability one might want to restrict the search space to decomposable models, cf. e.g. Whittaker (1990), Chap. 12, or Edwards (2000), Chap. 6. Otherwise a non-complete model search is necessary.

Model Search

The system of all possible models possesses itself a (directed) graph structure, corresponding to the partial ordering induced by set inclusion of the respective edge sets. A graph G_0, say, is a child of a graph G, if G has exactly one edge more than G_0. The fact that we know how to compare nested models, as described in Sect. 3.1, suggests a search along the edges of this lattice. A classic, simple search, known as *backward elimination*, is carried out as follows. Start with the saturated model, and in each step remove one edge. To determine which edge, compute all deviances between the current model and all models with exactly one edge less. The edge corresponding to the smallest deviance difference is deleted, unless all deviances are above the significance level, i.e. all edges are significant. Then the algorithm stops. The search in the opposite direction, starting from the empty graph and including significant edges, is also possible and known as *forward selection*. Although both schemes have been reported to produce similar results, there is a substantial conceptual difference that favors backward elimination. The latter searches among models consistent with the data, while forward selection steps through inconsistent models. The result of an LR test has no sensible interpretation if both models compared are actually invalid. On the other hand, forward selection is to be preferred for sparse graphs.

Of course, many variants exist, e.g., one may remove all non-significant edges at once, then successively include edges again, apply an alternative stopping rule (e.g. overall deviance against the saturated model) or generally alternate between elimination and selection steps. Another model search strategy in graphical models is known as the Edwards-Havránek procedure (Edwards and Havránek 1985, 1987; Smith 1992). It is a global search, but reduces the search space, similar to the branch-and-bound principle by making use of the lattice structure.

One Step Model Selection

The simplicity of a one step MSP is, of course, very appealing. They become increasingly desirable as there has been an enormous growth in the dimensionality of data sets, and several proposals have been made in the recent past (Drton and Perlman 2004, 2008; Meinshausen and Bühlmann 2006; Castelo and Roverato 2006). For instance, the SINful procedure by Drton and Perlman (2008) is a simple model selection scheme, which consists of setting all partial correlations to zero for which the absolute value of the sample partial correlation is below a certain threshold. This threshold is determined in such a way that overall probability of selecting an incorrect edge, i.e. the probability that the estimated model is too large, is controlled.

4 Robustness

Most of what has been presented in the previous section, the classical GGM theory, has been developed in the seventies and the eighties of the last century. Since then graphical models have become popular tools of data analysis, and the statistical theory of Gaussian graphical models remains an active field of research. Many authors have in particular addressed the $n < p$ problem (a weak point of the ML theory) as in recent years one often encounters huge data sets, where the number of variables exceeds by far the number of observations. Another line of research considers GGMs in the Bayesian framework. It is beyond the scope of a book chapter to give an exhaustive survey of the recent approaches—even if we restrict ourselves to undirected graphical models for continuous data. We want to focus on another weak point of the normal ML theory: its lack of robustness, which has been pointed out, e.g., by Kuhnt and Becker (2003) and Gottard and Pacillo (2007).

Robustness denotes the property of a statistical method to yield good results also if the assumptions for which it is designed are violated. Small deviations from the assumed model shall have only a small effect, and robustness can be seen as a continuity property. This includes the often implied meaning of robustness as invulnerability against outliers. For example, any neighborhood of a normal distribution (measured in the Kolmogorov metric) contains arbitrarily heavy-tailed distributions (measured in kurtosis, say). Outlier generating models with a small outlier fraction are actually very *close* to the pure data model.

There are two general conceptual approaches when it comes to robustifying a statistical analysis: identify the outliers and remove them, or use robust estimators that preferably nullify, but at least reduce the harmful impact of outliers. Graphical modeling—as an instance of the model selection problem—is a field where the advantages of the second approach become apparent. In its most general perception an outlier is a "very unlikely" observation under a given model, cf. Davies and Gather (1993). Irrespective of the particular rule applied to decide whether an observation is deemed an outlier or not, any sensible rule ought to give different answers for different models. An outlier in a specific GGM may be a quite likely observation in the saturated model.

This substantially complicates outlier detection in any type of graphical models, suggesting it must at least be applied iteratively, alternating with model selection steps. For Gaussian graphical models, however, we have the relieving fact that an outlier w.r.t. a normal distribution basically coincides with an *outlier* in its literal meaning: a point far away from the majority of the data. Hence, strongly outlying points tend to be ouliers w.r.t. any Gaussian model, no matter which—if any—conditional or marginal independences it obeys.

Our focus will therefore lie in the following on robust estimation. Note that Gaussian graphical modeling, as presented in the previous section, exclusively relies on $\hat{\Sigma}$. It is a promising approach to replace the initial estimate $\hat{\Sigma}$ by a robust substitute and hence robustify all subsequent analysis. We can make use of the well developed robust estimation theory of multivariate scatter.

4.1 Robust Estimation of Multivariate Scatter

Robust estimation in multivariate data analysis has long been recognized as a challenging task. Over the last four decades much work has been devoted to the problem and many robust alternatives of the sample mean and the sample covariance matrix have been proposed, e.g. M-estimators (Maronna 1976; Tyler 1987), Stahel-Donoho estimators (Stahel 1981; Donoho 1982; Maronna and Yohai 1995; Gervini 2002), S-estimators (Davies 1987; Lopuhaä 1989; Rocke 1996), MVE and MCD

(Rousseeuw 1985; Davies 1992; Butler et al. 1993; Croux and Haesbroeck 1999; Rousseeuw and Van Driessen 1999), τ-estimators (Lopuhaä 1991), CM-estimators (Kent and Tyler 1996), reweighted and data-depth based estimators (Lopuhaä 1999; Gervini 2003; Zuo and Cui 2005). Many variants exist, and the list is far from complete. For a more detailed account see e.g. the book Maronna et al. (2006) or the review article Zuo (2006).

The asymptotics and robustness properties of the estimators are to a large extent well understood. The computation often requires to solve challenging optimization problems, but improved search heuristics are nowadays available. What remains largely an open theoretical question is the exact distribution for small samples. Constants of finite sample approximations usually have to be assessed numerically.

There are several measures that quantify and thus allow to compare the robustness properties of estimators. We want to restrict our attention to the influence function, introduced by Hampel (1971). Toward this end we have to adopt the notion that estimators are functionals $S : \mathscr{F} \to \Theta$ defined on a class of distributions $\mathscr{F}$. In the case of matrix-valued scatter estimators S, the image space Θ is $\mathscr{S}_p$. The specific estimate computed from a data set $\mathbb{X}_n$ is the functional evaluated at the corresponding empirical distribution function $\mathbb{F}_n = \frac{1}{n}\sum_{i=1}^{n} \delta_{\mathbf{x}_i}$, where $\delta_{\mathbf{x}}$ denotes the Dirac-measure which puts unit mass at the point $\mathbf{x} \in \mathbb{R}^p$. For instance, the sample covariance matrix $\hat{\Sigma}$ is simply the functional $\mathrm{Var}(\cdot)$, which is defined on all distributions with finite second moments, evaluated at $\mathbb{F}_n$. The *influence function* of S at the distribution F is defined as

$$IF(\mathbf{x}; S, F) = \lim_{\varepsilon \searrow 0} \frac{1}{\varepsilon}\left(S(F_{\varepsilon,\mathbf{x}}) - S(F)\right), \quad \mathbf{x} \in \mathbb{R}^p,$$

where $F_{\varepsilon,\mathbf{x}} = (1-\varepsilon)F + \varepsilon\delta_{\mathbf{x}}$. In words, the influence function is the directional derivative of the functional S at the "point" $F \in \mathscr{F}$ in the direction of $\delta_{\mathbf{x}} \in \mathscr{F}$. It describes the *influence* of an infinitesimal contamination at point $\mathbf{x} \in \mathbb{R}^p$ on the functional S, when the latter is evaluated at the distribution F. Of course, in terms of robustness, the influence of any contamination is preferably small. A robust estimator has in particular a bounded influence function, i.e. the maximal absolute influence $\sup\{\, \|IF(\mathbf{x}; S, F)\| \mid \mathbf{x} \in \mathbb{R}^p\}$, also known as *gross-error sensitivity*, is finite.

The influence function is said to measure the *local robustness* of an estimator. Another important robustness measure, which in contrast measures the global robustness but which we will not pursue further here, is the *breakdown point* (asymptotic breakdown point (Hampel 1971), finite-sample breakdown point (Donoho and Huber 1983, see also Davies and Gather 2005)). Roughly, the finite-sample replacement breakdown point is the minimal fraction of contaminated data points that can drive the estimate to the boundary of the parameter space. For details on robustness measures see e.g. Hampel et al. (1986).

It is a very desirable property of scatter estimators to transform in the same way as the (population) covariance matrix—the quantity they aim to estimate—under affine linear transformations. A scatter estimator $\hat{S}$ is said to be *affine equivariant*, if it satisfies $\hat{S}(\mathbb{X}_n A^T + \mathbf{1}_n \mathbf{b}^T) = A\hat{S}(\mathbb{X}_n)A^T$ for any full rank matrix $A \in \mathbb{R}^{p\times p}$

and vector $\mathbf{b} \in \mathbb{R}^p$. We want to make a notational distinction between S, the functional working on distributions, and $\hat{S}$, the corresponding estimator working on data (strictly speaking a series of estimators indexed by n), i.e. $S(\mathbb{F}_n) = \hat{S}(\mathbb{X}_n)$. The equivariance is indeed an important property, due to various reasons. For instance, any statistical analysis based on such estimators is independent of any change of the coordinate system, may it be re-scaling or rotations of the data. Also, affine equivariance implies that at any elliptical population distribution (such as the Gaussian distribution) indeed a multiple of the covariance matrix is unbiasedly estimated, cf. Proposition 10 below. Furthermore the estimate obtained is usually positive definite with probability one, which is crucial for any subsequent analysis, e.g. we know that the derived partial correlation matrix estimator $-\text{Corr}(\hat{S}^{-1})$ actually reflects a "valid" dependence structure.

The classes of estimators listed above all possess this equivariance property—or at least the pseudo-equivariance described below. Historically though, affine equivariance for robust estimators is not a self-evident property. Contrary to univariate moment-based estimators (such as the sample variance), the highly robust quantile-based univariate scale estimators (such as the median absolute deviation, MAD) do not admit a straightforward affine equivariant generalization to higher dimensions.

In Gaussian graphical models we are interested in partial correlations and zero entries in the inverse covariance matrix, for which we need to know Σ only up to a constant. The knowledge of the overall scale is not relevant, and we require a slightly weaker condition than affine equivariance in the above sense, which we want to call *affine pseudo-equivariance* or *proportional affine equivariance*.

Condition C1. $\hat{S}(\mathbb{X}_n A^T + \mathbf{1}_n \mathbf{b}^T) = g(AA^T) A \hat{S}(\mathbb{X}_n) A^T$ for $\mathbf{b} \in \mathbb{R}^p$, $A \in \mathbb{R}^{p \times p}$ with full rank, and $g : \mathbb{R}^{p \times p} \to \mathbb{R}$ satisfying $g(I_p) = 1$.

This condition basically merges two important special cases, the proper affine equivariance described above, i.e. $g \equiv 1$, and the case of shape estimators in the sense of Paindaveine (2008), which corresponds to $g = 1/\det(\cdot)$. The following proposition can be found in a similar form in Bilodeau and Brenner (1999), p. 212.

Proposition 10. *In the MVN model, i.e.* $\mathbb{X}_n = (\mathbf{X}_1^T, \ldots, \mathbf{X}_n^T)^T$ *with* $\mathbf{X}_1, \ldots, \mathbf{X}_n \sim N_p(\boldsymbol{\mu}, \Sigma)$ *i.i.d., any affine pseudo-equivariant scatter estimator* $\hat{S} = \hat{S}(\mathbb{X}_n)$ *satisfies*

(1) $\mathbb{E}\hat{S} = a_n \Sigma$ *and*
(2) $\text{Var}(\text{vec}\hat{S}) = 2\, b_n\, M_p(\Sigma \otimes \Sigma) + c_n \text{vec}\Sigma (\text{vec}\Sigma)^T$,

where (a_n), (b_n) *and* (c_n) *are sequences of real numbers with* $a_n, b_n \geq 0$ *and* $c_n \geq -2b_n/p$ *for all* $n \in \mathbb{N}$.

Proposition 7 tells us that for $\hat{S} = \hat{\Sigma}$ we have $a_n = \frac{n}{n-1}$, $b_n = \frac{1}{n}$ and $c_n \equiv 0$. For root-n-consistent estimators the general form of variance re-appears in the asymptotic variance, and they fulfill

Condition C2. There exist constants $a, b > 0$ and $c \geq -2b/p$ such that

$$\sqrt{n}\text{vec}(\hat{S} - a\,\Sigma) \xrightarrow{\mathscr{L}} N_{p^2}\left(\mathbf{0},\ 2a^2 b M_p(\Sigma \otimes \Sigma) + a^2 c \text{vec}\Sigma (\text{vec}\Sigma)^T\right).$$

The continuous mapping theorem and the multivariate delta method yield the general form of the asymptotic variance of any partial correlation estimator derived from a scatter estimator satisfying Condition C2.

Proposition 11. *If $\hat{S}$ fulfills Condition* C2, *the corresponding partial correlation estimator $\hat{P}^S = -\mathrm{Corr}(\hat{S}^{-1})$ satisfies*

$$\sqrt{n}\,\mathrm{vec}(\hat{P}^S - P) \xrightarrow{\mathscr{L}} N_{p^2}(\mathbf{0}, 2b\Gamma M_p(K \otimes K)\Gamma^T), \tag{11}$$

where b is the same as in Condition C2 *and Γ is as in Proposition* 5.

Thus the comparison of the asymptotic efficiencies of partial correlation matrix estimators based on affine pseudo- equivariant scatter estimators reduces to the comparison of the respective values of the scalar b. For $\hat{S} = \hat{\Sigma}$ we have $b = 1$ by Proposition 5. Also, general results for the influence function of pseudo-equivariant estimators can be given, cf. Hampel et al. (1986), Chap. 5.3.

Proposition 12.

(1) *At the Gaussian distribution $F = N_p(\mu, \Sigma)$ the influence function of any functional S satisfying Condition* C1 *has, if it exists, the form*

$$IF(\mathbf{x}; S, F) = g(\Sigma)\left(\alpha(d(\mathbf{x}))(\mathbf{x}-\mu)(\mathbf{x}-\mu)^T - \beta(d(\mathbf{x}))\Sigma\right), \tag{12}$$

where $d(\mathbf{x}) = \sqrt{(\mathbf{x}-\mu)^T K(\mathbf{x}-\mu)}$, g is as in Condition C1 *and α and β are suitable functions* $[0,\infty) \to \mathbb{R}$.

(2) *Assuming that $\hat{S}$ is Fisher-consistent for $a\Sigma$, i.e. $S(F) = a\Sigma$, with $a > 0$, cf. Condition* C2, *the influence function of the corresponding partial correlation matrix functional $P^S = -\mathrm{Corr}(S^{-1})$ is*

$$IF(\mathbf{x}; P^S, F) = \frac{\alpha(d(\mathbf{x}))g(\Sigma)}{a} \times \left(\frac{1}{2}\left(\Pi_D K_D^{-1} P + (\Pi_D K_D^{-1} P)^T\right) - K_D^{-\frac{1}{2}} \Pi K_D^{-\frac{1}{2}}\right),$$

where $\Pi = K(\mathbf{x}-\mu)(\mathbf{x}-\mu)^T K$.

In the case of the sample covariance matrix $\hat{\Sigma}(\mathbb{X}_n) = \mathrm{Var}(\mathbb{F}_n)$ we have $a = 1$ and $\alpha = \beta \equiv 1$. Thus (12) reduces to $IF(\mathbf{x}; \mathrm{Var}, F) = (\mathbf{x}-\mu)(\mathbf{x}-\mu)^T - \Sigma$, which is not only unbounded, but even increases quadratically with $\|\mathbf{x}-\mu\|$. We will now give two examples of robust affine equivariant estimators, that have been proposed in the context of GGMs.

The Minimum Covariance Determinant (MCD) Estimator

The idea behind the MCD estimator is that outliers will increase the volume of the ellipsoid specified by the sample covariance matrix, which is proportional to the square root of its determinant. The MCD is defined as follows. A subset $\eta \subset \{1, \ldots, n\}$ of fixed size $h = \lfloor sn \rfloor$ with $\frac{1}{2} \le s < 1$ is determined such $\det(\hat{\Sigma}^{\eta})$ with

$$\hat{\Sigma}^{\eta} = \frac{1}{h} \sum_{i \in \eta} (\mathbf{X}_i - \bar{\mathbf{X}}^{\eta})(\mathbf{X}_i - \bar{\mathbf{X}}^{\eta})^T \quad \text{and} \quad \bar{\mathbf{X}}^{\eta} = \frac{1}{h} \sum_{i \in \eta} \mathbf{X}_i$$

is minimal. The mean μ_{MCD} and covariance matrix $\hat{\Sigma}_{\text{MCD}}$ computed from this subsample are called the *raw MCD location*, respectively *scatter estimate*. Based on the raw estimate $(\mu_{\text{MCD}}, \hat{\Sigma}_{\text{MCD}})$ a reweighted scatter estimator $\hat{\Sigma}_{\text{RMCD}}$ is computed from the whole sample:

$$\hat{\Sigma}_{\text{RMCD}} = \left(\sum_{i=1}^{n} w_i \right)^{-1} \sum_{i=1}^{n} w_i (\mathbf{X}_i - \mu_{\text{MCD}})(\mathbf{X}_i - \mu_{\text{MCD}})^T,$$

where $w_i = 1$ if $(\mathbf{X}_i - \mu_{\text{MCD}})^T \hat{\Sigma}_{\text{MCD}}^{-1} (\mathbf{X}_i - \mu_{\text{MCD}}) < \chi^2_{p,0.975}$ and zero otherwise. Usually the estimate is multiplied by a consistency factor (corresponding to $1/a$ in Condition C2) to achieve consistency for Σ at the MVN distribution. Since this is irrelevant for applications in GGMs we omit the details. The respective values of the constants b and c in Condition C2 as well as the function α and β in Proposition 12 are given in Croux and Haesbroeck (1999).

The reweighting step improves the efficiency and retains the high global robustness (breakdown point of roughly $1 - s$ for $s \ge 1/2$) of the raw estimate. Although the minimization over $\binom{n}{h}$ subsets is of non-polynomial complexity, the availability of fast search heuristics (e.g. Rousseeuw and Van Driessen 1999) along with the aforementioned good statistical properties have rendered the RMCD a very popular robust scatter estimator, and several authors (Becker 2005; Gottard and Pacillo 2008) have suggested its use for Gaussian graphical modeling.

The Proposal by Miyamura and Kano

Miyamura and Kano (2006) proposed another affine equivariant robust scatter estimator in the GGM framework. The idea is here a suitable adjustment of the ML equations. The Miyamura-Kano estimator $\hat{\Sigma}_{MK}$ falls into the class of M-estimators, as considered in Huber and Ronchetti (2009), and is defined as the scatter part Σ of the solution (μ, Σ) of

$$\frac{1}{n} \sum_{i=1}^{n} \exp\left(-\frac{\xi\, d^2(\mathbf{X}_i)}{2} \right) (\mathbf{X}_i - \mu) = \mathbf{0} \quad \text{and}$$

$$\frac{1}{n} \sum_{i=1}^{n} \exp\left(-\frac{\xi\, d^2(\mathbf{X}_i)}{2} \right) \left(\Sigma - (\mathbf{X}_i - \mu)(\mathbf{X}_i - \mu)^T \right) = \frac{\xi}{(\xi + 1)^{(p+2)/2}} \Sigma,$$

where $\xi \geq 0$ is a tuning parameter and $d(\mathbf{x})$ is, as in Proposition 12, the Mahalanobis distance of $\mathbf{x} \in \mathbb{R}^p$ w.r.t. $\boldsymbol{\mu}$ and Σ. Large values of ξ correspond to a more robust (but less efficient) estimate, i.e. less weight is given to outlying observations. The Gaussian likelihood equations are obtained for $\xi = 0$.

4.2 Robust Gaussian Graphical Modeling

The classical GGM theory is completely based on the sample covariance matrix $\hat{\Sigma}$: the ML estimates in Theorem 2, the deviance test statistic D_n in (10) and model selection procedures such as backward elimination, Edwards-Havránek or Drton-Perlman. Thus replacing the normal MLE by a robust, affine equivariant scatter estimator and applying the GGM methodology in analogous manner is an intuitive way of performing robust graphical modeling, insensitive to outliers in the data. Since the asymptotics of affine (pseudo-)equivariant estimators are well established (at the normal distribution), and, as described in Sect. 4.1, their general common structure is not much different from that of the sample covariance matrix, *asymptotic* statistical methods can rather easily be adjusted by means of standard asymptotic arguments.

Estimation under a Given Graphical Model

We have discussed properties of equivariant scatter estimators and indicated their usefulness for Gaussian graphical models. However they just provide alternatives for the unconstrained estimation. Whereas the ML paradigm dictates the solution of (9) as an optimal way of estimating a covariance matrix with a graphical model and exact normality, it is not quite clear what is the best way of robustly estimating a covariance matrix that obeys a zero pattern in its covariance. Clearly, Theorem 2 suggests to simply solve equations (9) with $\hat{\Sigma}$ replaced by any suitable robust estimator $\hat{S}$. This approach has the advantage that consistency of the estimator under the model is easily assessed. In case of a decomposable model the estimator can be computed by the decomposition of Proposition 8, or generally by the IPS algorithm, for which convergence has been shown and which comes at no additional computational cost. Becker (2005) has proposed to apply IPS to the reweighted MCD.

However, a thorough study of scatter estimators under graphical models is still due, and it might be that other possibilities are more appropriate in certain situations. Many robust estimators are defined as the solution of a system of equations. A different approach is to alter these estimation equations in a suitable way that forces a zero pattern on the inverse. This requires a new algorithm, the convergence of which has to be assessed individually. This route has been taken by Miyamura and Kano (2006). Their algorithm performs an IPS approximation at each step and is hence relatively slow.

A problem remains with both strategies. Scatter estimators, if they have not a structure as simple as the sample covariance, generally do not possess the "consistency property" that the estimate of a margin appears as a submatrix of the estimate of the whole vector. The ML estimate $\hat{\Sigma}_G$ in the decomposable as well as the general case is composed from the unrestricted estimates of the cliques, cf. Theorem 2 and Proposition 8, which makes it in particular possible to compute the MLE for $p \geq n$. One way to circumvent this problem is to drop the affine equivariance and resort to robust "pairwise" estimators, such as the Gnanadesikan-Kettenring estimator (Gnanadesikan and Kettenring 1972; Maronna and Zamar 2002) or marginal sign and rank matrices (Visuri et al. 2000; Vogel et al. 2008). Besides having the mentioned consistency property pairwise estimators are also very fast to compute.

Testing and Model Selection

The deviance test can be applied analogously with minor adjustments when based on an affine equivariant scatter estimator. Similarly to the partial correlation estimator $\hat{P}^S$ in Proposition 11, the asymptotic distribution of the generalized deviance D_n^S, computed from any root-n-consistent, equivariant estimate $\hat{S}$, differs from that of the ML-deviance (10) only by a factor, see Tyler (1983) or Bilodeau and Brenner (1999), Chap. 13, for details. However, as noted in Sect. 3.2, the χ^2 approximation of the uncorrected deviance may be rather inaccurate for small n. Generalizations of finite-sample approximations or the exact test in Proposition 9 are not equally straightforward. Since the exact distribution of a robust estimator is usually unknown, one will have to resort to Monte Carlo or bootstrap methods.

Model selection procedures that only require a covariance estimate can be robustified in the same way. Besides the classical search procedures this is also true for the SINful procedure by Drton and Perlman (2008), of which Gottard and Pacillo (2008) studied a robustified version based on the RMCD.

4.3 Concluding Remarks

The use of robust methods is strongly advisable, particularly in multivariate analysis, where the whole structure of the data is not immediately evident. Even if one refrains from relying solely on a robust analysis, it is in any case an important diagnostic tool. A single gross error or even mild deviations from the assumed model may render the results of a sample covariance based data analysis useless. The use of alternative, robust estimators provides a feasible safeguard, which comes at the price of a small loss in efficiency and a justifiable increase in computational costs.

Although there is an immense amount of literature on multivariate robust estimation and applications thereof (robust tests, regression, principal component analysis, discrimination analysis etc., see e.g. Zuo 2006 for references), the list of publications addressing robustness in graphical models is (still) rather short. We have described how GGMs can be robustified using robust, affine equivariant estimators. An in-depth study of this application of robust scatter estimation seems to be still open.

The main limitation of this approach is that it works well only for sufficiently large n, and on any account only for $n > p$, since, as pointed out above, usually an initial estimate of full dimension is required. Also note that, for instance, the computation of the MCD requires $h > p$. The finite-sample efficiency of many robust estimators is low, and with the exact distributions rarely accessible, methods based on such estimators rely even more on asymptotics than likelihood methods.

The processing of very high-dimensional data ($p \gg n$) becomes increasingly relevant, and in such situations it is unavoidable and even, if n is sufficiently large, dictated by computational feasibility, to assemble the estimate of Σ, restricted to a given model, from marginal estimates. A high dimensional, robust graphical modeling, combining robustness with applicability in large dimensions, remains a challenging topic of future research.

Acknowledgement

The authors gratefully acknowledge Alexander Dürre's assistance in preparing the figures and the financial support of the SFB 823.

References

Baba, K., Shibata, R., Sibuya, M.: Partial correlation and conditional correlation as measures of conditional independence. Aust. N. Z. J. Stat. **46**(4), 657–664 (2004)

Becker, C.: Iterative proportional scaling based on a robust start estimator. In: Weihs, C., Gaul, W. (eds.) Classification—The Ubiquitous Challenge, pp. 248–255. Springer, Heidelberg (2005)

Bilodeau, M., Brenner, D.: Theory of Multivariate Statistics. Springer Texts in Statistics. Springer, New York (1999)

Buhl, S.L.: On the existence of maximum likelihood estimators for graphical Gaussian models. Scand. J. Stat. **20**(3), 263–270 (1993)

Butler, R.W., Davies, P.L., Jhun, M.: Asymptotics for the minimum covariance determinant estimator. Ann. Stat. **21**(3), 1385–1400 (1993)

Castelo, R., Roverato, A.: A robust procedure for Gaussian graphical model search from microarray data with p larger than n. J. Mach. Learn. Res. **7**, 2621–2650 (2006)

Cox, D.R., Wermuth, N.: Multivariate Dependencies: Models, Analysis and Interpretation. Monographs on Statistics and Applied Probability, vol. 67. Chapman and Hall, London (1996)

Croux, C., Haesbroeck, G.: Influence function and efficiency of the minimum covariance determinant scatter matrix estimator. J. Multivar. Anal. **71**(2), 161–190 (1999)

Davies, P.L.: Asymptotic behaviour of S-estimates of multivariate location parameters and dispersion matrices. Ann. Stat. **15**, 1269–1292 (1987)

Davies, P.L.: The asymptotics of Rousseeuw's minimum volume ellipsoid estimator. Ann. Stat. **20**(4), 1828–1843 (1992)

Davies, P.L., Gather, U.: The identification of multiple outliers. J. Am. Stat. Assoc. **88**(423), 782–801 (1993)

Davies, P.L., Gather, U.: Breakdown and groups. Ann. Stat. **33**(3), 977–1035 (2005)

Dawid, A.P., Lauritzen, S.L.: Hyper Markov laws in the statistical analysis of decomposable graphical models. Ann. Stat. **21**(3), 1272–1317 (1993)

Deming, W.E., Stephan, F.F.: On a least squares adjustment of a sampled frequency table when the expected marginal totals are known. Ann. Math. Stat. **11**, 427–444 (1940)

Dempster, A.P.: Covariance selection. Biometrics **28**, 157–175 (1972)

Donoho, D.L.: Breakdown properties of multivariate location estimators. PhD thesis, Harvard University (1982)

Donoho, D.L., Huber, P.J.: The notion of breakdown point. In: Bickel, P.J., Doksum, K.A., Hodges, J.L. (eds.): Festschrift for Erich L. Lehmann, pp. 157–183. Wadsworth, Belmont (1983)

Drton, M., Perlman, M.D.: Model selection for Gaussian concentration graphs. Biometrika **91**(3), 591–602 (2004)

Drton, M., Perlman, M.D.: A SINful approach to Gaussian graphical model selection. J. Stat. Plan. Inference **138**(4), 1179–1200 (2008)

Edwards, D.: Introduction to graphical modeling. Springer Texts in Statistics. Springer, New York (2000)

Edwards, D., Havránek, T.: A fast procedure for model search in multidimensional contingency tables. Biometrika **72**, 339–351 (1985)

Edwards, D., Havránek, T.: A fast model selection procedure for large families of models. J. Am. Stat. Assoc. **82**, 205–213 (1987)

Eriksen, P.S.: Tests in covariance selection models. Scand. J. Stat. **23**(3), 275–284 (1996)

Gervini, D.: The influence function of the Stahel–Donoho estimator of multivariate location and scatter. Stat. Probab. Lett. **60**(4), 425–435 (2002)

Gervini, D.: A robust and efficient adaptive reweighted estimator of multivariate location and scatter. J. Multivar. Anal. **84**(1), 116–144 (2003)

Gnanadesikan, R., Kettenring, J.R.: Robust estimates, residuals, and outlier detection with multiresponse data. Biometrics **28**(1), 81–124 (1972)

Gottard, A., Pacillo, S.: On the impact of contaminations in graphical Gaussian models. Stat. Methods Appl. **15**(3), 343–354 (2007)

Gottard, A., Pacillo, S.: Robust concentration graph model selection. Comput. Stat. Data Anal. (2008). doi:10.1016/j.csda.2008.11.021

Hampel, F.R.: A general qualitative definition of robustness. Ann. Math. Stat. **42**, 1887–1896 (1971)

Hampel, F.R., Ronchetti, E.M., Rousseeuw, P.J., Stahel, W.A.: Robust Statistics. The Approach Based on Influence Functions. Wiley Series in Probability and Mathematical Statistics. Wiley, New York (1986)

Huber, P.J., Ronchetti, E.M.: Robust Statistics, 2nd edn. Wiley Series in Probability and Statistics. Wiley, Hoboken (2009)

Kent, J.T., Tyler, D.E.: Constrained M-estimation for multivariate location and scatter. Ann. Stat. **24**(3), 1346–1370 (1996)

Kuhnt, S., Becker, C.: Sensitivity of graphical modeling against contamination. In: Schader, Martin, et al. (eds.) Between Data Science and Applied Data Analysis, Proceedings of the 26th Annual Conference of the Gesellschaft für Klassifikation e. V., Mannheim, Germany, July 22–24, 2002, pp. 279–287. Springer, Berlin (2003)

Lauritzen, S.L.: Graphical Models. Oxford Statistical Science Series, vol. 17. Oxford University Press, Oxford (1996)

Lopuhaä, H.P.: On the relation between S-estimators and M-estimators of multivariate location and covariance. Ann. Stat. **17**(4), 1662–1683 (1989)

Lopuhaä, H.P.: Multivariate τ-estimators for location and scatter. Can. J. Stat. **19**(3), 307–321 (1991)

Lopuhaä, H.P.: Asymptotics of reweighted estimators of multivariate location and scatter. Ann. Stat. **27**(5), 1638–1665 (1999)

Magnus, J.R., Neudecker, H.: Matrix Differential Calculus with Applications in Statistics and Econometrics, 2nd edn. Wiley Series in Probability and Statistics. Wiley, Chichester (1999)

Maronna, R.A.: Robust M-estimators of multivariate location and scatter. Ann. Stat. **4**, 51–67 (1976)

Maronna, R.A., Yohai, V.J.: The behavior of the Stahel-Donoho robust multivariate estimator. J. Am. Stat. Assoc. **90**(429), 330–341 (1995)

Maronna, R.A., Zamar, R.H.: Robust estimates of location and dispersion for high-dimen sional datasets. Technometrics **44**, 307–317 (2002)

Maronna, R.A., Martin, D.R., Yohai, V.J.: Robust Statistics: Theory and Methods. Wiley Series in Probability and Statistics. Wiley, Chichester (2006)

Meinshausen, N., Bühlmann, P.: High-dimensional graphs and variable selection with the Lasso. Ann. Stat. **34**(3), 1436–1462 (2006)

Miyamura, M., Kano, Y.: Robust Gaussian graphical modeling. J. Multivar. Anal. **97**(7), 1525–1550 (2006)

Paindaveine, D.: A canonical definition of shape. Stat. Probab. Lett. **78**(14), 2240–2247 (2008)

Porteous, B.T.: Stochastic inequalities relating a class of log-likelihood ratio statistics to their asymptotic χ^2 distribution. Ann. Stat. **17**(4), 1723–1734 (1989)

Rocke, D.M.: Robustness properties of S-estimators of multivariate location and shape in high dimension. Ann. Stat. **24**(3), 1327–1345 (1996)

Rousseeuw, P.J.: Multivariate estimation with high breakdown point. In: Grossmann, W., Pflug, G.C., Vincze, I., Wertz, W. (eds.) Mathematical Statistics and Applications, Vol. B, Proc. 4th Pannonian Symp. Math. Stat., Bad Tatzmannsdorf, Austria, September 4–10, 1983, pp. 283–297. Reidel, Dordrecht (1985)

Rousseeuw, P.J., Van Driessen, K.: A fast algorithm for the minimum covariance determinant estimator. Technometrics **41**, 212–233 (1999)

Roverato, A.: Cholesky decomposition of a hyper inverse Wishart matrix. Biometrika **87**(1), 99–112 (2000)

Smith, P.W.F.: Assessing the power of model selection procedures used when graphical modeling. In: Dodge, Y., Whittaker, J. (eds.) Computational Statistics, vol. I, pp. 275–280. Physica, Heidelberg (1992)

Speed, T.P., Kiiveri, H.T.: Gaussian Markov distributions over finite graphs. Ann. Stat. **14**, 138–150 (1986)

Srivastava, M., Khatri, C.: An Introduction to Multivariate Statistics. North Holland, New York (1979)

Stahel, W.: Robust estimation: Infinitesimal optimality and covariance matrix estimation. PhD thesis, ETH Zürich (1981)

Tyler, D.E.: Robustness and efficiency properties of scatter matrices. Biometrika **70**, 411–420 (1983)

Tyler, D.E.: A distribution-free M-estimator of multivariate scatter. Ann. Stat. **15**, 234–251 (1987)

Visuri, S., Koivunen, V., Oja, H.: Sign and rank covariance matrices. J. Stat. Plan. Inference **91**(2), 557–575 (2000)

Vogel, D., Köllmann, C., Fried, R.: Partial correlation estimates based on signs. In Heikkonen, J. (ed.) Proceedings of the 1st Workshop on Information Theoretic Methods in Science and Engineering. TICSP series # 43 (2008)

Vogel, D.: On generalizing Gaussian graphical models. In: Ciumara, R., Bădin, L. (eds.) Proceedings of the 16th European Young Statisticians Meeting, University of Bucharest, pp. 149–153 (2009)

Whittaker, J.: Graphical Models in Applied Multivariate Statistics. Wiley Series in Probability and Mathematical Statistics. Wiley, Chichester (1990)

Zuo, Y.: Robust location and scatter estimators in multivariate analysis. In: Fan, J., Koul, H. (eds.) Frontiers in Statistics. Dedicated to Peter John Bickel on Honor of his 65th Birthday, pp. 467–490. Imperial College Press, London (2006)

Zuo, Y., Cui, H.: Depth weighted scatter estimators. Ann. Stat. **33**(1), 381–413 (2005)

Strong Laws of Large Numbers and Nonparametric Estimation

Harro Walk

Abstract Elementary approaches to classic strong laws of large numbers use a monotonicity argument or a Tauberian argument of summability theory. Together with results on variance of sums of dependent random variables they allow to establish various strong laws of large numbers in case of dependence, especially under mixing conditions. Strong consistency of nonparametric regression estimates of local averaging type (kernel and nearest neighbor estimates), pointwise as well as in L_2, can be considered as a generalization of strong laws of large numbers. Both approaches can be used to establish strong universal consistency in the case of independence and, mostly by sharpened integrability assumptions, consistency under ρ-mixing or α-mixing. In a similar way Rosenblatt-Parzen kernel density estimates are treated.

1 Introduction

The classic strong law of large numbers of Kolmogorov deals with independent identically distributed integrable real random variables. An elementary approach has been given by Etemadi (1981). He included the arithmetic means of nonnegative (without loss of generality) random variables truncated at the number equal to the index, between fractions with the first $\lceil a^{N+1} \rceil$ summands in the numerator and the denominator a^N and fractions with the first $\lfloor a^N \rfloor$ summands in the numerator and the denominator a^{N+1} ($a > 1$, rational), investigated the almost sure (a.s.) convergence behavior of the majorant sequence and the minorant sequence by use of Chebyshev's inequality and let then go $a \downarrow 1$. This method was refined by Etemadi (1983) himself, Csörgő et al. (1983) and Chandra and Goswami (1992, 1993) and extended to the investigation of nonparametric regression and density estimates un-

Harro Walk, Department of Mathematics, Universität Stuttgart, Pfaffenwaldring 57, 70569 Stuttgart, Germany
e-mail: harro.walk@mathematik.uni-stuttgart.de

L. Devroye et al. (eds.), *Recent Developments in Applied Probability and Statistics*,
DOI 10.1007/978-3-7908-2598-5_8,

der mixing conditions by Irle (1997) and to the proof of strong universal pointwise consistency of nearest neighbor regression estimates under independence by Walk (2008a).

Another approach to strong laws of large numbers was proposed by Walk (2005b). Classic elementary Tauberian theorems (Lemma 1a,b) in summmability theory allow to conclude convergence of a sequence (s_n) of real numbers from convergence of their arithmetic means (C_1 summability, Cesàro summability of (s_n)) together with a so-called Tauberian condition on variation of the original sequence (s_n). If (s_n) itself is a sequence of basic arithmetic means $(a_1 + \cdots + a_n)/n$, as the realization in the strong law of large numbers, then the Tauberian condition simply means that (a_n) is bounded from below. In this context the other assumption (C_1-summability of the sequence of basic arithmetic means) is usually replaced by the more practicable, but equivalent, C_2-summability of (a_n), see Lemma 1a. For the sequence of nonnegative truncated random variables centered at their expectations which are bounded by the finite expectation in Kolmogorov's strong law of large numbers, the simple Tauberian condition is obviously fulfilled. To show almost sure (a.s.) C_2-summability of the sequence to 0, it then suffices to show a.s. convergence of a series of nonnegative random variables by taking expectations, see Theorem 1a. The summmability theory approach has been extended by Walk (2005a, 2008b) to establish strong universal L_2-consistency of Nadaraya-Watson type kernel regression estimates (under independence) and strong consistency under mixing conditions and sharpened moment conditions. Both described approaches have different areas of application and will be used in this paper.

In Sect. 2 strong laws of large numbers under conditions on the covariance and more generally under conditions on the variance of sums of random variables (Theorem 1) and under mixing conditions (Theorem 2) are stated. For the two latter situations proofs via the summability theory approach are given. We shall deal with ρ-mixing and α-mixing conditions. Theorem 2a specialized to the case of independence states Kolmogorov's strong law of large numbers and is a consequence of Theorem 1a, which itself is an immediate consequence of the Tauberian Lemma 1a.

Section 3 deals with strong pointwise consistency of Nadaraya-Watson kernel regression estimates under ρ-mixing and α-mixing (Theorem 3), Devroye-Wagner semirecursive kernel regression estimates under ρ-mixing (Theorem 4) and k_n-nearest neighbor regression estimates under independence (Theorem 5). In the proof of Theorem 3 truncation of the response variables is justified by a monotonicity argument of Etemadi type, asymptotic unbiasedness is established by a generalized Lebesgue density theorem of Greblicki et al. (1984), and a.s. convergence after truncation and centering at expectations is shown by exponential inequalities of Peligrad (1992) and Rhomari (2002). Theorem 4 is a result on strong universal pointwise consistency, i.e., strong pointwise consistency for each distribution of (X, Y) with $\mathbf{E}|Y| < \infty$ (X d-dimensional prediction random vector, Y real response random variable); it is an extension from independence Walk (2001) to ρ-mixing due to Frey (2007) by use of the Tauberian Lemma 2 on weighted means. Theorem 5 is a strong universal consistency result of Walk (2008a). Its proof uses Etemadi's monotonicity argument and will be omitted. Irle (1997) uses mixing and boundedness assumptions (Remark 5). Section 4 first points out strong universal L_2-consistency (strong L_2-consistency for all distributions of (X, Y) with $\mathbf{E}\{|Y|^2\} < \infty$ under independence) of k_n-nearest neighbor, semirecursive Devroye-Wagner kernel and Nadaraya-Watson type kernel estimates (Devroye et al. 1994; Györfi et al. 1998 and Walk 2005a, respectively), see Theorem 6 (without proof). Under ρ- and α-mixing and sharpened moment conditions, Theorem 7 (Walk 2008b) states strong L_2- consistency of Nadaraya-Watson regression estimates. Its proof uses the summability theory approach and will be omitted.

The final Sect. 5 deals with Rosenblatt-Parzen kernel density estimates under ρ- and α-mixing. L_1-consistency (Theorem 8) is proven by use of a monotonicity argument of Etemadi type.

2 Strong Laws of Large Numbers

The following lemma states elementary classical Tauberian theorems of Landau (1910) and Schmidt (1925). They allow to conclude Cesàro summability (C_1-summability, i.e., convergence of arithmetic means) from C_2-summability or to conclude convergence from C_1-summability, in each case under an additional assumption (so-called Tauberian condition). A corresponding result of Szász (1929) and Karamata (1938) concerns weighted means (Lemma 2). References for these and related results are Hardly (1949), pp. 121, 124–126, 145, Zeller and Beekmann (1970), pp. 101, 103, 111–113, 117, Korevaar (2004), pp. 12–16, 58, compare also Walk (2005b, 2007).

Lemma 1. (a) *Let the sequence $(a_n)_{n\in\mathbb{N}}$ of real numbers satisfy*

$$\frac{1}{\binom{n+1}{2}} \sum_{j=1}^{n} \sum_{i=1}^{j} a_i \to 0, \tag{1}$$

i.e., C_2-summability of $(a_n)_{n\in\mathbb{N}}$ to 0, or sharper

$$\sum_{n=1}^{\infty} \frac{1}{n^3} \left(\sum_{i=1}^{n} a_i \right)^2 < \infty, \tag{2}$$

and the Tauberian condition

$$\inf_n a_n > -\infty. \tag{3}$$

Then

$$\frac{1}{n} \sum_{i=1}^{n} a_i \to 0. \tag{4}$$

(b) *Let the sequence $(s_n)_{n\in\mathbb{N}}$ of real numbers satisfy*

$$\frac{1}{n}\sum_{k=1}^{n} s_k \to 0 \tag{5}$$

and the Tanberian condition

$$\liminf\, (s_N - s_M) \geq 0 \quad \textit{for } M \to \infty,\ M < N,\ N/M \to 1, \tag{6}$$

i.e.,

$$\liminf\, (s_{N_n} - s_{M_n}) \geq 0$$

for each sequence $((M_n, N_n))$ *in* $\mathbb{N}^2$ *with* $M_n \to \infty$, $M_n < N_n$, $N_n/M_n \to 1$ $(n \to \infty)$.
Then

$$s_n \to 0. \tag{7}$$

To make the paper more self-contained we shall give direct proofs of Lemma 1a and Lemma 1b. Remark 1b states (with proof) that Lemma 1b implies Lemma 1a. The notations $\lfloor s \rfloor$ and $\lceil s \rceil$ for the integer part and the upper integer part of the nonnegative real number s will be used.

Proof (of Lemma 1).

(a) (2) implies (1), because

$$\left| \frac{1}{n^2} \sum_{j=1}^{n} \sum_{i=1}^{j} a_i \right|^2 \leq \frac{1}{n^4} n \sum_{j=1}^{n} \left(\sum_{i=1}^{j} a_i \right)^2 \to 0 \quad (n \to \infty)$$

by the Cauchy-Schwarz inequality, (2) and the Kronecker lemma. (3) means $a_n \geq -c$, $n \in \mathbb{N}$, for some $c \in \mathbb{R}_+$. With

$$t_n := \sum_{i=1}^{n} a_i, \qquad w_n := \sum_{j=1}^{n} t_j, \quad n \in \mathbb{N},$$

for $k \in \{1, \ldots, n\}$ one obtains

$$w_{n+k} - w_n = t_n k + \sum_{j=n+1}^{n+k} (t_j - t_n) \geq k t_n - k^2 c,$$

$$w_{n-k} - w_n = t_n(-k) + \sum_{j=n-k+1}^{n} (t_n - t_j) \geq -k t_n - k^2 c$$

(compare Taylor expansion), thus

$$\frac{w_n - w_{n-k}}{nk} - \frac{kc}{n} \leq \frac{t_n}{n} \leq \frac{w_{n+k} - w_n}{nk} + \frac{kc}{n}.$$

(1) implies

$$\sigma_n := \max\{|w_l|; l = 1, \ldots, 2n\} = o(n^2),$$
$$k(n) := \min\{1 + \lfloor\sqrt{\sigma_n}\rfloor, n\} = o(n).$$

Therefore

$$\frac{|t_n|}{n} \leq \frac{2\sigma_n}{nk(n)} + \frac{k(n)}{n}c$$
$$= \frac{k(n)}{n}\frac{2\sigma_n}{k(n)^2} + \frac{k(n)}{n}c \leq (2+c)\frac{k(n)}{n} \to 0 \quad (n \to \infty),$$

i.e. (4).

(b) With

$$z_n := \sum_{k=1}^{n} s_k, \quad n \in \mathbb{N},$$

for $k \in \{1, \ldots, \lceil\frac{n}{2}\rceil\}$ we obtain as before

$$z_{n+k} - z_n = s_n k + \sum_{j=n+1}^{n+k} (s_j - s_n),$$

$$z_{n-k} - z_n = s_n(-k) + \sum_{j=n-k+1}^{n} (s_n - s_j),$$

thus for $n \geq 2$

$$-\frac{2}{k/n} \sup_{j\in\{\lfloor\frac{n}{2}\rfloor, \lfloor\frac{n}{2}\rfloor+1,\ldots\}} \frac{|z_j|}{j} + (s_n - s_{j(n,k)})$$
$$\leq \frac{n}{k}\frac{z_n}{n} - \frac{n-k}{k}\frac{z_{n-k}}{n-k} + \min_{j\in\{n-k+1,\ldots,n\}} (s_n - s_j)$$
$$\leq s_n$$
$$\leq \frac{n+k}{k}\frac{z_{n+k}}{n+k} - \frac{n}{k}\frac{z_n}{n} - \min_{j\in\{n+1,\ldots,n+k\}} (s_j - s_n)$$
$$\leq 2\frac{1+k/n}{k/n} \sup_{j\in\{n,n+1,\ldots\}} \frac{|z_j|}{j} - (s_{j^*(n,k)} - s_n)$$

with suitable $j(n,k) \in \{n-k+1, \ldots, n\}$, $j^*(n,k) \in \{n+1, \ldots, n+k\}$. Now choose $k = k(n) \in \{1, \ldots, \lceil\frac{n}{2}\rceil\}$ such that $k(n)/n \to 0$ so slowly that, besides $\sup_{j\in\{\lfloor\frac{n}{2}\rfloor, \lfloor\frac{n}{2}\rfloor+1,\ldots\}} \frac{|z_j|}{j} \to 0$ $(n \to \infty)$ (by (5)), even

$$\frac{1}{\frac{k(n)}{n}} \sup_{j\in\{\lfloor\frac{n}{2}\rfloor, \lfloor\frac{n}{2}\rfloor+1,\ldots\}} \frac{|z_j|}{j} \to 0 \quad (n \to \infty).$$

Therefore and by $k(n)/n \to 0$ (once more) together with (6) we obtain

$$
\begin{aligned}
0 &\le \liminf(s_n - s_{j(n,k(n))}) \\
&\le \liminf s_n \le \limsup s_n \\
&\le -\liminf(s_{j^*(n,k(n))} - s_n) \le 0,
\end{aligned}
$$

which yields (7). □

Remark 1.

(a) Assumption (6) in Lemma 1b is fulfilled if

$$
s_{n+1} - s_n \ge u_n + v_n + w_n
$$

with $u_n = O(\frac{1}{n})$, convergence of $(\frac{1}{n}\sum_{i=1}^{n} i v_i)$, $\sum_{n=1}^{\infty} n w_n^2 < \infty$.
For

$$
\left| \sum_{n=M+1}^{N} u_n \right| \le \text{const} \sum_{n=M+1}^{N} \frac{1}{n} \to 0,
$$

$$
\begin{aligned}
\left| \sum_{n=M+1}^{N} v_n \right| &= \left| \sum_{n=M+1}^{N} \frac{1}{n}(n v_n) \right| \\
&= \left| \frac{1}{N+1} \sum_{n=1}^{N} n v_n - \frac{1}{M+1} \sum_{n=1}^{M} n v_n + \sum_{n=M+1}^{N} \frac{1}{n(n+1)} \sum_{i=1}^{n} i v_i \right| \\
&\quad \text{(by partial summation)} \\
&\le o(1) + \sup_{n \in \mathbb{N}} \left(\frac{1}{n+1} \left| \sum_{i=1}^{n} i v_i \right| \right) \sum_{n=M+1}^{N} \frac{1}{n} \\
&\to 0,
\end{aligned}
$$

and (by the Cauchy-Schwarz inequality)

$$
\begin{aligned}
\left| \sum_{n=M+1}^{N} w_n \right| &= \left| \sum_{n=M+1}^{N} n^{-\frac{1}{2}} (n^{\frac{1}{2}} w_n) \right| \\
&\le \left(\sum_{n=M+1}^{N} \frac{1}{n} \right)^{\frac{1}{2}} \left(\sum_{n=1}^{\infty} n w_n^2 \right)^{\frac{1}{2}} \to 0
\end{aligned}
$$

for $M \to \infty$, $M < N$, $N/M \to 1$.

(b) Lemma 1b implies Lemma 1a. For, under the assumptions of Lemma 1a, with $s_n := (a_1 + \cdots + a_n)/n$ one has

$$\frac{1}{n}\sum_{k=1}^{n} s_k = \frac{1}{n(n+1)}\sum_{k=1}^{n}(a_1+\cdots+a_k) + \frac{1}{n}\sum_{k=1}^{n}\frac{1}{k(k+1)}\sum_{j=1}^{k}(a_1+\cdots+a_j)$$

(by partial summation)

$$\to 0 \quad (n\to\infty)$$

by (1), i.e., (5) is fulfilled. Further, with suitable $c \in \mathbb{R}_+$,

$$\begin{aligned} s_{n+1} - s_n &= \frac{a_{n+1}}{n} - \left(\frac{1}{n} - \frac{1}{n+1}\right)(a_1+\cdots+a_{n+1}) \\ &\geq -\frac{c}{n} - \frac{1}{n}s_{n+1} \\ &=: u_n + v_n \end{aligned}$$

(by (3)), where $u_n = O(\frac{1}{n})$ and

$$\frac{1}{n}\sum_{i=1}^{n} i v_i = -\frac{1}{n}\sum_{i=1}^{n} s_{i+1} \to 0 \quad (n\to\infty),$$

by (5). Thus, by (a), (6) is fulfilled. Now Lemma 1b yields (7), i.e., (4).

(c) Analogously to (b) one shows that Lemma 1b implies the variant of Lemma 1a where assumption (3) is replaced by $a_n \geq -c_n$, $n \in \mathbb{N}$, for some sequence (c_n) in $\mathbb{R}_+$ with convergence of $(\frac{1}{n}\sum_{i=1}^{n} c_i)$.

Part (a) of the following Theorem 1 immediately follows from Lemma 1a, compare Walk (2005b), see the proof below. Part (b) is due to Chandra and Goswami (1992, 1993) and has been shown by a refinement of Etemadi's (1981, 1983) argument. Part (c) contains the classic Rademacher-Menchoff theorem and is obtained according to Serfling (1970b), proof of Theorem 2.1 there; its condition can be slightly weakend (see Walk 2007). $\mathbf{Cov}_+$ denotes the nonnegative part of $\mathbf{Cov}$, i.e., $\max\{0, \mathbf{Cov}\}$.

Theorem 1. *Let (Y_n) be a sequence of square integrable real random variables. If*

(a) $Y_n \geq 0$, $\sup_n \mathbf{E}Y_n < \infty$ *and*

$$\sum_{n=1}^{\infty}\frac{1}{n^3}\mathbf{Var}\left\{\sum_{i=1}^{n} Y_i\right\} < \infty, \tag{8}$$

or

(b) $Y_n \geq 0$, $\sup_n \frac{1}{n}\sum_{k=1}^{n}\mathbf{E}Y_k < \infty$ *and*

$$\sum_{n=1}^{\infty}\frac{1}{n^2}\sum_{i=1}^{n}\mathbf{Cov}_+(Y_i, Y_n) < \infty \quad \left(\Leftrightarrow \sum_{n=1}^{\infty}\frac{1}{n^3}\sum_{j=1}^{n}\sum_{i=1}^{j}\mathbf{Cov}_+(Y_i, Y_j) < \infty\right)$$

or

(c) $\sum_{n=1}^{\infty} \frac{(\log(n+1))^2}{n^2} \sum_{i=1}^{n} \mathbf{Cov}_+(Y_i, Y_n) < \infty$,

then

$$\frac{1}{n}\sum_{i=1}^{n}(Y_i - \mathbf{E}Y_i) \to 0 \quad a.s. \tag{9}$$

Proof (of Theorem 1a). Obviously $(Y_n - \mathbf{E}Y_n)$ is bounded from below. Equation (8) yields

$$\sum_{n=1}^{\infty} \frac{1}{n^3}\left|\sum_{i=1}^{n}(Y_i - \mathbf{E}Y_i)\right|^2 < \infty \quad \text{a.s.}$$

Thus (9) follows by Lemma 1a. □

Remark 2. Analogously one can show that in Theorem 1a the condition $\sup_n \mathbf{E}Y_n < \infty$ may be replaced by convergence of the sequence $(\frac{1}{n}\sum_{i=1}^{n}\mathbf{E}Y_i)$. Instead of Lemma 1a one uses Remark 1c which is based on Lemma 1b.

Theorem 1a and a corresponding theorem for weighted means based on Lemma 2 below allow to apply results on the variance of sums of dependent random variables (see Theorem 2a and Theorem 4, respectively, with proofs). In the special case of independence, Theorem 2a is Kolmogorov's strong law of large numbers, and its proof by Theorem 1a is elementary.

Remark 3. If the square integrable real random variables Y_n satisfy

$$|\mathbf{Cov}(Y_i, Y_j)| \le r(|i - j|),$$

then

$$\sum_{n=1}^{\infty} \frac{1}{n} r(n) < \infty$$

or in the case of weak stationarity the weakest possible condition

$$\sum_{n=3}^{\infty} \frac{\log\log n}{n \log n} r(n) < \infty$$

imply

$$\frac{1}{n}\sum_{i=1}^{n}(Y_i - \mathbf{E}Y_i) \to 0 \quad \text{a.s.}$$

(see Walk 2005b and Gaposhkin 1977, respectively).

Lemma 2 generalizes Lemma 1a and will be applied in Sect. 3.

Lemma 2. *Let* $0 < \beta_n \uparrow \infty$ *with* $\beta_{n+1}/\beta_n \to 1$ *and set* $\gamma_n := \beta_n - \beta_{n-1}$ $(n \in \mathbb{N})$ *with* $\beta_0 := 0$. *Let* (a_n) *be a sequence of real numbers bounded from below. If*

$$\frac{1}{\beta_n}\sum_{k=1}^{n}\frac{\gamma_k}{\beta_k}\sum_{j=1}^{k}\gamma_j a_j \to 0$$

or sharper

$$\sum_{n=1}^{\infty}\frac{\gamma_n}{\beta_n^3}\left(\sum_{k=1}^{n}\gamma_k a_k\right)^2 < \infty,$$

then

$$\frac{1}{\beta_n}\sum_{i=1}^{n}\gamma_i a_i \to 0.$$

Also Chandra and Goswami (1992, 1993) gave their result in a more general form with $1/n$ and $1/j^2$ (and $1/n^3$) replaced by $1/\beta_n$ and $1/\beta_j^2$ (and $(\beta_n - \beta_{n-1})/\beta_n^3$), respectively, in Theorem 1b above, where $0 < \beta_n \uparrow \infty$.

The following theorem establishes validity of the strong law of large numbers under some mixing conditions. Part (a) comprehends Kolmogorov's classic strong law of large numbers for independent identically distributed integrable real random variables and, as this law, can be generalized to the case of random variables with values in a real separable Banach space.

We shall use the ρ-mixing and the α-mixing concept of dependence of random variables. Let $(Z_n)_{n\in\mathbb{N}}$ be a sequence of random variables on a probability space $(\Omega, \mathcal{A}, P)$. $\mathcal{F}_m^n$ denotes the σ-algebra generated by $(Z_m, \ldots, Z_n)$ for $m \le n$. The ρ-mixing and α-mixing coefficients are defined by

$$\rho_n := \sup_{k\in\mathbb{N}} \sup\{|\operatorname{corr}(U,V)|;\ U \in L_2(\mathcal{F}_1^k),\ V \in L_2(\mathcal{F}_{k+n}^\infty),\ U, V \text{ real-valued}\},$$

$$\alpha_n := \sup_{k\in\mathbb{N}} \sup\{|P(A\cap B) - P(A)P(B)|;\ A \in \mathcal{F}_1^k,\ B \in \mathcal{F}_{k+n}^\infty\},$$

respectively. The sequence (Z_n) is called ρ-mixing, if $\rho_n \to 0$, and α-mixing, if $\alpha_n \to 0$ $(n \to \infty)$. It holds $4\alpha_n \le \rho_n$ (see, e.g., Györfi et al. 1989, p. 9, and Doukhan 1994, p. 4). $\log_+$ below denotes the nonnegative part of log, i.e., $\max\{0, \log\}$.

Theorem 2. *Let the real random variables Y_n be identically distributed.*

(a) *If* $\mathbf{E}|Y_1| < \infty$ *and if* (Y_n) *is independent or, more generally,* ρ*-mixing with*

$$\sum_{n=1}^{\infty}\frac{1}{n}\rho_n < \infty,$$

$$\textit{e.g., if } \rho_n = O\left(\frac{1}{(\log n)^{1+\delta}}\right) \quad \textit{for some } \delta > 0,$$

then

$$\frac{1}{n}\sum_{i=1}^{n} Y_i \to \mathbf{E}Y_1 \quad \textit{a.s.}$$

(a$_1$) *If* $\mathbf{E}\{|Y_1|\log_+|Y_1|\} < \infty$ *and if* (Y_n) *is* ρ*-mixing, then*

$$\frac{1}{n}\sum_{i=1}^{n} Y_i \to \mathbf{E}Y_1 \quad a.s.$$

(b) *If* $\mathbf{E}\{|Y_1|\log_+|Y_1|\} < \infty$ *and if* (Y_n) *is* α*-mixing with* $\alpha_n = O(n^{-\alpha})$ *for some* $\alpha > 0$*, then*

$$\frac{1}{n}\sum_{i=1}^{n} Y_i \to \mathbf{E}Y_1 \quad a.s.$$

Proof. Let $Y_n \geq 0$ without loss of generality. We set $Y_n^{[c]} := Y_n 1_{\{Y_n \leq c\}},\ c > 0$.

(a) We use a well-known truncation argument. Because of $\mathbf{E}Y_1 < \infty$, we have a.s. $Y_n = Y_n^{[n]}$ for some random index on. Therefore and because of $\mathbf{E}Y_n^{[n]} \to \mathbf{E}Y$, it suffices to show

$$\frac{1}{n}\sum_{i=1}^{n}\left(Y_i^{[i]} - \mathbf{E}Y_i^{[i]}\right) \to 0 \quad \text{a.s.}$$

Because of $Y_n^{[n]} \geq 0$, $\mathbf{E}Y_n^{[n]} \leq \mathbf{E}Y < \infty$, by Theorem 1a it is enough to show

$$\sum_{n=1}^{\infty}\frac{1}{n^3}\mathbf{Var}\left\{\sum_{i=1}^{n} Y_i^{[i]}\right\} < \infty.$$

Application of Lemma 3a below for real random variables yields

$$\mathbf{Var}\left\{\sum_{i=1}^{n} Y_i^{[i]}\right\} \leq Cn\mathbf{E}\left\{\left(Y_n^{[n]}\right)^2\right\} = Cn\mathbf{E}\left\{\left(Y_1^{[n]}\right)^2\right\}$$

with some constant $C \in \mathbb{R}_+$. In the special case of independence one immediately obtains the inequality with $C = 1$. From this and the well-known relation

$$\begin{aligned}
\sum_{n=1}^{\infty}\frac{1}{n^2}\mathbf{E}\left\{\left(Y_1^{[n]}\right)^2\right\} &= \sum_{n=1}^{\infty}\sum_{i=1}^{n}\frac{1}{n^2}\int_{(i-1,i]} t^2 P_{Y_1}(dt) \\
&= \sum_{i=1}^{\infty}\int_{(i-1,i]} t^2 P_{Y_1}(dt)\sum_{n=i}^{\infty}\frac{1}{n^2} \\
&\leq \sum_{i=1}^{\infty}\frac{2}{i}\int_{(i-1,i]} t^2 P_{Y_1}(dt) \\
&\leq 2\mathbf{E}Y_1 < \infty
\end{aligned}$$

we obtain the assertion.

(a$_1$) Let $\epsilon = \frac{1}{4}$, $\kappa = \frac{1}{8}$. From the integrability assumption we obtain as in the first step of the proof of Theorem 3 below (specialization to X_n = const) that

$$\frac{1}{n}\sum_{i=1}^{n}\left(Y_i - Y_i^{[i^\kappa]}\right) \to 0 \quad \text{a.s.}$$

As in part (a) it is enough to show

$$\sum_{n=1}^{\infty}\frac{1}{n^3}\mathbf{Var}\left\{\sum_{i=1}^{n} Y_i^{[i^\kappa]}\right\} < \infty.$$

Application of Lemma 3a below for real random variables yields

$$\mathbf{Var}\left\{\sum_{i=1}^{n} Y_i^{[i^\kappa]}\right\} \le C(\epsilon)n^{1+\epsilon}\mathbf{E}\left\{\left(Y_n^{[n^\kappa]}\right)^2\right\} \le C(\epsilon)n^{1+\epsilon+2\kappa}$$

for some $C(\epsilon) < \infty$ and thus the assertion.

(b) Let $\kappa = \frac{1}{4}\min\{1, \alpha\}$. As in (a$_1$) it is enough to show

$$\sum_{n=1}^{\infty}\frac{1}{n^3}\mathbf{Var}\left\{\sum_{i=1}^{n} Y_i^{[i^\kappa]}\right\} < \infty.$$

Application of Lemma 3b below for real random variables yields

$$\mathbf{Var}\left\{\sum_{i=1}^{n} Y_i^{[i^\kappa]}\right\} = O\left(n^{2\kappa+2-\min\{1,\alpha\}}\log(n+1)\right)$$

and thus the assertion. □

Part (a) of the following lemma is due to Peligrad (1992), Proposition 3.7 and Remark 3.8. Part (b) is an immediate consequence of an inequality of Dehling and Philipp (1982), Lemma 2.2. Parts (c) and (d) are due to Liebscher (1996), Lemma 2.1, and Rio (1993), pp. 592, 593, respectively.

Lemma 3. (a) *Let (Z_n) be a ρ-mixing sequence of square integrable variables with values in a real separable Hilbert space. Then for each $\epsilon > 0$*

$$\mathbf{Var}\left\{\sum_{i=1}^{n} Z_i\right\} \le C(\epsilon)n^{1+\epsilon}\max_{i=1,\dots,n}\mathbf{Var}Z_i$$

for some $C(\epsilon) < \infty$. If additionally

$$\sum_{n=1}^{\infty}\rho_{2^n} < \infty \quad \textit{or, equivalently,} \quad \sum_{n=1}^{\infty}\frac{1}{n}\rho_n < \infty,$$

then

$$\mathbf{Var}\left\{\sum_{i=1}^{n} Z_i\right\} \le Cn \max_{i=1,\dots,n} \mathbf{Var} Z_i$$

for some $C < \infty$.

(b) *Let* (Z_n) *be an* α*-mixing sequence of essentially bounded random variables with values in a real separable Hilbert space with* $\alpha_n = O(n^{-\alpha})$ *for some* $\alpha > 0$, *then*

$$\mathbf{Var}\left\{\sum_{i=1}^{n} Z_i\right\} \le Cn^{2-\min\{1,\alpha\}} \log(n+1) \max_{i=1,\dots,n} (\text{ess sup} \|Z_i\|)^2$$

for some $C < \infty$. *In the case* $\alpha \ne 1$ *the assertion holds without the factor* $\log(n+1)$.

(c) *Let* (Z_n) *be an* α*-mixing sequence of real random variables with* $\alpha_n = O(n^{-\alpha})$ *for some* $\alpha > 1$ *and* $\mathbf{E}\{|Z_n|^{2\alpha/(\alpha-1)}\} < \infty$, $n \in \mathbb{N}$. *Then*

$$\mathbf{Var}\left\{\sum_{i=1}^{n} Z_i\right\} \le Cn \log(n+1) \max_{i=1,\dots,n} \left(\mathbf{E}\left\{|Z_i|^{2\alpha/(\alpha-1)}\right\}\right)^{\frac{\alpha-1}{\alpha}}$$

for some $C < \infty$.

(d) *Let* (Z_n) *be a weakly stationary* α*-mixing sequence of identically distributed real random variables with* $\alpha_n = O(\delta^n)$ *for some* $\delta \in (0, 1)$ *and* $\mathbf{E}\{Z_1^2 \log_+ |Z_1|\} < \infty$, *then*

$$\mathbf{Var}\left\{\sum_{i=1}^{n} Z_i\right\} \le Cn\mathbf{E}\left\{Z_1^2 \log_+ |Z_1|\right\}$$

for some $C < \infty$.

3 Pointwise Consistent Regression Estimates

In regression analysis, on the basis of an observed d-dimensional random predictor vector X one wants to estimate the non-observed real random response variable Y by $f(X)$ with a suitable measurable function $f : \mathbb{R}^d \to \mathbb{R}$. In case of a square integrable Y one is often interested to minimize the L_2 risk or mean squared error $\mathbf{E}\{|f(X) - Y|^2\}$. As is well known the optimal f is then given by the regression function m of Y on X defined by $m(x) := \mathbf{E}\{Y|X = x\}$. This follows from

$$\mathbf{E}\{|f(X) - Y|^2\} = \int_{\mathbb{R}^d} |f(x) - m(x)|^2 \mu(dx) + \mathbf{E}\{|m(X) - Y|^2\},$$

where μ denotes the distribution of X. Usually the distribution $P_{(X,Y)}$ of (X, Y), especially m, is unknown. If there is the possibility to observe a training sequence $(X_1, Y_1), (X_2, Y_2), \dots$ of $(d + 1)$-dimensional random vectors distributed like

(X, Y) up to the index n, one now wants to estimate m by $m_n(x) := m_n(X_1, Y_1, \ldots, X_n, Y_n; x)$ in such a way that

$$\int |m_n(x) - m(x)|^2 \mu(dx) \to 0 \quad (n \to \infty)$$

almost surely (a.s.) or at least in probability. Inspired by $m(x) = \mathbf{E}(Y|X = x)$, $x \in \mathbb{R}^d$, one uses local averaging methods, where $m(x)$ is estimated by the average of those Y_i where X_i is close to x. Inspired by the above minimum property one also uses least squares methods, which minimize the empirical L_2 risk over a general set $\mathcal{F}_n$ of functions. The classic partitioning regression estimate (regressogram) is a local averaging method as well as a least squares method where $\mathcal{F}_n$ consists of the functions which are constant on each set belonging to a partition $\mathcal{P}_n$ of $\mathbb{R}^d$.

A frequently used local averaging estimate is the regression kernel estimate of Nadaraya and Watson. It uses a kernel function $K : \mathbb{R}^d \to \mathbb{R}_+$, usually with $0 < \int K(x)\lambda(dx) < \infty$ (λ denoting the Lebesgue-Borel measure on $\mathcal{B}_d$), e.g., $K = 1_{S_{0,1}}$ (naive kernel), $K(x) = (1 - \|x\|^2)1_{S_{0,1}}(x)$ (Epanechnikov kernel), $K(x) = (1 - \|x\|^2)^2 1_{S_{0,1}}(x)$ (quartic kernel) and $K(x) = e^{-\|x\|^2/2}$ (Gaussian kernel), with $x \in \mathbb{R}^d$, and bandwidth $h_n \in (0, \infty)$, usually satisfying $h_n \to 0$, $nh_n^d \to \infty$ $(n \to \infty)$, e.g., $h_n = cn^{-\gamma}$ $(c > 0,\ 0 < \gamma d < 1)$. ($S_{x,h}$ for $x \in \mathbb{R}^d$, $h > 0$ denotes the sphere in $\mathbb{R}^d$ with center x and radius h.) The estimator m_n is defined by

$$m_n(x) := \frac{\sum_{i=1}^n Y_i K(\frac{x-X_i}{h_n})}{\sum_{i=1}^n K(\frac{x-X_i}{h_n})}, \quad x \in \mathbb{R}^d \tag{10}$$

with $0/0 := 0$. The k_n-nearest neighbor $(k_n - NN)$ regression estimate m_n of m is defined by

$$m_n(x) := \frac{1}{k_n}\sum_{i=1}^n Y_i 1_{\{X_i \text{ is among the } k_n \text{NNs of } x \text{ in } (X_1,\ldots,X_n)\}} \tag{11}$$

with $k_n \in \{1, \ldots, n-1\}$, $n \geq 2$, usually satisfying $k_n/n \to 0$, $k_n \to \infty$ $(n \to \infty)$. Ambiguities in the definition of NNs (on the basis of the Euclidean distance in $\mathbb{R}^d$) can be solved by random tie-breaking. As to references see Györfi et al. (2002).

A regression estimation sequence is called strongly universally (L_2-)consistent (usually in the case that the sequence of identically distributed $(d+1)$-dimensional random vectors $(X_1, Y_1), (X_2, Y_2), \ldots$ is independent), if

$$\int |m_n(x) - m(x)|^2 \mu(dx) \to 0 \quad \text{a.s.} \tag{12}$$

for all distributions of (X, Y) with $\mathbf{E}\{Y^2\} < \infty$. It is called strongly universally pointwise consistent, if

$$m_n(x) \to m(x) \quad \text{a.s. mod } \mu$$

for all distributions of (X, Y) with $\mathbf{E}|Y| < \infty$. (mod μ means that the assertion holds for μ-almost all $x \in \mathbb{R}^d$.) Correspondingly one speaks of weak consistency if one has convergence in first mean (or in probability).

Results on strong universal pointwise or L_2-consistency will be stated which generalize Kolmogorov's strong law of large numbers. If the independence condition there is relaxed to a mixing condition, mostly the moment condition $\mathbf{E}|Y| < \infty$ or $\mathbf{E}\{|Y|^2\} < \infty$ for pointwise or L_2-consistency, respectively, has to be strengthened to $\mathbf{E}\{|Y| \log_+ |Y|\} < \infty$ or higher moment conditions. We shall use ρ-mixing and α-mixing conditions. No continuity assumptions on the distribution of X will be made.

This section and the next section deal with strong pointwise consistency and with strong L_2-consistency, respectively.

In this section, more precisely, strong pointwise consistency of Nadaraya-Watson estimates (Theorem 3), strong universal pointwise consistency of semi-recursive Devroye-Wagner estimates (Theorem 4), both under mixing conditions, and strong universal pointwise consistency of k_n-nearest neighbor estimates under independence (Theorem 5) are stated.

Theorem 3. *Let $(X, Y), (X_1, Y_1), (X_2, Y_2), \ldots$ be identically distributed $(d+1)$-dimensional random vectors with $\mathbf{E}\left\{|Y| \log_+ |Y|\right\} < \infty$. Let K be a measurable function on $\mathbb{R}^d$ satisfying $c_1 H(\|x\|) \leq K(x) \leq c_2 H(\|x\|)$, $x \in \mathbb{R}^d$, for some $0 < c_1 < c_2 < \infty$ and a nondecreasing function $H : \mathbb{R}_+ \to \mathbb{R}_+$ with $H(+0) > 0$ and $t^d H(t) \to 0$ $(t \to \infty)$, e.g., naive, Epanechnikov, quartic and Gaussian kernel. For $n \in \mathbb{N}$, with bandwidth $h_n > 0$, define m_n by* (10).

(a) *If the sequence $((X_n, Y_n))$ is ρ-mixing with $\rho_n = O(n^{-\rho})$ for some $\rho > 0$ and if h_n is chosen as $h_n = cn^{-\gamma}$ with $c > 0$, $0 < \gamma d < 2\rho/(1+2\rho)$, then*

$$m_n(x) \to m(x) \quad \textit{a.s.} \bmod \mu.$$

(b) *If the sequence $((X_n, Y_n))$ is α-mixing with $\alpha_n = O(n^{-\alpha})$ for some $\alpha > 1$ and if h_n is chosen as $h_n = cn^{-\gamma}$ with $c > 0$, $0 < \gamma d < (2\alpha - 2)/(2\alpha + 3)$, then*

$$m_n(x) \to m(x) \quad \textit{a.s.} \bmod \mu.$$

Remark 4. Theorem 3 in both versions (a) and (b) comprehends the case of independent identically distributed random vectors with choice $h_n = cn^{-\gamma}$ satisfying $0 < \gamma d < 1$ treated in Kozek et al. (1998), Theorem 2, with a somewhat more general choice of h_n, but with a somewhat stronger integrability condition such as $\mathbf{E}\{|Y| \log_+ |Y| (\log_+ \log_+ |Y|)^{1+\delta}\} < \infty$ for some $\delta > 1$. In the proof of Theorem 3 exponential inequalities of Peligrad (1992) and Rhomari (2002) together with the above variance inequalities and a generalized Lebesgue density theorem due to Greblicki et al. (1984) together with a covering lemma for kernels are used. In the independence case the classic Bernstein exponential inequality, see Györfi et al. (2002), Lemma A.2, can be used.

Regarding Lemma 3a,c we can state the exponential inequalities of Peligrad (1992) and Rhomari (2002) for ρ-mixing and α-mixing sequences, respectively, of bounded real random variables in the following somewhat specialized form.

Lemma 4. *Let Z_n, $n \in \mathbb{N}$, be bounded real random variables and set*

$$L_n := \max_{i=1,\dots,n} \operatorname{ess\,sup} |Z_i|.$$

(a) *Let (Z_n) be ρ-mixing with $\rho_n = O(n^{-\rho})$ for some $\rho > 0$. Then there are constants $c_1, c_2 \in (0,\infty)$ such that for all $n \in \mathbb{N}$, $\epsilon^* > 0$*

$$P\left\{\left|\sum_{i=1}^{n}(Z_i - \mathbf{E}Z_i)\right| > \epsilon^*\right\} \le c_1 \exp\left(-\frac{c_2\epsilon^*}{n^{1/2}\max_{i=1,\dots,n}(\mathbf{E}\{|Z_i|^2\})^{1/2} + L_n n^{1/(1+2\rho)}}\right).$$

(b) *Let (Z_n) be α-mixing with $\alpha_n = O(n^{-\alpha})$ for some $\alpha > 1$. Then there are constants $c_1, c_2 \in (0,\infty)$ such that for all $n \in \mathbb{N}$, $\epsilon^* > 0$, $\beta \in (0,1)$*

$$P\left\{\left|\sum_{i=1}^{n}(Z_i - \mathbf{E}Z_i)\right| > \epsilon^*\right\} \le 4\exp\left(-\frac{c_1(\epsilon^*)^2}{n\log(n+1)\max_{i=1,\dots,n}(\mathbf{E}\{|Z_i|^{2\alpha/(\alpha-1)}\})^{(\alpha-1)/\alpha} + \epsilon^* L_n n^{\beta}}\right) + c_2 \max\left\{\left(\frac{L_n n}{\epsilon^*}\right)^{\frac{1}{2}}, 1\right\} n^{1-\beta-\beta\alpha}.$$

The following generalized Lebesgue density theorem is due to Greblicki et al. (1984) (see also Györfi et al. 2002, Lemma 24.8).

Lemma 5. *Let K as in Theorem 3, $0 < h_n \to 0$ $(n \to \infty)$, and let μ be a probability measure on $\mathcal{B}_d$. Then for all μ-integrable functions $f : \mathbb{R}^d \to \mathbb{R}$,*

$$\frac{\int K(\frac{x-t}{h_n}) f(t)\mu(dt)}{\int K(\frac{x-t}{h_n})\mu(dt)} \to f(x) \mod \mu.$$

The next lemma is due to Devroye (1981) (see also Györfi et al. 2002, Lemma 24.6).

Lemma 6. *Let μ be a probability measure on $\mathcal{B}_d$ and $0 < h_n \to 0$ $(n \to \infty)$. Then*

$$\liminf \frac{\mu(S_x, h_n)}{h_n^d} > 0 \mod \mu.$$

It follows a covering lemma. It can be proven as Lemma 23.6 in Györfi et al. (2002) where $K = \widetilde{K}$ is used.

Lemma 7. *Let H, $\widetilde{H}$ and K, $\widetilde{K}$ be functions as H and K, respectively, in Theorem 3. There exists $\tilde{c} \in (0, \infty)$ depending only on K and $\widetilde{K}$ such that for all $h > 0$ and $u \in \mathbb{R}^d$*

$$\int \frac{\widetilde{K}(\frac{x-u}{h})}{\int K(\frac{x-t}{h})\mu(dt)}\mu(dx) \le \tilde{c}.$$

Proof (of Theorem 3). It suffices to show

$$\bar{m}_n(x) := \frac{\sum_{i=1}^n Y_i K(\frac{x-X_1}{h_n})}{n \int K(\frac{x-t}{h_n})\mu(dt)} \to m(x) \quad \text{a.s. mod } \mu, \tag{13}$$

because this result together with its special case for $Y_i = \text{const} = 1$ yields the assertion. Let $Y_i \ge 0$, $0 \le K \le 1$, without loss of generality.

In the first step, for an arbitrary fixed $\kappa > 0$ and $Y_i^* := Y_i^{[i^\kappa]} := Y_i 1_{[Y_i \le i^\kappa]}$, we show

$$\frac{\sum_{i=1}^n (Y_i - Y_i^*) K(\frac{x-X_i}{h_n})}{n \int K(\frac{x-t}{h_n})\mu(dt)} \to 0 \quad \text{a.s. mod } \mu, \tag{14}$$

which together with (16) below yields the assertion. The notation $K_h(x)$ for $K(\frac{x}{h})$ $(x \in \mathbb{R}^d, h > 0)$ will be used.

According to a monotonicity argument of Etemadi (1981), for (14) it suffices to show

$$V_n(x) := \frac{\sum_{i=1}^{2^{n+1}} (Y_i - Y_i^*) K_{h_{2^n}}(x - X_i)}{2^n \int K_{h_{2^{n+1}}}(x-t)\mu(dt)} \to 0 \quad \text{a.s. mod } \mu.$$

We notice

$$h_{2^n} / h_{2^{n+1}} = 2^\gamma,$$

thus

$$K_{h_{2^n}} = K_{h_{2^{n+1}}}\left(\frac{\cdot}{2^\gamma}\right) =: \widetilde{K}_{h_{2^{n+1}}}$$

and, because of Lemma 7,

$$\begin{aligned} &\int \frac{K_{h_{2^n}}(x-z)}{\int K_{h_{2^{n+1}}}(x-t)\mu(dt)}\mu(dx) \\ &= \int \frac{\widetilde{K}_{h_{2^{n+1}}}(x-z)}{\int K_{h_{2^{n+1}}}(x-t)\mu(dt)}\mu(dx) \le \tilde{c} < \infty \end{aligned}$$

for all $z \in \mathbb{R}^d$ and all n. Therefore, with suitable constants c_3, $c_4(\kappa)$,

$$
\begin{aligned}
\mathbf{E}\sum_{n=1}^{\infty}\int V_n(x)\mu(dx) &\le \tilde{c}\sum_{n=1}^{\infty}2^{-n}\sum_{i=1}^{2^{n+1}}\mathbf{E}\{Y_i - Y_i^*\}\\
&\le \tilde{c}\sum_{i=1}^{\infty}\left(\sum_{n=\max\{1,\lfloor(\log i)/(\log 2)\rfloor-1\}}^{\infty}2^{-n}\right)\mathbf{E}Y_i 1_{[Y_i>i^\kappa]}\\
&\le c_3\sum_{i=1}^{\infty}\frac{1}{i}\sum_{j=\lfloor i^\kappa\rfloor}^{\infty}\int_{(j,j+1]}vP_Y(dv)\\
&\le c_3\sum_{j=1}^{\infty}\left(\sum_{i=1}^{\lceil (j+1)^{\frac{1}{\kappa}}\rceil}\frac{1}{i}\right)\int_{(j,j+1]}vP_Y(dv)\\
&\le c_4(\kappa)\mathbf{E}\left\{Y\log_+ Y\right\} < \infty.
\end{aligned}
$$

This yields (14). In the second step we show

$$
\frac{\sum_{i=1}^{n}\mathbf{E}\{Y_i^* K(\frac{x-X_i}{h_n})\}}{n\int K(\frac{x-t}{h_n})\mu(dt)} \to m(x) \quad \bmod \mu. \tag{15}
$$

We have

$$
\frac{\sum_{i=1}^{n}\mathbf{E}\{Y_i^* K(\frac{x-X_i}{h_n})\}}{n\int K(\frac{x-t}{h_n})\mu(dt)} \le \frac{\mathbf{E}\{YK(\frac{x-X}{h_n})\}}{\int K(\frac{x-t}{h_n})\mu(dt)} = \frac{\int m(t)K(\frac{x-t}{h_n})\mu(dt)}{\int K(\frac{x-t}{h_n})\mu(dx)} \to m(x) \quad \bmod \mu
$$

by Lemma 5. Because of Lemma 6 we have

$$
n\int K\left(\frac{x-t}{h_n}\right)\mu(dt) \ge d^*(x)n^{1-\gamma d} \to \infty \quad \bmod \mu
$$

(compare (18) below), thus

$$
\begin{aligned}
&\liminf_{n\to\infty}\frac{\sum_{i=1}^{n}\mathbf{E}\{Y_i^* K(\frac{x-X_i}{h_n})\}}{n\int K(\frac{x-t}{h_n})\mu(dt)}\\
&\ge \lim_{n\to\infty}\frac{n\mathbf{E}\{Y1_{[Y\le N]}K(\frac{x-X}{h_n})\}}{n\int K(\frac{x-t}{h_n})\mu(dt)}\\
&= \lim_{n\to\infty}\frac{\int \mathbf{E}\{Y1_{[Y\le N]}|X=t\}K(\frac{x-t}{h_n})\mu(dt)}{\int K(\frac{x-t}{h_n})\mu(dt)}\\
&= \mathbf{E}(Y1_{[Y\le N]}|X=x) \quad \bmod \mu \text{ (for each } N\in\mathbb{N}\text{, by Lemma 5)}\\
&\to \mathbf{E}(Y|X=x) = m(x) \quad (N\to\infty),
\end{aligned}
$$

which leads to (15). Together with (17) below we shall have

$$\frac{\sum_{i=1}^n Y_i^* K(\frac{x-X_i}{h_n})}{n\int K(\frac{x-t}{h_n})\mu(dt)} \to m(x) \quad \text{a.s. mod } \mu, \tag{16}$$

which together with (14) yields (13). In the third step we show

$$\frac{\sum_{i=1}^n [Y_i^* K(\frac{x-X_i}{h_n}) - \mathbf{E}Y_i^* K(\frac{x-X_i}{h_n})]}{n\int K(\frac{x-t}{h_n})\mu(dt)} \to 0 \quad \text{a.s. mod } \mu \tag{17}$$

distinguishing the cases of (a) ρ-mixing and (b) α-mixing.

(a) According to Lemma 6

$$\mu\left(\left\{x \in \mathbb{R}^d;\ \liminf \frac{\mu(S_{x,h_n})}{h_n^d} = 0\right\}\right) = 0.$$

Neglecting this set we have

$$\begin{aligned}\int K\left(\frac{x-t}{h_n}\right)\mu(dt) &\geq c^* \int 1_{S_{0,r^*}}\left(\frac{x-t}{h_n}\right)\mu(dt)\\ &\geq d(x)h_n^d = d^*(x)n^{-\gamma d}\end{aligned} \tag{18}$$

for all n with suitable $c^* > 0$, $r^* > 0$, $d^*(x) > 0$. Choose an arbitrary $\epsilon > 0$. Noticing

$$\mathbf{E}\left\{\left(Y_i^* K\left(\frac{x-X_i}{h_n}\right)\right)^2\right\} \leq n^{2\kappa}\int K\left(\frac{x-t}{h_n}\right)\mu(dt) \quad (i = 1, \ldots, n),$$

by Lemma 4a with $\epsilon^* = \epsilon n \int K(\frac{x-t}{h_n})\mu(dt)$ we obtain for suitable $c_1, c_2 \in (0, \infty)$

$$\begin{aligned}
&\sum_{n=1}^\infty P\left\{\frac{1}{n\int K(\frac{x-t}{h_n})\mu(dt)}\left|\sum_{i=1}^n\left[Y_i^* K\left(\frac{x-X_i}{h_n}\right) - \mathbf{E}Y_i^* K\left(\frac{x-X_i}{h_n}\right)\right]\right| > \epsilon\right\}\\
&\leq c_1 \sum_{n=1}^\infty \exp\left(-\frac{c_2 \epsilon n \int K(\frac{x-t}{h_n})\mu(dt)}{n^{1/2}n^\kappa(\int K(\frac{x-t}{h_n})\mu(dt))^{1/2} + n^\kappa n^{1/(1+2\rho)}}\right)\\
&\leq c_1 \sum_{n=1}^\infty \exp\left(-\frac{1}{2}c_2\epsilon\right.\\
&\qquad \left.\times \min\left\{n^{\frac{1}{2}-\kappa}\left(\int K\left(\frac{x-t}{h_n}\right)\mu(dt)\right)^{1/2}, n^{1-\kappa-\frac{1}{1+2\rho}}\int K\left(\frac{x-t}{h_n}\right)\mu(dt)\right\}\right)\\
&\leq c_1 \sum_{n=1}^\infty \exp\left(-\frac{1}{2}c_2\epsilon \min\left\{d^*(x)^{\frac{1}{2}} n^{\frac{1}{2}-\kappa-\frac{1}{2}\gamma d}, d^*(x) n^{1-\kappa-\frac{1}{1+2\rho}-\gamma d}\right\}\right)
\end{aligned}$$

(by 18)

$$= c_1 \sum_{n=1}^{\infty} \exp\left(-\frac{1}{2} c_2 \epsilon \min\left\{d^*(x)^{\frac{1}{2}}, d^*(x)\right\} n^{\min\{\frac{1}{2}-\kappa-\frac{1}{2}\gamma d, 1-\kappa-\frac{1}{1+2\rho}-\gamma d\}}\right)$$
$$< \infty \bmod \mu,$$

if $1-\gamma d - 2\kappa > 0$ and $1 - 1/(1+2\rho) - \gamma d - \kappa > 0$. Both conditions are fulfilled under the assumptions on ρ and γ, if $\kappa > 0$ is chosen sufficiently small. Thus (17) is obtained by the Borel-Cantelli lemma.

(b) As in (a) we have (18). Choose an arbitrary $\epsilon > 0$. For suitable constants c_1, c_2 by Lemma 4b with $\epsilon^* = \epsilon n \int K(\frac{x-t}{h_n})\mu(dt)$ and $\beta \in (0,1)$ we obtain for ϵ sufficiently small

$$\sum_{n=1}^{\infty} P\left\{\frac{1}{n\int K(\frac{x-t}{h_n})\mu(dt)}\left|\sum_{i=1}^{n}\left[Y_i^* K\left(\frac{x-X_i}{h_n}\right) - \mathbf{E}Y_i^* K\left(\frac{x-X_i}{h_n}\right)\right]\right| > \epsilon\right\}$$
$$\leq 4\sum_{n=1}^{\infty} \exp\left(\frac{-c_1\epsilon^2 n^2 (\int K(\frac{x-t}{h_n})\mu(dt))^2}{n^{2\kappa+1}\log(n+1)(\int K(\frac{x-t}{h_n})\mu(dt))^{1-1/\alpha} + \epsilon n^{1+\kappa+\beta}\int K(\frac{x-t}{h_n})\mu(dt)}\right)$$
$$+ c_2 \left(\frac{n^{\kappa}}{\epsilon\int K(\frac{x-t}{h_n})\mu(dt)}\right)^{\frac{1}{2}} n^{1-\beta-\beta\alpha}$$
$$\leq 4\sum_{n=1}^{\infty}\exp\left(-\frac{c_1\epsilon}{2}\int K\left(\frac{x-t}{h_n}\right)\mu(dt)\right.$$
$$\left.\times \min\left\{\epsilon n^{1-2\kappa}(\log(n+1))^{-1}\left(\int K\left(\frac{x-t}{h_n}\right)\mu(dt)\right)^{1/\alpha}, n^{1-\kappa-\beta}\right\}\right)$$
$$+ c_2\epsilon^{-\frac{1}{2}} d^*(x)^{-\frac{1}{2}} \sum_{n=1}^{\infty} n^{\frac{\kappa}{2}+\frac{\gamma d}{2}+1-(1+\alpha)\beta}$$
$$\leq 4\sum_{n=1}^{\infty}\exp\left(-\frac{c_1\epsilon}{2} d^*(x) n^{-\gamma d}\right.$$
$$\left.\times\min\left\{\epsilon d^*(x)^{1/\alpha} n^{1-2\kappa-\gamma d/\alpha}(\log(n+1))^{-1}, n^{1-\kappa-\beta}\right\}\right)$$
$$+ c_2\epsilon^{-\frac{1}{2}} d^*(x)^{-\frac{1}{2}} \sum_{n=1}^{\infty} n^{\frac{\kappa}{2}+\frac{\gamma d}{2}+1-(1+\alpha)\beta} < \infty \quad \bmod \mu,$$

if $1-\gamma d(\alpha+1)/\alpha - 2\kappa > 0$, $1-\beta-\gamma d-\kappa > 0$ and $-4+2(1+\alpha)\beta-\kappa-\gamma d > 0$. These conditions are fulfilled under the assumptions on α and γ, if one chooses $\beta = 5/(3+2\alpha)$ and $\kappa > 0$ sufficiently small. Thus (17) is obtained. □

If the above Nadaraya-Watson kernel regression estimate is replaced by the semi-recursive Devroye-Wagner (1980b) kernel regression estimate, then strong universal pointwise consistency in the case of independent identically distributed random vectors (X_n, Y_n) can be stated, i.e., under the only condition $\mathbf{E}|Y_1| < \infty$ strong consistency P_{X_1}-almost everywhere (see Walk 2001). This result has been extended to the ρ-mixing case under the condition $\sum \rho_n < \infty$ by Frey (2007). The case of bounded Y was treated by Ferrario (2004) under more general α-mixing and ρ-mixing conditions on the basis of the generalized Theorem 1b mentioned in context of Lemma 2.

In the following the result of Frey (2007) and its proof will be given.

Theorem 4. *Let $(X, Y), (X_1, Y_1), (X_2, Y_2), \ldots$ be identically distributed $(d+1)$-dimensional random vectors with $\mathbf{E}|Y| < \infty$. Let K be a symmetric measurable function on $\mathbb{R}^d$ satisfying $c_1 1_{S_{0,R}} \le K \le c_2 1_{S_{0,R}}$ for some $0 < R < \infty$, $0 < c_1 < c_2 < \infty$ (so-called boxed kernel with naive kernel $K = 1_{S_{0,1}}$ as a special case). With $n \in \mathbb{N}$ and $h_n > 0$ set*

$$m_n(x) := \frac{\sum_{i=1}^n Y_i K(\frac{x-X_i}{h_i})}{\sum_{i=1}^n K(\frac{x-X_i}{h_i})}, \quad x \in \mathbb{R}^d$$

where $\frac{0}{0} := 0$. If the sequence $((X_n, Y_n))$ is ρ-mixing with $\sum \rho_n < \infty$ (e.g., $\rho_n = O(n^{-\rho})$ for some $\rho > 1$) and if h_n is chosen as $h_n = cn^{-\gamma}$ with $c > 0$, $0 < \gamma d < \frac{1}{2}$, then

$$m_n(x) \to m(x) \quad \textit{a.s.} \bmod \mu.$$

Theorem 4 comprehends Kolmogorov's strong law of large numbers (special case that μ is a Dirac measure). The semirecursive kernel estimate has the numerical advantage that a new observation leads only to an addition of a new summand in the numerator and in the denominator, but the observations obtain different weights. In the proof we give in detail only the part which differs from the proof in Walk (2001).

Proof (of Theorem 4). Without loss of generality assume $Y_i \ge 0$. The case of bounded Y, also with denominator replaced by its expectation, is comprehended by Ferrario (2004). Therefore in the case $\mathbf{E}|Y| < \infty$ it is enough to show existence of a $c \in (0, \infty)$ independent of the distribution of (X, Y) with

$$\limsup_{n\to\infty} \frac{\sum_{i=1}^n Y_i K(\frac{x-X_i}{h_i})}{1 + \sum_{i=1}^n \int K(\frac{x-t}{h_n})\mu(dt)} \le cm(x) \quad \text{a.s.} \bmod \mu \tag{19}$$

(compare Lemma 8 below). Let the compact support of K by covered by finitely many closed spheres $S_1, \ldots, S_N$ with radius $R/2$. Fix $k \in \{1, \ldots, N\}$. For all $t \in \mathbb{R}^d$ and all $n \in \mathbb{N}$, from $x \in t + h_n S_k$ it can be concluded

$$K\left(\frac{\cdot - x}{h_i}\right) \ge \frac{c_1}{c_2} K\left(\frac{\cdot - t}{h_i}\right) 1_{S_k}\left(g\left(\frac{\cdot - t}{h_i}\right)\right) \tag{20}$$

for all $i = \{1, \ldots, n\}$. It suffices to show

$$\limsup_{n\to\infty} \frac{\sum_{i=1}^n Y_i K(\frac{x-X_i}{h_i}) 1_{S_k}(\frac{x-X_i}{h_i})}{1 + \sum_{i=1}^n \int K(\frac{x-t}{h_i})\mu(dt)} \le c'm(x) \quad \text{a.s.} \bmod \mu \tag{21}$$

for some $c' < \infty$. With

$$r_n := r_n(t) := \frac{1}{c_2} \int K\left(\frac{x-t}{h_n}\right) 1_{t+h_n S_k}(x)\mu(dx),$$
$$R_n := r_1 + \cdots + r_n, \quad n \in \mathbb{N},$$

for $t \in \mathbb{R}^d$ we can choose indices $p_i = p(t,k,i) \uparrow \infty$ $(i \to \infty)$ such that

$$R_{p_i} \leq i+1, \tag{22}$$

$$\sum_{j=p_i}^{\infty} \frac{r_j}{(1+R_j)^2} < \frac{1}{i} \tag{23}$$

holds. For $p(t,k,\cdot)$ we define the inverse function $q(t,k,\cdot)$ on $\mathbb{N}$ by

$$q(t,k,n) := \max\{i \in \mathbb{N};\ p(t,k,i) \leq n\}.$$

Set

$$Z_i := Y_i 1_{[Y_i \leq q(X_i,k,i)]}, \quad i \in \mathbb{N}.$$

Now it will be shown

$$\frac{\sum_{i=1}^n [Z_i K(\frac{x-X_i}{h_i}) 1_{S_k}(\frac{x-X_i}{h_i}) - \mathbf{E}\{Z_i K(\frac{x-X_i}{h_i}) 1_{S_k}(\frac{x-X_i}{h_i})\}]}{c_1 + \sum_{i=1}^n \int K(\frac{x-t}{h_i})\mu(dt)} \to 0 \quad \text{a.s. mod } \mu \tag{24}$$

by an application of Lemma 2. We notice $\int K(\frac{x-t}{h_n})\mu(dt) \leq c_2$,

$$\sum_{i=1}^n \int K\left(\frac{x-t}{h_i}\right)\mu(dt) \uparrow \infty \quad \text{mod } \mu \tag{25}$$

because of $\int K(\frac{x-t}{h_i})\mu(dt) \geq c_1 \mu(x + h_i S_{0,1}) \geq c_1 c(x) h_i^d$ by Lemma 6 with $c(x) > 0$ mod μ and $\sum h_n^d = \infty$ by $0 < \gamma d < 1$. Further $Z_n \geq 0$ and

$$\limsup \frac{\mathbf{E} Z_n K(\frac{x-X_n}{h_n}) 1_{S_k}(\frac{x-X_n}{h_n})}{\int K(\frac{x-t}{h_n})\mu(dt)} \leq \lim \frac{\int m(t) K(\frac{x-t}{h_n})\mu(dt)}{\int K(\frac{x-t}{h_n})\mu(dt)} = m(x) \quad \text{mod } \mu$$

by Lemma 5. With $W_j(x) := Z_j K(\frac{x-X_j}{h_j}) 1_{S_k}(\frac{x-X_j}{h_j})$ we obtain

$$\begin{aligned}
&\mathbf{Var}\left\{\sum_{j=1}^n W_j(x)\right\} \\
&\leq \sum_{j=1}^n \mathbf{Var}\{W_j(x)\} + \sum_{j=1}^n \sum_{l=1, l\neq j}^n \rho_{|j-l|} \left(\mathbf{Var}\{W_j(x)\}\right)^{\frac{1}{2}} \left(\mathbf{Var}\{W_l(x)\}\right)^{\frac{1}{2}}
\end{aligned}$$

$$\leq \sum_{j=1}^{n} \mathbf{Var}\left\{W_j(x)\right\} + \frac{1}{2}\sum_{j=1}^{n}\sum_{l=1,l\neq j}^{n} \rho_{|j-l|}\left[\mathbf{Var}\left\{W_j(x)\right\} + \mathbf{Var}\left\{W_l(x)\right\}\right]$$

$$\leq \left(1+2\sum_{j=1}^{\infty}\rho_j\right)\sum_{j=1}^{n}\mathbf{Var}\left\{W_j(x)\right\}$$

$$= c^* \sum_{j=1}^{n} \mathbf{Var}\left\{W_j(x)\right\}$$

with $c^* < \infty$ by the assumption on (ρ_n), thus

$$\sum_{n=1}^{\infty} \frac{\int K(\frac{x-t}{h_n})\mu(dt)\mathbf{Var}\{\sum_{j=1}^{n} W_j(x)\}}{(c_1 + \sum_{i=1}^{n}\int K(\frac{x-t}{h_i})\mu(dt)\})^3}$$

$$\leq c^* \sum_{n=1}^{\infty} \frac{\int K(\frac{x-t}{h_n})\mu(dt)\sum_{j=1}^{n}\mathbf{Var}\{W_j(x)\}}{(c_1 + \sum_{i=1}^{n}\int K(\frac{x-t}{h_i})\mu(dt))^3}$$

$$= c^* \sum_{j=1}^{\infty} \mathbf{Var}\left\{W_j(x)\right\} \sum_{n=j}^{\infty} \frac{\int K(\frac{x-t}{h_n})\mu(dt)}{(c_1 + \sum_{i=1}^{n}\int K(\frac{x-t}{h_i})\mu(dt))^3}$$

$$\leq c^{**} \sum_{j=1}^{\infty} \frac{\mathbf{Var}\{W_j(x)\}}{(c_1 + \sum_{i=1}^{j}\int K(\frac{x-t}{h_i})\mu(dt))^2}$$

$$\leq c^{**} c_2 \sum_{n=1}^{\infty} \frac{\mathbf{E} Z_n^2 K(\frac{x-X_n}{h_n}) 1_{S_k}(\frac{x-X_n}{h_n})}{(c_1 + \sum_{i=1}^{n}\int K(\frac{x-s}{h_i})\mu(ds))^2}$$

with suitable $c^{**} < \infty$. Now, by (20),

$$\int \sum_{n=1}^{\infty} \frac{\mathbf{E} Z_n^2 K(\frac{x-X_n}{h_n}) 1_{S_k}(\frac{x-X_n}{h_n})}{(c_1 + \sum_{i=1}^{n}\int K(\frac{x-s}{h_i})\mu(ds))^2}\mu(dx)$$

$$\leq \sum_{n=1}^{\infty} \int \left[\int \frac{\mathbf{E}\{Z_n^2 | X_n = t\} K(\frac{x-t}{h_n}) 1_{S_k}(\frac{x-t}{h_n})}{(c_1 + \sum_{i=1}^{n}\int \frac{c_1}{c_2} K(\frac{s-t}{h_i}) 1_{S_k}(\frac{s-t}{h_i})\mu(ds))^2}\mu(dx)\right]\mu(dt)$$

$$= \frac{1}{c_1^2}\sum_{n=1}^{\infty} \int \left[\int \frac{\int v^2 P_{Z_n|X_n=t}(dv) K(\frac{x-t}{h_n}) 1_{S_k}(\frac{x-t}{h_n})}{(1 + \sum_{i=1}^{n}\frac{1}{c_2}\int K(\frac{s-t}{h_i}) 1_{S_k}(\frac{s-t}{h_i})\mu(ds))^2}\mu(dx)\right]\mu(dt)$$

$$= \frac{c_2}{c_1^2}\sum_{n=1}^{\infty} \int \left[\int \frac{\sum_{i=1}^{q(t,k,n)} (\int_{(i-1,i]} v^2 P_{Z_n|X_n=t}(dv)) \frac{1}{c_2} K(\frac{x-t}{h_n}) 1_{S_k}(\frac{x-t}{h_n})}{(1 + \sum_{i=1}^{n}\frac{1}{c_2}\int K(\frac{s-t}{h_i}) 1_{S_k}(\frac{s-t}{h_i})\mu(ds))^2}\right.$$

$$\left. \times\ \mu(dx)\right]\mu(dt)$$

$$= \frac{c_2}{c_1^2}\sum_{n=1}^{\infty}\int\left[\int \frac{\sum_{i=1}^{q(t,k,n)}(\int_{(i-1,i]} v^2 P_{Y|X=t}(dv))\frac{1}{c_2}K(\frac{x-t}{h_n})1_{S_k}(\frac{x-t}{h_n})}{(1+\sum_{i=1}^{n}\frac{1}{c_2}\int K(\frac{s-t}{h_i})1_{S_k}(\frac{s-t}{h_i})\mu(ds))^2}\right.$$

$$\left.\times\ \mu(dx)\right]\mu(dt)$$

$$= \frac{c_2}{c_1^2}\int\left[\sum_{i=1}^{\infty}\int_{(i-1,i]} v^2 P_{Y|X=t}(dv)\right.$$

$$\left.\times \sum_{n=p(t,k,i)}^{\infty}\frac{\frac{1}{c_2}\int K(\frac{x-t}{h_n})1_{S_k}(\frac{x-t}{h_n})\mu(dx)}{(1+\sum_{i=1}^{n}\frac{1}{c_2}\int K(\frac{s-t}{h_i})1_{S_k}(\frac{s-t}{h_i})\mu(ds))^2}\right]\mu(dt)$$

$$\leq \frac{c_2}{c_1^2}\int\left[\sum_{i=1}^{\infty}\int_{(i-1,i]} v^2 P_{Y|X=t}(dv)\frac{1}{i}\right]\mu(dt)$$

(by 23)

$$\leq \frac{c_2}{c_1^2}\int\left[\sum_{i=1}^{\infty}\int_{(i-1,i]} v P_{Y|X=t}(dv)\right]\mu(dt)$$

$$\leq \frac{c_2}{c_1^2}\mathbf{E}Y < \infty.$$

Therefore

$$\sum_{n=1}^{\infty}\frac{\int K(\frac{x-t}{h_n})\mu(dt)\mathbf{Var}\{\sum_{j=1}^{n} Z_j K(\frac{x-X_j}{h_j})1_{S_k}(\frac{x-X_j}{h_j})\}}{(c_1+\sum_{i=1}^{n}\int K(\frac{x-t}{h_i})\mu(dt))^3} < \infty \quad \text{mod } \mu,$$

and (24) follows by Lemma 2.

In the next step we notice

$$\limsup \frac{\sum_{i=1}^{n}\mathbf{E}Z_i K(\frac{x-X_i}{h_i})1_{S_k}(\frac{x-X_i}{h_i})}{c_1+\sum_{i=1}^{n}\int K(\frac{x-t}{h_i})\mu(dt)} \leq \lim \frac{\sum_{i=1}^{n}\int m(t)K(\frac{x-t}{h_i})\mu(dt)}{c_1+\sum_{i=1}^{n}\int K(\frac{x-t}{h_i})\mu(dt)}$$

$$= m(x) \quad \text{mod } \mu$$

because of (25) and Lemma 5. This together with (24) yields

$$\limsup \frac{\sum_{i=1}^{n} Z_i K(\frac{x-X_i}{h_i})1_{S_k}(\frac{x-X_i}{h_i})}{c_1+\sum_{i=1}^{n}\int K(\frac{x-t}{h_i})\mu(dt)} \leq m(x) \quad \text{a.s. mod } \mu. \tag{26}$$

In the last step one obtains (21) from (26) and (25) by noticing

$$\sum_{n=1}^{\infty} P\left[Z_n 1_{S\cap S_k}\left(\frac{x-X_n}{h_n}\right) \neq Y_n 1_{S\cap S_k}\left(\frac{x-X_n}{h_n}\right)\right] < \infty \quad \text{mod } \mu, \tag{27}$$

where $S := S_{0,R}$, together with the Borel-Cantelli lemma, and (27) follows from

$$\int \sum_{n=1}^{\infty} P\left[Y_n > q(X_n, k, n), X_n \in x - h_n (S \cap S_k)\right] \mu(dx)$$

$$= \int \sum_{n=1}^{\infty} \left(\int P\left[Y > q(t, k, n) | X = t\right] 1_{x - h_n (S \cap S_k)}(t) \mu(dt) \right) \mu(dx)$$

$$= \sum_{n=1}^{\infty} \int P\left[Y > q(t, k, n) | X = t\right] \mu\left(t + h_n (S \cap S_k)\right) \mu(dt)$$

$$\leq \int \sum_{i=1}^{\infty} P\left[Y \in (i, i+1] | X = t\right] \sum_{n=1}^{p(t,k,i+1)} \mu\left(t + h_n (S \cap S_k)\right) \mu(dt)$$

$$\leq \frac{c_2}{c_1} \int \sum_{i=1}^{\infty} P\left[Y \in (i, i+1] | X = t\right] (i+2) \mu(dt) \quad \text{(by 22)}$$

$$\leq 3 \frac{c_2}{c_1} \mathbf{E} Y < \infty.$$

□

For k_n-nearest neighbor regression estimation with integrable response random variable Y and d-dimensional predictor random vector X on the basis of independent data, the following theorem states strong universal pointwise consistency, i.e., strong consistency P_X-almost everywhere for general distribution of (X, Y) with $\mathbf{E}|Y| < \infty$. The estimation is symmetric in the data, does not use truncated observations and contains Kolmogorov's strong law of large numbers as the special case that P_X is a Dirac measure. It can be considered as a universal strong law of large numbers for conditional expectations. Let for the observed copies of (X, Y) the k_n-nearest neighbor (k_n-NN) regression estimate $m_n(x)$ of $m(x) := \mathbf{E}(Y|X = x)$ be defined by (11).

Theorem 5. *Let* $(X, Y), (X_1, Y_1), (X_2, Y_2), \ldots$ *be independent identically distributed* $(d+1)$*-dimensional random vectors with* $\mathbf{E}|Y| < \infty$. *Choose* $k_n \in \min\{\lceil cn^{\beta} \rceil, n-1\}$ *with* $c > 0$, $\beta \in (0, 1)$ *for* $n \in \{2, 3, \ldots\}$. *Then*

$$m_n(x) \to m(x) \quad a.s. \bmod \mu.$$

As to the proof (with somewhat more general k_n) and related results we refer to Walk (2008a). The proof uses Etemadi's (1981) monotonicity argument, a generalized Lebesgue density theorem concerning $\mathbf{E}m_n(x) \to m(x) \bmod \mu$, a covering lemma for nearest neighbors and Steele's (1986) version of the Efron-Stein inequality for the variance of a function of independent identically distributed random variables.

Whether at least in the case of independence strong universal pointwise consistency for Nadaraya-Watson kernel regression estimates or for classic partitioning regression estimates holds, is an open problem.

Remark 5. Let the situation in Theorem 5 be modified by assuming that the sequence $(X_1, Y_1), (X_2, Y_2), \ldots$ of identically distributed $(d+1)$-dimensional random vectors is α-mixing with $\alpha_n = O(n^{-\alpha})$ such that $0 < 1 - \beta < \min\{\alpha/2, \alpha/(\alpha+1)\}$ and that Y_i is bounded. Let tie-breaking be done by enlarging the dimension d of the predictor vectors to $d+1$ by use of new (independent!) random variables equidistributed on $[0, 1]$ as additional components (see Györfi et al. 2002, pp. 86, 87). Then Theorem 2 in Irle (1997) states

$$m_n(x) \to m(x) \quad \text{a.s. mod } \mu.$$

Analogously, by use of Lemma 3a, one obtains the same convergence assertion under ρ-mixing where $0 < 1 - \beta < 1$.

4 L_2-Consistent Regression Estimates

The pioneering paper on universal consistency of nonparametric regression estimates is Stone (1977). It contains a criterion of weak universal L_2-consistency of local averaging estimates under independence. The conditions for k_n-nearest neighbor estimates and for Nadaraya-Watson kernel estimates were verified by Stone (1977) and by Devroye and Wagner (1980a) and Spiegelman and Sacks (1980), respectively. The following theorem concerns strong universal L_2-consistency of k_n-nearest neighbor estimates Devroye et al. (1994), of semirecursive Devroye-Wagner kernel estimates (Györfi et al. 1998) and modified Nadaraya-Watson kernel estimates (Walk 2005a).

Theorem 6. *Let (X, Y), (X_1, Y_1), $(X_2, Y_2), \ldots$ be independent identically distributed $(d+1)$-dimensional random vectors with* $\mathbf{E}\{Y^2\} < \infty$.

(a) *Let the k_n-NN regression estimates m_n of m be defined by* (11) *with $k_n \in \{1, \ldots, n-1\}$, $n \geq 2$, satisfying $k_n/n \to 0$, $k_n/\log n \to \infty$ $(n \to \infty)$ and random tie-breaking. Then* (12) *holds.*

(b) *Let the semirecursive Devroye-Wagner kernel regression estimates m_n, $n \geq 2$, be defined by*

$$m_n(x) := \frac{Y_1 K(0) + \sum_{i=2}^{n} Y_i K(\frac{x-X_i}{h_i})}{K(0) + \sum_{i=2}^{n} K(\frac{x-X_i}{h_i})}, \quad x \in \mathbb{R}^d,$$

with symmetric λ-integrable kernel $K : \mathbb{R}^d \to \mathbb{R}_+$ satisfying

$$\alpha H(\|x\|) \leq K(x) \leq \beta H(\|x\|), \quad x \in \mathbb{R}^d,$$

for some $0 < \alpha < \beta < \infty$ and nonincreasing $H : \mathbb{R}_+ \to \mathbb{R}_+$ with $H(+0) > 0$ and with bandwidths $h_n > 0$ satisfying

$$h_n \downarrow 0 \quad (n \to \infty), \qquad \sum_{n=2}^{\infty} h_n^d = \infty,$$

e.g., $h_n = cn^{-\gamma}$ *with* $c > 0$, $0 < \gamma d < 1$. *Then* (12) *holds.*

(c) *Let the Nadaraya-Watson type kernel regression estimates* m_n, $n \in \mathbb{N}$, *be defined by*

$$m_n(x) := \frac{\sum_{i=1}^{n} Y_i K(\frac{x-X_i}{h_n})}{\max\{\delta, \sum_{i=1}^{n} K(\frac{x-X_i}{h_n})\}}, \qquad x \in \mathbb{R}^d,$$

with an arbitrary fixed $\delta > 0$, *a smooth kernel* $K : \mathbb{R}^d \to \mathbb{R}_+$ *(see below) and bandwidths* $h_n > 0$ *satisfying*

$$h_n \downarrow 0, \qquad nh_n^d \to \infty \quad (n \to \infty), \qquad h_n - h_{n+1} = O(h_n/n),$$

e.g., $h_n = cn^{-\gamma}$ *with* $c > 0$, $0 < \gamma d < 1$. *Then* (12) *holds.*

In Theorem 6c the modification of Nadaraya-Watson estimates consists of a truncation of the denominator from below by an arbitrary positive constant, see Spiegelman and Sacks (1980). Smooth kernel means a kernel K of the form $K(x) = H(\|x\|)$, where H is a continuously differentiable nonincreasing function on $\mathbb{R}_+$ with $0 < H(0) \leq 1$, $\int H(s)s^{d-1}ds < \infty$ such that R with $R(s) := s^2 H'(s)^2/H(s)$, $s \geq 0$ $(0/0 := 0)$, is bounded, piecewise continuous and, for s sufficiently large, nonincreasing with $\int R(s)s^{d-1}ds < \infty$. Examples are the quartic and the Gaussian kernel. In the proof of Theorem 6a one shows

$$\limsup_{n\to\infty} \frac{n}{k_n} \max_{i=1,\dots,n} \int 1_{\{X_i \text{is among the } k_n \text{nearest neighbors of } x \text{in } (X_1,\dots,X_n)\}} \mu(dx)$$
$$\leq \text{const} < \infty \quad \text{a.s.}$$

and uses Kolmogorov's strong law of large numbers for $Y_1^2, Y_2^2, \dots$ and Lemma 8 (with $p = 2$, $\delta = 0$) below. Theorem 6b is proven by martingale theory, a covering argument and Lemmas 5, 6 and 8 (with $p = 2$, $\delta = 0$). In both cases, for details and further literature we refer to Györfi et al. (2002). In the proof of Theorem 6c strong consistency for bounded Y (due to Devroye and Krzyżak 1989), Lemma 8 (with $p = 2$, $\delta = 0$) and summability theory (Lemma 1b), together with Lemmas 5, 6, 7 and Steele's (1986) version of the Efron-Stein inequality for variances are used.

The following lemma (see Györfi 1991, Theorem 2 with proof, and Györfi et al. 2002, Lemma 23.3; compare also the begin of the proof of Theorem 4) allows to reduce problems of strong consistency of kernel or nearest neighbor regression estimates to two simpler problems. It holds more generally for local averaging estimation methods.

Lemma 8. *Let* $p \geq 1$ *and* $\delta \geq 0$ *be fixed. Denote the Nadaraya-Watson or semirecursive Devroye-Wagner or* k_n*-NN regression estimates in the context of* $(d+1)$*-dimensional identically distributed random vectors* (X, Y), (X_1, Y_1), (X_2, Y_2), $\dots$ *by* m_n. *The following statement* (a) *is implied by statement* (b):

(a) *for all* Y *with* $\mathbf{E}\{|Y|^{(1+\delta)p}\} < \infty$

$$\int |m_n(x) - m(x)|^p \mu(dx) \to 0 \quad a.s.;$$

(b) *for all bounded* Y

$$\int |m_n(x) - m(x)|\mu(dx) \to 0 \quad a.s.,$$

and there exists a constant $c < \infty$ *such that for all* $Y \geq 0$ *with* $\mathbf{E}\{Y^{1+\delta}\} < \infty$

$$\limsup \int m_n(x)\mu(dx) \leq c\mathbf{E}Y \quad a.s. \tag{28}$$

For fixed $\delta \geq 0$ *the statement* (a) *for* $p > 1$ *follows from* (a) *for* $p = 1$.

If we allow stronger moment conditions on Y (and X) we can relax the independence assumption for kernel estimates. Here Nadaraya-Watson kernel estimates m_n are considered with kernels $K : \mathbb{R}^d \to \mathbb{R}_+$ of the form $K(x) = H(\|x\|)$, $x \in \mathbb{R}^d$, where $H : \mathbb{R}_+ \to \mathbb{R}_+$ is a Lipschitz continuous nonincreasing function with $0 < H(0) \leq 1$ and $\int H(s)s^{d-1}ds < \infty$ such that the function $s \to s|H'(s)|$ (defined λ-almost everywhere on $\mathbb{R}_+$) is nonincreasing for s sufficiently large (e.g., Epanechnikov, quartic and Gaussian kernel). The following result Walk (2008b) concerns L_2-consistency.

Theorem 7. *Let* $(X, Y), (X_1, Y_1), (X_2, Y_2), \ldots$ *be identically distributed* $(d + 1)$-*dimensional random vectors with* $\mathbf{E}\{|Y|^p\} < \infty$ *for some* $p > 4$, $\mathbf{E}\{\|X\|^q\} < \infty$ *for some* $q > 0$. *Choose bandwidths* $h_n = cn^{-\gamma}$ $(c > 0,\ 0 < \gamma d < 1)$. *If the sequence* $((X_n, Y_n))$ *is* ρ-*mixing and* $0 < \gamma d < 1 - \frac{4}{p} - \frac{2d}{pq}$ *or if it is* α-*mixing with* $\alpha_n = O(n^{-\alpha})$, $\alpha > 0$, *with* $0 < \gamma d < \min\{1, \alpha\} - \frac{4}{p} - \frac{2d}{pq}$, *then*

$$\int |m_n(x) - m(x)|^2 \mu(dx) \to 0 \quad a.s.$$

If Y is essentially bounded, then no moment condition on X is needed and the conditions on γ are $0 < \gamma d < 1$ in the ρ-mixing case and $0 < \gamma d < \min\{1, \alpha\}$ in the α-mixing case. If X is bounded, then the conditions on γ are $0 < \gamma d < 1 - \frac{4}{p}$ in the ρ-mixing case and $0 < \gamma d < \min\{1, \alpha\} - \frac{4}{p}$ in the α-mixing case. In this context we mention that a measurable transformation of X to bounded X does not change the L_2 risk $\mathbf{E}\{|Y - m(X)|^2\}$.

In the proof of Theorem 7, by Lemma 8 we treat the corresponding L_1-consistency problem with $p > 2$. The integrability assumption on Y allows to truncate Y_i (≥ 0) at $i^{1/p}$. Because of

$$\left| \frac{\sum_{i=1}^n Y_i 1_{[Y_i \le i^{1/p}]} K(\frac{x-X_i}{h_n})}{\sum_{i=1}^n K(\frac{x-X_i}{h_n})} - \frac{\sum_{i=1}^n Y_i 1_{[Y_i \le i^{1/p}]} K(\frac{x-X_i}{h_n})}{n \int K(\frac{x-t}{h_n}) \mu(dt)} \right|$$
$$\le n^{\frac{1}{p}} \left| \frac{\sum_{i=1}^n K(\frac{x-X_i}{h_n})}{n \int K(\frac{x-t}{h_n}) \mu(dt)} - 1 \right|,$$

it suffices to investigate the convergence behavior of the latter term and of the simplified estimator

$$\frac{\sum_{i=1}^n Y_i 1_{[Y_i \le i^{1/p}]} K(\frac{\cdot - X_i}{h_n})}{n \int K(\frac{\cdot - t}{h_n}) \mu(dt)}.$$

considered as a random variable with values in the real separable Hilbert space $L_2(\mu)$. This can be done by use of the Tauberian Lemma 1b, the covering Lemma 7 and Lemma 3a,b on the variance of sums of Hilbert space valued random variables under mixing together with a result of Serfling (1970a), Corollary A3.1, on maximum cumulative sums.

5 Rosenblatt-Parzen Density Estimates under Mixing Conditions

In this section we investigate the Rosenblatt-Parzen kernel density estimates in view of strong L_1-consistency under mixing conditions, namely ρ- and α-mixing. In the latter case the α-mixing condition in Theorem 4.2.1(iii) in Györfi et al. (1989) (see also Györfi and Masry 1990) is weakened, essentially to that in Theorem 4.3.1 (iii) there on the Wolverton-Wagner and Yamato recursive density estimates. In the proof we use Etemadi's concept and not Tauberian theory, because in the latter case a Lipschitz condition on the kernel should be imposed.

Theorem 8. *Let the d-dimensional random vectors X_n, $n \in \mathbb{N}$, be identically distributed with density f, and assume that (X_n) is ρ-mixing or α-mixing with $\alpha_n = O(n^{-\alpha})$ for some $\alpha > 0$. If for the Rosenblatt-Parzen density estimates*

$$f_n(x) := f_n(X_1, \ldots, X_n; x) := \frac{1}{nh_n^d} \sum_{i=1}^n K\left(\frac{x - X_i}{h_n}\right), \quad x \in \mathbb{R}^d \tag{29}$$

the kernel K is chosen as a square λ-integrable density on $\mathbb{R}^d$ with $K(rx) \ge K(x)$ for $0 \le r \le 1$ and the bandwidths are of the form $h_n = cn^{-\gamma}$, $0 < c < \infty$, with $0 < \gamma d < 1$ in the ρ-mixing case and $0 < \gamma d < \min\{1, \alpha\}$ in the α-mixing case, then

$$\int |f_n(x) - f(x)| \lambda(dx) \to 0 \quad a.s. \tag{30}$$

Proof. Because of the simple structure of the denominator in (29) we can use Etemadi's monotonicity argument. Let $(\Omega, \mathcal{A}, P)$ be the underlying probability space. For rational $a > 1$ and $n \in \mathbb{N}$ set

$$q(a,n) := \min\{a^N; a^N > n, N \in \mathbb{N}\}, \qquad p(a,n) := q(a,n)/a.$$

Then for f_n one has the majorant

$$g(a,n,\cdot) := \frac{1}{p(a,n)h^d_{q(a,n)}} \sum_{i=1}^{\lceil q(a,n)\rceil} K\left(\frac{\cdot - X_i}{h_{p(a,n)}}\right)$$

and a corresponding minorant $b(a,n,\cdot)$. Let $\|\cdot\|_1$ and $\|\cdot\|_2$ denote the norms in $L_1(\lambda)$ and $L_2(\lambda)$, respectively. In order to show

$$\left\|g(a,n,\cdot) - a^{1+\gamma d} f\right\|_1 \to 0 \quad (n \to \infty) \text{ a.s.},$$

i.e.,

$$\left\|\sum_{i=1}^{\lceil a^{N+1}\rceil} V_{N,i} - a^{1+\gamma d} f\right\|_1 \to 0 \quad (N \to \infty) \text{ a.s.},$$

with

$$V_{N,i}(x) := \frac{1}{a^N h^d_{a^{N+1}}} K\left(\frac{x - X_i}{h_{a^N}}\right),$$

it suffices, according to Györfi et al. (1989), pp. 76, 77, to show

$$\left\|\sum_{i=1}^{\lceil a^{N+1}\rceil} \mathbf{E}V_{N,i} - a^{1+\gamma d} f\right\|_1 \to 0 \tag{31}$$

and

$$\left\|\sum_{i=1}^{\lceil a^{N+1}\rceil} (V_{N,i} - \mathbf{E}V_{N,i})\right\|_2 \to 0 \quad \text{a.s.} \tag{32}$$

(31) follows from Theorem 1 in Chap. 2 of Devroye and Györfi (1985). Noticing

$$\|V_{N,i}\|_2 = \frac{1}{a^N h^d_{a^{N+1}}} h^{\frac{d}{2}}_{a^N} \left(\int K(s)^2 \lambda(ds)\right)^{\frac{1}{2}},$$

one obtains

$$\sum_{N=1}^{\infty} \mathbf{E} \left\|\sum_{i=1}^{\lceil a^{N+1}\rceil} (V_{N,i} - \mathbf{E}V_{N,i})\right\|_2^2 < \infty$$

by Lemma 3a,b and thus (32). Analogously one has

$$\|b(a,n,\cdot) - a^{-1-\gamma d} f\|_1 \to 0 \quad \text{a.s.}$$

Thus for P-almost all realizations $b^*(a,n,\cdot) \le f_n^* \le g^*(a,n,\cdot)$, one obtains that for all rational $a > 1$

$$\|g^*(a,n,\cdot) - a^{1+\gamma d} f\|_1 \to 0, \qquad \|b^*(a,n,\cdot) - a^{-1-\gamma d} f\|_1 \to 0.$$

Let (n_k) be an arbitrary sequence of indices in $\mathbb{N}$. Then a subsequence (n_{k_l}) exists such that for all rational $a > 1$ (by Cantor's diagonal method)

$$g^*(a, n_{k_l}, \cdot) \to a^{1+\gamma d} f, \qquad b^*(a, n_{k_l}, \cdot) \to a^{-1-\gamma d} f$$

λ-almost everywhere, thus $f_{n_{k_l}}^* \to f$ λ-almost everywhere and, by the Riesz-Vitali-Scheffé lemma, $\|f_{n_{k_l}}^* - f\|_1 \to 0$. Therefore $\|f_n^* - f\|_1 \to 0$, i.e., (30) is obtained. □

In order to establish in the situation of Theorem 8 strong consistency λ-almost everywhere for bounded K, in the α-mixing case one needs to strengthen the condition on γ to $0 < \gamma d < \min\{\frac{\alpha}{2}, \frac{\alpha}{\alpha+1}\}$, according to Irle (1997). Another result, where the freedom of choice in Theorem 8 is preserved, is given in the following corollary. The proof is similar to that of Theorem 8 and will be omitted. It uses variance inequalities of Peligrad (1992) and Rìo (1993), respectively (see Lemma 3a,d).

Corollary 1. *Let the density K be as in Theorem* 3. *Assume further the conditions of Theorem* 8 *with (X_n) ρ-mixing or (X_n) weakly stationary and α-mixing with $\alpha_n = O(\delta^n)$ for some $\delta \in (0,1)$ and $\int K(x)^2 \log_+ K(x)\lambda(dx) < \infty$, further $0 < \gamma d < 1$. Then*

$$f_n(x) \to f(x) \quad a.s. \bmod \lambda.$$

References

Chandra, T.K., Goswami, A.: Cesàro uniform integrability and the strong law of large numbers. Sankhyā, Ser. A **54**, 215–231 (1992)

Chandra, T.K., Goswami, A.: Corrigendum: Cesàro uniform integrability and the strong law of large numbers. Sankhyā, Ser. A **55**, 327–328 (1993)

Csörgő, S., Tandori, K., Totik, V.: On the strong law of large numbers for pairwise independent random variables. Acta Math. Hung. **42**, 319–330 (1983)

Dehling, H., Philipp, W.: Almost sure invariance principles for weakly dependent vector-valued random variables. Ann. Probab. **10**, 689–701 (1982)

Devroye, L.: On the almost everywhere convergence of nonparametric regression function estimates. Ann. Stat. **9**, 1310–1319 (1981)

Devroye, L., Györfi, L.: Nonparametric Density Estimation: The L_1-View. Wiley, New York (1985)

Devroye, L., Krzyżak, A.: An equivalence theorem for L_1 convergence of the kernel regression estimate. J. Stat. Plan. Inference **23**, 71–82 (1989)

Devroye, L., Wagner, T.J.: Distribution-free consistency results in nonparametric discrimination and regression function estimation. Ann. Stat. **8**, 231–239 (1980a)

Devroye, L., Wagner, T.J.: On the L_1-convergence of kernel estimators of regression functions with applications in discrimination. Z. Wahrscheinlichkeitstheor. Verw. Geb. **51**, 15–25 (1980b)

Devroye, L., Györfi, L., Krzyżak, A., Lugosi, G.: On the strong universal consistency of nearest neighbor regression function estimates. Ann. Stat. **22**, 1371–1385 (1994)
Doukhan, P.: Mixing: Properties and Examples. Springer, New York (1994)
Etemadi, N.: An elementary proof of the strong law of large numbers. Z. Wahrscheinlichkeitstheor. Verw. Geb. **55**, 119–122 (1981)
Etemadi, N.: On the law of large numbers for nonnegative random variables. J. Multivar. Anal. **13**, 187–193 (1983)
Ferrario, P.: The Strong law of large numbers and applications. Laurea Thesis, Politecnico di Milano (2004)
Frey, S.: Konvergenzverhalten von Regressionsschätzungen unter Mischungsbedingungen. Wissenschaftliche Arbeit, Universität Stuttgart (2007)
Gaposhkin, V.F.: Criteria for the strong law of large numbers for some classes of second-order stationary processes and homogeneous random fields. Theory Probab. Appl. **22**, 286–310 (1977)
Greblicki, W., Krzyżak, A., Pawlak, M.: Distribution-free pointwise consistency of kernel regression estimate. Ann. Stat. **12**, 1570–1575 (1984)
Györfi, L.: Universal consistencies of a regression estimate for unbounded regression functions. In: Roussas, G. (ed.) Nonparametric Functional Estimation and Related Topics. NATO ASI Series, pp. 329–338. Kluwer Academic, Dordrecht (1991)
Györfi, L., Härdle, W., Sarda, P., Vieu, P.: Nonparametric Curve Estimation for Mixing Time Series. Springer, New York (1989)
Györfi, L., Masry, E.: The L_1 and L_2 strong consistency of recursive kernel density estimation from dependent samples. IEEE Trans. Inf. Theory **IT 36**, 531–539 (1990)
Györfi, L., Kohler, M., Walk, H.: Weak and strong universal consistency of semi-recursive partitioning and kernel regression estimates. Stat. Decis. **16**, 1–18 (1998)
Györfi, L., Kohler, M., Krzyżak, A., Walk, H.: A Distribution-Free Theory of Nonparametric Regression. Springer Series in Statistics. Springer, New York (2002)
Hardly, G.H.: Divergent Series. Oxford University Press, London (1949)
Irle, A.: On consistency in nonparametric estimation under mixing conditions. J. Multivar. Anal. **60**, 123–147 (1997)
Karamata, J.: Einige Sätze über die Rieszschen Mittel. Acad. R. Serbe, Bull. Acad. Sci. Math. Nat.(Belgrade) A **4**, 121–137 (1938)
Korevaar, J.: Tauberian Theory: A Century of Developments. Springer, Berlin (2004)
Kozek, A.S., Leslie, J.R., Schuster, E.F.: On a univeral strong law of large numbers for conditional expectations. Bernoulli **4**, 143–165 (1998)
Landau, E.: Über die Bedeutung einiger neuer Grenzwertsätze der Herren Hardy und Axer. Pr. Mat.-Fiz. **21**, 91–177 (1910)
Liebscher, E.: Strong convergence of sums of α-mixing random variables with applications to density estimation. Stoch. Process. Appl. **65**, 69–80 (1996)
Peligrad, M.: Properties of uniform consistency of the kernel estimators of density and of regression functions under dependence assumptions. Stoch. Stoch. Rep. **40**, 147–168 (1992)
Rhomari, N.: Approximation et inégalités exponentielles pour les sommes de vecteurs aléatoires dépendants. C. R. Acad. Sci. Paris, Ser. I **334**, 149–154 (2002)
Rio, E.: Covariance inequalities for strongly mixing processes. Ann. Inst. H. Poincaré, Sect. B **29**, 587–597 (1993)
Schmidt, R.: Über divergente Folgen und lineare Mittelbildungen. Math. Z. **22**, 89–152 (1925)
Serfling, R.J.: Moment inequalities for the maximum cumulative sum. Ann. Math. Stat. **41**, 1227–1234 (1970a)
Serfling, R.J.: Convergence properties of S_n under moment restrictions. Ann. Math. Stat. **41**, 1235–1248 (1970b)
Spiegelman, C., Sacks, J.: Consistent window estimation in nonparametric regression. Ann. Stat. **8**, 240–246 (1980)
Steele, J.M.: An Efron-Stein inequality for nonsymmetric statistics. Ann. Stat. **14**, 753–758 (1986)
Stone, C.J.: Consistent nonparametric regression. Ann. Stat. **5**, 595–645 (1977)
Szász, O.: Verallgemeinerung und neuer Beweis einiger Sätze Tauberscher Art. Sitzungsber. Bayer. Akad. Wiss. München, Math.-Phys. Kl. **59**, 325–340 (1929)

Walk, H.: Strong universal pointwise consistency of recursive regression estimates. Ann. Inst. Stat. Math. **53**, 691–707 (2001)

Walk, H.: Strong universal consistency of smooth kernel regression estimates. Ann. Inst. Stat. Math. **57**, 665–685 (2005a)

Walk, H.: Strong laws of large numbers by elementary Tauberian arguments. Monatshefte Math. **144**, 329–346 (2005b)

Walk, H.: Almost sure Cesàro and Euler summability of sequences of dependent random variables. Arch. Math. **89**, 466–480 (2007)

Walk, H.: A universal strong law of large numbers for conditional expectations via nearest neighbors. J. Multivar. Anal. **99**, 1035–1050 (2008a)

Walk, H.: Strong consistency of kernel estimates under dependence. Stat. Probab. Lett. (2008b). doi:10.1016/j.spl.2010.03.010

Zeller, K., Beekmann, W.: Theorie der Limitierungsverfahren, 2 Aufl. Springer, Berlin (1970)

Institute of Applied Mathematics at Middle East Technical University, Ankara (Panel Discussion Contribution)

Bülent Karasözen

Abstract The foundation and development of the Institute of Applied Mathematics (IAM) at the Middle East Technical University (METU) Ankara is briefly outlined. The impact of the Institute on the mathematical based interdisciplinary graduate education and research is discussed.

1 Introduction

Mathematics and computational science are utilized in almost every discipline of science, engineering and industry. New application areas of mathematics are being discovered constantly and older techniques are applied in new ways and in emerging fields. Industry relies on applied mathematics and computational sciences for the design and optimization of aircraft, automobiles, computers, communications systems, prescription of drugs etc. Computation is now regarded as the third approach along with theory and experiment to advance the scientific knowledge and industrial research. Simulations enable the study of complex systems that would be too expensive or dangerous by experiment Petzold et al. (2001). Simulations based on computation have become a crucial part of the present infrastructure of the sciences, engineering and industry. The importance of mathematics in the development of new technologies, in financial and economical sectors are well recognized by the developed countries like USA, Europe, Japan and others and is well documented by several reports in Glotzer et al. (2009), Wright and Chorin (1999).

Starting in 1990's several interdisciplinary mathematical research institutes and graduate programs emerged around the world. Examples of such institutes supported by the governments are *Institute for Mathematics and Applications (IMA)*, Univer-

Bülent Karasözen, Institute of Applied Mathematics & Department of Mathematics, Middle East Technical University, 06531 Ankara, Turkey
e-mail: bulent@metu.edu.tr

L. Devroye et al. (eds.), *Recent Developments in Applied Probability and Statistics*,
DOI 10.1007/978-3-7908-2598-5_9, © Springer-Verlag Berlin Heidelberg 2010

sity of Minnesota, *Mathematical Biosciences Institute*, Columbus, Ohio, *DFG Research Center Matheon*, Berlin.

Two important aspects of contemporary research are closely connected with the establishment of these applied mathematics research institutes and graduate programs:

- *Interdisciplinary teamwork:* The successful solution of complex problems in science and in industry requires the collaboration of experts from diverse disciplines. It is generally recognized by industry that mathematics is the lingua franca of science and engineering and many industries include mathematicians in industrial research teams. However, not all mathematicians have the right skills to collaborate in an industrial environment, and not all industries appreciate the long-term approach favored by mathematicians. Industry and academia have to engage with each other in new ways that encourage innovation through interdisciplinary teamwork.
- *Transfer of Mathematical Knowledge:* The mathematical research community has developed a host of techniques that could be of significant benefit to science, technology and thus to the society as a whole. Translating these techniques into practical terms and implementing them in applicable paradigms are, however, not straightforward.

The Institute of Applied Mathematics(IAM) at the Middle East Technical University is the first kind of an interdisciplinary Mathematics Institute in Turkey. It was founded in 2002 with three programs; Cryptography, Financial Mathematics, Scientific Computing. Actuarial Sciences was an option within Financial Mathematics Program and became a separate one in 2009. The Cryptography, Financial Mathematics, Actuarial Sciences programs offer master degrees with and without thesis option, the Scientific Computing program only master program with thesis option. All programs except Actuarial Sciences have PhD options.

Objectives of the Institute were stated in the foundation document as:

- Coordination of mathematics based research at METU in fostering interdisciplinary collaboration between the department of Mathematics and the other departments.
- Training graduates from different disciplines at the Master's level in theoretical and practical aspects of mathematical sciences with the aim of developing their skills in solving real life problems and applying them to science, engineering and industry in order to address the interdisciplinary needs of both the public and private sectors.
- Organizing and conducting international workshops, summer schools in chosen research areas and short courses to industrial partners.
- Cultivation of the collaboration among research groups in mathematics, science and engineering departments at METU and with other universities.
- Providing a platform for active participation of research groups from METU in the international research community by establishing research networks and participating in international projects.

2 Structure of the Programs

The specific aims, main research areas and graduate courses are given below:

Scientific Computing Program:

Parallel to the emergence of Scientific Computing (SC) as an interdisciplinary research area, many graduate programs were developed in recent years at the leading universities of the world (see for an overview of the SC programs in USA, Petzold et al. 2001 and in Schäfer 2010). The SC graduate programs are designed according to the multidisciplinary nature of SC and include the areas of Applied Mathematics, Numerical Analysis and Mathematical Modeling. In addition to a background in mathematics and computer science, a SC graduate must have a thorough education in an application area. Knowledge of computer science, and in particular numerical algorithms, software design and visualization, enable the SC graduate to make efficient use of computers and the graduates of the Program are expected to communicate within a team of engineers, computer scientists and mathematicians to solve difficult practical problems.

The main research fields are: Continuous optimization with emphasis on derivative free optimization, semi-infinite optimization, non-smooth optimization, robust optimization, conic quadratic programming, optimal control of partial differential equations, application of finite elements in fluid dynamics, metabolic and gene regulation networks.

Core courses: Scientific Computing I, II, Mathematical Modeling, Numerical Optimization.

Elective courses: Inverse Problems, Statistical Learning and Simulation, Applied Nonlinear Dynamics, Hybrid Systems, Advanced Continuous Optimization, Optimal Control and Adaptive Finite Elements, Finite Elements: Theory and Practice, Game Theory, Basic Algorithms and Programming.

Cryptography Program:

The Cryptography Group (CG) of IAM conducts research in areas such as design, evaluation, and implementation of cryptographic algorithms and protocols. CG's theoretical work on cryptographic algorithms and protocols are based on discrete mathematics. The major focus of the research will be in applied and theoretical cryptography. CG's research areas can be broadly categorized as follows: Design and analysis of pseudorandom sequences, elliptic and hyperelliptic curve cryptography, computational number theory, coding theory, computational methods in quadratic fields, algorithms for finite Abelian groups.

Core courses: Introduction to Cryptography, Stream Ciphers, Applications of Finite Fields, Public Key Cryptography, Block Ciphers.

Elective Courses: Elliptic Curves in Cryptography, Combinatorics, Algorithmic Graph Theory, Computer Algebra, Algebraic Aspects of Cryptography, Quantum Cryptography, Algebraic Geometric Codes.

Financial Mathematics Program:

The last decades witnessed the projection of sophisticated mathematical techniques to the center of the finance industry. In the 80's many investment banks hired mathematicians, physicists and engineers to become financial engineers (see for an overview of the development finance mathematics, Korn 2010). Gradually, the main skills defining this professional category are being clarified and today many universities all over the world are designing programs to develop modeling and mathematical expertise in financial applications. In Turkey, the financial sector has enjoyed unparalleled expansion in the last decades and more sophisticated financial instruments are expected to be introduced into the sector in the forthcoming ones. Already there are serious attempts to integrate derivative securities and markets into the Turkish financial system. These developments will lead to a demand for talented people trained in the field of financial mathematics. One of the fields connected to Financial Mathematics is Actuarial Sciences. By analyzing uncertainty of real life events, actuaries create and apply programs aimed at managing risks. The option of *Financial Mathematics in Life and Pension Insurance* focuses on life contingencies.

Core courses are: Financial Derivatives, Financial Management, Stochastic Calculus for Finance, Financial Economics, Time Series Applied to Finance, Probability Theory, Simulation.

Elective courses: Markov Decisions Processes, Stochastic Processes, Regulation and Supervision of Financial Risk, Interest Rate Models, Financial Economics, Risk Management and Insurance, Energy Trade and Risk Management, Decision-Making under Uncertainty, Financial Modeling with Jump Processes, Numerical Methods with Financial Applications, Portfolio Optimization, Pension Fund Mathematics.

3 Development of the Institute

The number of permanent members of the Institute was kept small in order to have an active and lively interdisciplinary research atmosphere. There are currently seven permanent members of the Institute. Institute has 48 affiliated faculty members from METU, 22 affiliated faculty members from other universities and from the industry and 19 research assistants.

Cooperation agreements with the foreign universities and scientific institutions: University of Technology, Darmstadt (Germany), Kaiserslautern University (Germany), University of the Aegean (Greece), The Institute of Mathematics of The Polish Academy of Sciences (Poland), Laboratoire de Mathématiques et Applications Université de La Rochelle (France), University of Ballarat (Australia).

Partnership with the Industry: Information and Technologies Authority (Turkey), ValuePrice AG, Frankfurt am Main (Germany).

Number of publications increased within the years steadily. Since 2003, 83 journal articles, 14 proceedings papers were published with IAM affiliation (Table 1).

Table 1 Publications

Year	2004	2005	2006	2007	2008	2009
Journal articles	3	8	10	10	23	29
Conference papers	–	1	4	2	7	3

The number of the students in 2008–2009 among the programs totally 168; 71 of them were in Cryptography Program, 67 in Financial Mathematics, 24 in Scientific Computing, 6 in Actuarial Sciences. The distribution of the registered student over the years was given in Table 2. The big majority of the students are still from Mathematics.

Table 2 Students according to their graduating departments

	Mathematics	Statistics	Engineering	Bus. & Econ.	Other
2003	52	19	19	10	3
2004	34	11	11	11	4
2005	30	12	14	9	1
2006	28	5	11	3	1
2007	28	6	9	10	1
2008	31	7	12	8	–
2009	47	6	9	6	2

Each year, the Institute produces each year a sufficient number of graduates and in the recent years first graduates with PhD degrees have been appeared (Table 3).

Table 3 Graduates

	MSc with thesis	MSc without thesis	PhD
2004	10	5	–
2005	13	21	–
2006	10	11	–
2007	10	6	2
2008	17	13	3
2009	9	11	6

The Institute was very active in organizing international and national conferences. Since 2003, 25 conferences and workshops were organized by the Institute. Some of them are:

- EUROPT (Continuous Optimization Working Group of "The Association of European Operational Societies") Conferences on Continuous Optimization: 2003 in Istanbul, 2004 in Rhodes, 2005 in Pec, EURO ( ) Summer Institute 2004 in Ankara.
- Turkish-German Summer Academy in Advanced Engineering, 2003–2006 in Kuşadası in cooperation with TU Darmstadt, Department of Mechanical Engineering.

- Annual National Cryptography Conferences since 2005.
- Advanced Mathematical Methods in Finance, 2006 in Antalya.
- Networks in Computational Biology, 2006 in Ankara.
- Workshop on Recent Developments in Financial Mathematics and Stochastic Calculus: in Memory of Hayri Körezlioğlu, 2008 in Ankara.
- Hybrid Systems, 2008 in Istanbul.
- Workshop on Recent Developments in Applied Probability and Statistics: in Memory of Jürgen Lehn, 2009 in Ankara.
- Complex Systems—Theory and Applications in Sciences and Engineering, with Max Planck Institutes, Leipzig, Magdeburg, Stuttgart, 2009 in Ankara.

Within these conferences several special Issues in journals were published:

- Hybrid Systems: Modeling, Simulation, Optimization, in Journal of Process Control, Vol. 19, 2009.
- Special issue on Networks in Computational Biology, in Discrete Applied Mathematics Vol. 157, 2009.
- Special Issue on Challenges of continuous optimization in theory and applications, in European Journal of Operational Research, Vol: 181, 2007.
- Advances in Continuous Optimization, European Journal of Operational Research, Vol. 169, 2006.
- Optimization in Data Mining, European Journal of Operational Research, Vol. 173, 2006.

One of the mechanisms of conducting research at the Institute was by establishing and participating in projects. IAM members participated in 21 different in projects since the foundation of the Institute. International and national projects coordinated by IAM members are:

- Volkswagen Foundation Project, "Optimization of Stirrer Configurations by Numerical Simulation" 2003–2005, with TU Darmstadt, Department of Mechanical Engineering.
- DAAD Project, Cooperation in the Field of Financial and Insurance Mathematics, 2003–2006, with TU Kaiserslautern and TU Darmstadt, Department of Mathematics.
- NSF-TÜBITAK Project, Development of Modeling and Optimization Tools for Hybrid systems, 2005–2008.
- DAAD Project, 2007–2011, with TU Darmstadt, Department of Mathematics.
- Turkish State Planning Organization: Research and Development in Cryptography: Design, Analysis and Implementation of Algorithms, 2004–2007.
- TÜBITAK (Turkish Research and Technical Council) Industry Project, Developments and Applications of Open Key Algorithms, 2005–2007.
- TÜBITAK Integrated PhD Program in Continuous Optimization, 2003–2007.
- TÜBITAK Research Project, Numerical Solution of Nuclear Fusion Problems by Boundary and Finite Element Methods, 2005–2007.

4 Conclusions

IAM is now well established within the university and within the scientific community in and outside of Turkey. IAM has shown through its research activities, the possibility of interdisciplinary research conducted through several projects. Especially in the areas like Cryptography and Financial Mathematics, the Institute attracted many graduate students. Also new research areas in mathematics, like optimization, data mining, cryptography, stochastic became popular.

References

Glotzer, S.C., et al.: International assessment of research and development in simulation-based engineering and science. WTEC Panel Report (2009). http://www.wtec.org/sbes/SBES-GlobalFinalReport.pdf

Korn, R.: Financial mathematics between stochastic differential equations and financial crisis. In: Devroye, L., Karasözen, B., Kohler, M., Korn, R. (eds.) Recent Developments in Applied Probability and Statistics: Dedicated to Jürgen Lehn, pp. 231–236. Springer/Physica-Verlag, Berlin (2010)

Petzold, L., et al.: Graduate education in computational science and engineering. SIAM Rev. **43**, 163–177 (2001)

Schäfer, M.: Computational science and engineering education programs in Germany. In: Devroye, L., Karasözen, B., Kohler, M., Korn, R. (eds.) Recent Developments in Applied Probability and Statistics: Dedicated to Jürgen Lehn, pp. 237–243. Springer/Physica-Verlag, Berlin (2010)

Wright, M., Chorin, A.: Mathematics and science. NSF Report (1999). http://www.nsf.gov/mps/dms/start.htm

Financial Mathematics: Between Stochastic Differential Equations and Financial Crisis (Panel Discussion Contribution)

Ralf Korn

Abstract We survey the role, scope and subject of modern financial mathematics. Besides its impact on scientific research in recent years and on the financial market (including the current crisis), we will also comment on the possibility to study financial mathematics and related areas at German universities.

1 Modern Financial Mathematics—A New Area of Mathematical Education and Research

Modern financial mathematics is a subject that emerged in the recent 30–40 years. We can mainly distinguish four areas (which have some overlap with each other):

- **Modeling**. The modeling of the dynamical evolution of various price processes such as stock prices, goods prices, interest rates, exchange rates (just to name a few) is a central topic of modern financial mathematics. However, it is very important that modeling does not necessarily mean forecasting.
- **Optimal Investment**. The task to determine an optimal strategy for investing a given amount of money at a financial market is maybe the first problem that comes to one‘s mind when thinking of financial mathematics. Here, we consider the possibility of continuously investing, consumption and rebalancing the holdings. This subject is also known as portfolio optimization.
- **Option pricing**. Here, finding the correct price of a derivative contract which yields a non-negative payment at some future time depending on the performance of an underlying good is the task to solve. The most famous result of financial mathematics, the Black-Scholes-formula for the price of European call and put options on a stock, is the cornerstone of modern financial mathematics.

Ralf Korn, Center for Mathematical and Computational Modeling (CM)² and Department of Mathematics, University of Kaiserslautern, 67653 Kaiserslautern, Germany
e-mail: korn@mathematik.uni-kl

L. Devroye et al. (eds.), *Recent Developments in Applied Probability and Statistics*,
DOI 10.1007/978-3-7908-2598-5_10,

- **Risk management**. As (nearly) all investments at financial markets contain some risky part (such as market risk (price risk) or credit risk). It is therefore on one hand necessary to know the risk of the current position (measured in terms of risk measures such as Value at Risk (a high (typically 95 or 99%) quantile of the loss distribution for some fixed time horizon)) and to manage it by buying insurance against losses such as suitable option contracts.

As uncertainty is one of the main modeling ingredients in financial mathematics, it is clear that probabilistic methods (in particular continuous-time stochastic processes, Itô calculus and Monte Carlo methods) are central. Further, partial differential equations of parabolic type play an important role when it comes to characterize option prices. Finally, numerical methods such as finite difference methods and methods of statistics are needed. On top of that the use of stochastic control theory to obtain optimal portfolios is an advanced mathematical ingredient. And last but not least statistical methods (such as least-squares, maximum likelihood or Bayes methods) are necessary to obtain reasonable input parameters.

2 Modern Financial Mathematics 1973–2007: A Success Story

For at least the last two decays, financial mathematics has been one of the most popular and successful areas of mathematics, both in practise and in theory. There have been three decisive events that helped in their ways:

- 1973: The Black-Scholes formula for pricing European call and put options was the first advanced result of modern financial mathematics that had a big impact on both theory and application (see Black and Scholes 1973).
- 1981: Harrison and Pliska (1981) introduced martingale methods and stochastic calculus in a detailed way to financial mathematics. This introduction created a lot of interest on both sides of the academic subject, finance and mathematics.
- End of the 80s: The fall of the Iron Curtain lead to the fact that a real wave of excellent probabilists from the former communist states concentrated onto financial mathematics which gave the subject a further boost.

The now rapidly evolving theory gained a big **impact on financial markets and the financial industry**. The acceptance of the new methods by the industry were impressively demonstrated by an enormous demand for mathematicians, a new type of jobs—the Quants, new financial products (such as Exotic options, credit derivatives, equity linked notes) that exploited the advanced mathematical theory and last but not least by the fact that now mathematicians gained top positions in banks.

Also, there was a demand for new research with computational but also theoretical aspects in areas such as Monte Carlo methods and random number generation, parallelization and computer architecture, Malliavin calculus, copulas, to name a few popular examples.

On the down side, there were already warning signs such as an extensive, sometimes blind believe in models which was caused by practitioners sometimes not understanding but using them and by mathematicians only interested in the model but not the application. On top of that, there was a tendency of arrogance towards statistical and econometric problems which did not produce great mathematical theorems but which are indispensable for a successful application of financial mathematics at real markets.

The **impact on the academic world** of financial mathematics could be seen by the creation of new study programmes such as *Master in Financial Mathematics* or *Master in Financial and Actuarial Mathematics*. Whole research directions were introduced such as *Computational Finance*, *Quantitative Finance* or *Phynance*, a somewhat strange term describing the physicists approach to financial mathematics. The diverse nature of the subject also required interdisciplinarity among mathematicians. New theoretical problems and areas such as the valuation of exotic options (where the market always produced new and more complicated payoff structures), the theory of risk measures and the arbitrage theory with its many variants of the fundamental asset pricing theorem were just some popular examples. The demand for the new subject by students lead to the introduction of new chairs in financial mathematics, maybe more than the subject actually needed.

Modern financial mathematics even has an **impact on the political world**. This can be seen in the rules introduced with **Basel II** where it is in particular required that internal risk models have to be based on recent statistical methods. **Solvency II**, the currently created equivalent of Basel II for the insurance industry, has clear relations to recent mathematical research in risk measurement and management.

3 Financial Mathematics in the Crisis: 2008–Today

The credit crisis that we are currently facing is certainly the biggest financial crisis since the introduction of methods of modern financial mathematics. Of course, there remains the question if the use of financial mathematics is connected or even responsible for the crisis. The answer to this question is not easy, but it can definitely be said that financial mathematics is not the main reason of the crisis. Financial mathematics has been a driving force behind the introduction of complex derivatives. Its use has also very often given traders the feeling that the models they are using **are** the reality.

However, wrong use of a model or a bad model can cause severe losses and damages to a certain branch of a company. But it cannot be blamed for what we observe now. Already the use of the term *credit crisis* gives a hint on what happened, although it is not simply that a lot of credits have defaulted. It is actually the way the credits were disguised as credit derivatives, often as so-called *CDOs* (collateralized debt obligations). By this innovation, participating in the credit business became an interesting subject for investors who usually trade in stocks and equally risky investment opportunities.

The basic idea behind a CDO is that a package of credits granted by a bank is put together as a bundle and form the basis of a firm, a so-called special purpose vehicle

(SPV). Then, parts of this company are sold as if the company would have issued shares. To make this more interesting (as otherwise the parts would only be parts of a credit portfolio), an action called **tranching** was introduced. Via this procedure, shares of different quality of the SPV could be offered:

- shares from the *equity tranche* which are receiving a high rate of return but which also have to take the risk of the first defaulting credits,
- shares from the *mezzanine tranche* that has to take the risk of the defaulting credits after the equity tranche is fully used up by already defaulted credits. It therefore receives a smaller rate of return than the equity tranche but a bigger one than
- shares from the *senior tranche* that should only consist of high class credits, preferably of a AAA-rating. For the shares a comparably low rate of return is paid.

So this repackaging and tranching of the collection of credits allowed investors of different risk adversity to invest in their preferred tranche of the SPV. Thus, the innovation was that a comparably standard business, the trading of credits, had been transformed into a new type (actually many types) of an—at least at first sight—attractive investment. As there was an enormous demand for CDOs in 2006 and 2007, there was pressure on the banks to satisfy this demand. This translated directly into the problem of the availability of credits that are needed as underlyings for a CDO. However, it—at least—seems that often those credits did not have the quality that was needed to give the different tranches of a CDO the quality they were advertised. Thus, the CDOs were overvalued and when it was first discovered that the credit quality was inferior, the correction of the CDO values lead to serious (liquidity) problems of their owner, one of the main reasons of the credit crisis.

So what has been the role of the financial mathematician and financial mathematics in this whole affair? Let me first state that a bad model can be a cause for serious losses but definitely not for losses that we have observed. The credit quality is the most important input parameter for valuing a CDO. If this information is totally wrong then the mathematical model used to calculate the value of a CDO cannot produce a reasonable number. However, the success of financial mathematics in other areas of finance gave traders the feeling that they could (maybe blindly) trust mathematical models. They did therefore not question the figures they came up with. It will thus be a serious future task for financial mathematicians to clarify what a model can do and what it cannot. And definitely, the reputation of modern methods of financial mathematics has suffered by the crisis although its contribution to the crisis itself is surely a minor one.

However, we will need even more financial mathematics in the future to

- obtain a better understanding of risks, maybe supported by legal frameworks such as Basel III, IV, ... or even better **World I** to give the financial markets a common legal basis.
- educate decision makers in banks and politics in financial mathematics, especially with regard to the risks and the consequences of new products and new regulations.

- be able to manage (and not only measure) risks in a sophisticated way.

Modern trading will not be possible without quantification and modeling of risks by financial mathematics. Even if a consequence of the current crisis would be to abandon some of the more exotic products that created whole branches of research in recent years, there remains a lot to do. The impact of financial mathematics will thus not decrease, but maybe the topics will change.

4 Financial Mathematics: How to Become a Quant?

While in former times mathematicians working in banks were more or less trained on the job, the practical success of modern financial mathematics has lead to the introduction of the specialist job of a quantitative analyst (for short: quant). This is nowadays the job for mathematicians that relies most on advanced financial mathematics methods. It is the main reason for the success of master programmes in financial mathematics as offered by many universities. Also, in Germany it is often possible to study mathematics with a focus on financial mathematics. While there are many master programmes that allow such a study programme, there are so far only very few bachelor programmes with a specialization in financial mathematics. The reason for this lies in the complexity of the methods needed in financial mathematics and it remains doubtful that they can be introduced on a sufficient level in a bachelor programme.

A second type of job that is closely related, but has a longer tradition than the quant is the actuary. Actuaries are mathematicians that deal with financial and actuarial risks in insurance companies. Although, their specialization is much more focused on insurance mathematics (such as life-insurance mathematics, risk theory, ...), it contains a significant part related to financial mathematics. With the growing complexity of financial products that are attractive for insurance companies, actuaries also have to be familiar with modern financial mathematics. To become an actuary in Germany, one has to pass a post-university education (including examinations) by the German actuarial society DAV.

Both types of jobs and also both the study programmes and/or focuses in financial and actuarial mathematics are currently much requested by students, a fact that lead to the introduction of such kind of study programmes at numerous German universities in recent years.

References

Black F., Scholes M.S.: The pricing of options and corporate liabilities. J. Polit. Econ. **81**, 637–654 (1973)

Harrison J.C., Pliska S.R.: Martingales and stochastic integrals in the theory of continuous trading. Stoch. Process. Appl. **11** (3), 215–260 (1981)

Computational Science and Engineering Education Programs in Germany (Panel Discussion Contribution)

Michael Schäfer

Abstract In recent years a remarkable variety of study programs in the field of Computational Science and Engineering emerged in Germany covering Bachelor, Master, and PhD education. A review on these developments is provided. Similarities and differences in curricula and formal organization issues are discussed.

1 Introduction

It is nowadays well recognized that computational science and/or engineering (CSE/CE) can be viewed as a key technology that will play a more and more important role in the future (e.g., Glotzer et al. 2009; Oden et al. 2006). Engineering applications are becoming increasingly complex. Consequently, the theory required to analyse corresponding advanced technical systems is becoming more and more complicated or even intractable. Experimental investigations are often too complex, too dangerous, too costly, or the experimental conditions are irreproducible. Methods of CSE/CE, including computer based modeling, analysis, simulation, and optimization, are a cost effective and efficient alternative to investigate engineering applications and to engineer new technical solutions. It can give insights into phenomena at a level that is unobtainable with traditional scientific approaches. CSE/CE contributes to finding optimal strategies addressing key issues in future technical developments for the economy and society, in areas such as energy, health, safety, and mobility.

To meet and master the corresponding challenges, it is beyond dispute that there is a strong need of an adequate education in the field of CSE/CE. The question is, what is the best way to do so? Thoughts on this can be found in two reports of the SIAM Working Group on CSE Education (Petzold et al. 2001 and Turner et

Michael Schäfer, Institute of Numerical Methods in Mechanical Engineering, Technische Universität Darmstadt, Dolivostr. 15, 64293 Darmstadt, Germany
e-mail: schaefer@fnb.tu-darmstadt.de

L. Devroye et al. (eds.), *Recent Developments in Applied Probability and Statistics*,
DOI 10.1007/978-3-7908-2598-5_11, © Springer-Verlag Berlin Heidelberg 2010

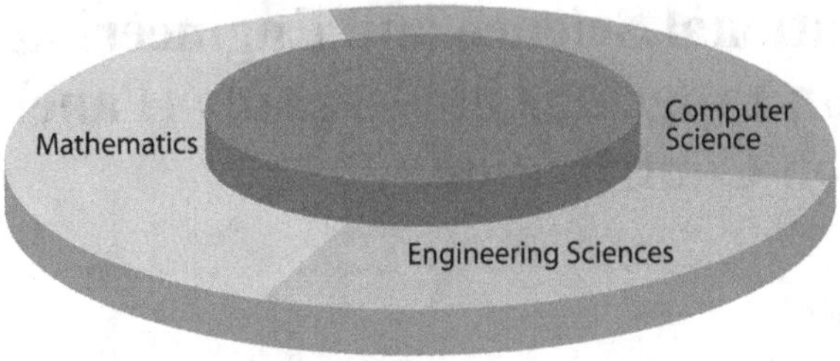

Fig. 1 Interdisciplinarity of Computational Engineering

al. 2006). It is common understanding that CSE/CE somehow is a combination of mathematics, computer science, and engineering sciences (see Fig. 1):

- Computer Science—to handle large simulation tools and software, to deal with large amounts of data in technical systems, visualization, virtual and augmented reality, and high-performance computing.
- Mathematics—to develop and investigate mathematical modeling and simulation methods for solving engineering problems as well as methods for optimization of components, systems, and networks.
- Engineering Sciences—to specify digital models and to develop and apply algorithms and software tools for use in future engineering environments for virtual prototyping, lifecycle simulation, network simulation, and simulated reality.

In science and industry, people are needed, who are able combine these skills and knowledge in an efficient way. There are several possibilities at different levels for organizing education in CSE/CE, i.e., as:

- minor subject within an existing study program (e.g., Mathematics, Computer Science, Mechanical/Civil/Electrical Engineering),
- postgraduate MSc program,
- consecutive BSc + MSc program,
- PhD study program.

All of the above options have been realized in the last years in Germany. Nowadays there are nearly 20 corresponding programs that have been established at different universities:

- Postgraduate MSc programs at U Bochum, TU Braunschweig, U Duisburg, TU München (2), U Rostock, and U Stuttgart.
- Consecutive BSc/MSc programs at RWTH Aachen, TU Darmstadt, U Erlangen, and U Hannover.
- PhD programs at RWTH Aachen, TU Darmstadt, U Erlangen, U Hannover, and TU München.
- Further related programs: Computational Science (e.g., U Bremen), Computational Physics (e.g., U Dortmund) Computational Biology, and more.

In particular, this illustrates that there is a strong activity in the field in Germany. In the present contribution we will give a brief overview of these developments and discuss several aspects of the programs.

2 Postgraduate MSc Programs

Postgraduate MSc programs provide a specialization in CSE/CE for students having a BSc degree in a classical discipline, i.e., mathematics, computer science, or engineering science.

Such programs that can be classified as more general ones are listed in Table 1 indicating important organizational issues and mandatory courses. It can be seen that the programs are organized quite differently and that there also is a significant variety concerning the content of teaching.

Table 1 More general MSc programs

	TU Braunschweig Computational Sciences in Engineering	TU München Computational Science and Engineering	U Rostock Computational Engineering
Semester	4	4	4
Language	50% English 50% German	English	80% English 20% German
Departments	Math+CS, CivE, ME, EE	CS, Math, CivE, ME, EE, Phs Chem	CS+EE, Math, CS
In charge	Interdept. committee	Interdept. committee	CS+EE
Enrollment	Participating Depts	CS	CS+EE
Mathematics	Intro Num. PDEs	Num. Analysis I, II Parallel Numerics	Num. Math, Num. PDEs, Num. Lin. Alg.
Comp. Science	Algorithms+Programs	Algorithms, Programming, Architecture+Networks, Visualization (selectable), OO-Programming	Software concepts, Data Managenent,
Engineering	(selectable) Fluid Mech. Thermodyn, EM Fields, Systems Theory, …	Comp. Mechanics, Comp. Fluid Dynamics, Parallel Architectures	(selectable) Robotics, Thermodyn.+Fluid Mech. Circuit Design, Control
Tailored	Intro Scientific Comp.	Scientific Computing	Computational Methods
Projects, Labs	Seminar Student Research Project	Seminar Scientific Computing Lab	Project Seminar Comp. Electromagnetics

Mandatory courses only

In Table 2 a summary of more specialized MSc programs is given. As can be observed these programs are dominated by civil engineering departments what is also reflected in the course programs. Of course, this is mainly due to the origin of the people pushing forward the programs.

Table 2 More specialized MSc programs

	U Bochum Computational Engineering	U Duisb.-Essen Computational Mechanics	U Stuttgart Computational Mechanics of Mat.+Structures	TU München Computational Mechanics
Semester	4	4	3	4
Language	English	50% English 50% German	English	English
Departments	CivE	CivE	CivE, CS, ME	CivE, CS, Math
In charge	CivE	CivE	CivE	CivE
Enrollment	CivE	CivE	CivE	CivE
Mathematics	ODE Numerics FEM	Intro Num. Meth. FEM	Discrete Math Num. Program.	FEM I, II
Comp. Science	Mod. Programming Concepts in Eng.	Comp. Lang. for Eng.		Software Lab I, II Par. Computing
Engineering	Mech. of Solids, Fluid Mech.	Tensor Calculus Continuum Mech. Test. of Met. Mat. Thermodyn.	Continuum Mech. Theory of Mat. I, II Struc.+Dyn.Sys. I, II Eng. Materials and Smart Systems I, II	Continuum Mech. Tensor Analysis Theory of Plates Theory of Shells Struct. Dyn. Hydromech.
Tailored	Comput. Modeling OO Modeling			Mod.+Sim. I, II, III Adv. Comp. Meth.
Projects, Labs	Case Studies in CE	Soft Skills I, II Project		Software Lab I, II

Mandatory courses only

In summary, one can observe a rather large variety in the programs in both organizational and teaching issues. However, the combination of courses from mathematics, computer science, and engineering science, possibly augmented by tailored courses, is a common feature.

3 Consecutive BSc/MSc Programs

In Table 3 an overview of consecutive BSc/MSc programs is given together with some organizational issues. Concerning the mandatory courses these programs are more similar as for the postgraduate MSc programs, although they differ in the overall number of semesters, i.e., 6–7 for BSc and 2–4 for MSc. All four programs focus on the engineering side of CSE/CE. They contain fundamentals of (Applied) Mathematics, Computer Science, Engineering as well as tailored courses to CE/CSE needs and characteristics. There are practical courses and several areas of application in the 5th and 6th semesters. Three programs are connected with dedicated interdepartmental research (and study) units for CSE/CE.

As an example, in Fig. 2 the structure of the program of the TU Darmstadt is indicated. In the Bachelor program there is a common set of basic courses for all

Table 3 Consecutive BSc/MSc programs

	RWTH Aachen Computational Engineering	U Erlangen Computational Engineering	U Hannover Computer based Engineering Sciences	TU Darmstadt Computational Engineering
Semester	7+3	6+4	6+2	6+4
Language	BSc German MSc English	BSc German MSc English	German	German
Departments	Math+CS, ME, EE MatE	EE	CivE, ME, EE, Math	CS, CivE, EE, ME, Math
In charge	ME	CS	CivE	Study Center
Enrollment	ME	CS	CivE	Study Center

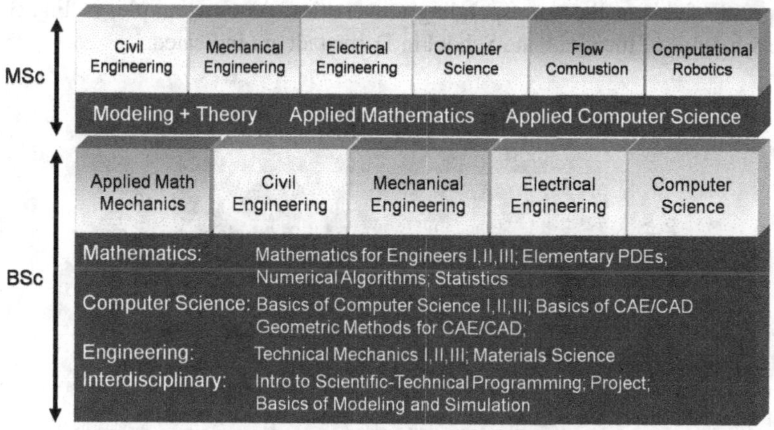

Fig. 2 Consecutive BSc/MSc program Computational Engineering at TU Darmstadt

students in the first four semesters covering fundamentals of mathematics, computer science, engineering, as well as interdisciplinary subjects. Afterwards the students must select one out of five different specialization areas. The Master program consists of compulsory elective courses. These cover general methodical subjects as well as—depending on the chosen specialization area—discipline specific methodical and application specific subjects. The specialization areas cover classical disciplines (i.e., civil, mechanical, or electrical engineering or computer science) as well as (possibly varying) interdisciplinary areas like flow and combustion or computational robotics.

4 PhD Programs

A strong impact on the field of CSE/CE originated from the recent German Excellence Initiative. In this framework a variety of Graduate Schools for PhD education in the field have been established:

- Heidelberg Graduate School of Mathematical and Computational Methods for the Sciences at U Heidelberg
- Aachen Institute for Advanced Study in Computational Engineering Science at RWTH Aachen
- Graduate School of Computational Engineering at TU Darmstadt
- International Graduate School of Science and Engineering at TU München

These programs offer a structured PhD education and support of excellent PhD students within an interdisciplinary environment. This way they provide a sound basis for high-level research and for educating highly qualified experts in CSE/CE.

The main research areas in the Graduate Schools vary according to the strengths in specific fields at the corresponding universities. As an example, in Fig. 3 the research fields of the Graduate School in Darmstadt are indicated.

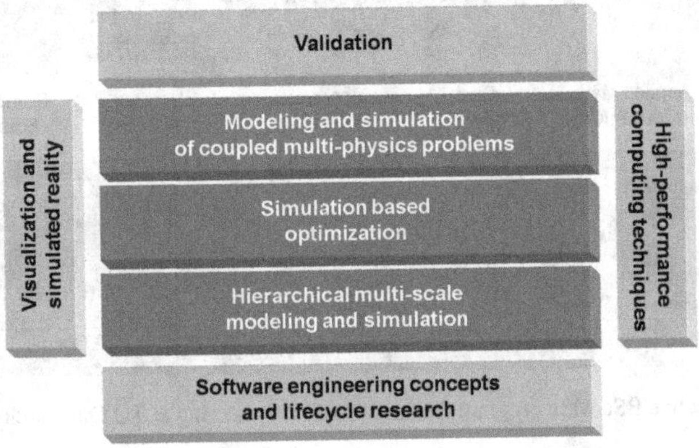

Fig. 3 Research areas of the Graduate School of Computational Engineering at TU Darmstadt

The Graduate Schools significantly contribute to further strengthen the field of CSE/CE in Germany.

5 Conclusions

One can observe many activities in CSE/CE education in Germany. Quite different approaches at different levels have been implemented. These cover general and specialized MSc programs, consecutive BSc/MSc programs, and structured PhD programs. With this, the foundations are laid for the expected future needs in the field. However, for all concepts there appears to be still room for improvements. Thus, further developments, taking into account feedback from science and industry, are necessary.

Acknowledgement

The author would like to thank Oskar von Stryk for collecting the data of the study programs.

References

Glotzer, S.C., et al.: International assessment of research and development in simulation-based engineering and science. WTEC Panel Report (2009)
Oden, J.T., et al.: Simulation-based engineering science. NSF Report, (2006)
Petzold, L., et al.: Graduate education in CSE. SIAM Working Group on CSE Education (2001)
Turner, P., et al.: Undergraduate computational science and engineering education. SIAM Working Group on CSE Education (2006)